ENGINEERING DESIGN AND TECHNICAL APPLICATIONS OF PHYSICAL SCIENCE

An Integrated Approach for Sustainable Development

ENGINEERING DESIGN AND TECHNICAL APPLICATIONS OF PHYSICAL SCIENCE

An Integrated Approach for Sustainable Development

Edited by

Chin Hua Chia, PhD

Sarani Zakaria, PhD

Ruey Shan Chen, PhD

Md. Shahariar Chowdhury, PhD

AAP | APPLE ACADEMIC PRESS

First edition published 2025

Apple Academic Press Inc.
1265 Goldenrod Circle, NE,
Palm Bay, FL 32905 USA

760 Laurentian Drive, Unit 19,
Burlington, ON L7N 0A4, CANADA

CRC Press
2385 NW Executive Center Drive,
Suite 320, Boca Raton FL 33431

4 Park Square, Milton Park,
Abingdon, Oxon, OX14 4RN UK

Apple Academic Press exclusively co-publishes with CRC Press, an imprint of Taylor & Francis Group, LLC

Library and Archives Canada Cataloguing in Publication

CIP data on file with Canada Library and Archives

Library of Congress Cataloging-in-Publication Data

CIP data on file with US Library of Congress

ISBN: 978-1-77491-752-7 (hbk)
ISBN: 978-1-77491-753-4 (pbk)
ISBN: 978-1-00353-681-9 (ebk)

About the Editors

Chin Hua Chia, PhD
Professor, Materials Science Program, School of Applied Physics, Universiti Kebangsaan Malaysia, Selangor, Malaysia

Chin Hua Chia, PhD, is a full Professor at the Department of Applied Physics, Faculty of Science and Technology at Universiti Kebangsaan Malaysia (National University of Malaysia), Malaysia. He is a recipient of the Young Scientist Award from the National University of Malaysia and the Malaysian Solid State Science and Technology (MASS) in 2012 and 2014, respectively. He also received the Distinguished Lectureship Award from the Chemical Society of Japan (CSJ) in 2017. He is a member of several professional organizations and has published several books and book chapters, as well as more than 230 articles in professional journals. His core research focuses on nanomaterials and polymer nanocomposite syntheses, with a particular emphasis on structure control to tailor the properties of the materials in determining their performances for environmental remediation applications, such as wastewater treatment, supercapacitors, and hydrogen production.

Sarani Zakaria, PhD
Professor, Materials Science Program, School of Applied Physics, Universiti Kebangsaan Malaysia, Selangor, Malaysia

Sarani Zakaria, PhD, is a Professor at the Department of Applied Physics, Faculty of Science and Technology, Universiti Kebangsaan Malaysia (also known as The National University of Malaysia). She started her career as a Research Assistant at the Chemistry Department, UPM (1987–1988) and a Research Officer at the Drug Center, USM (1988–1989). Then, she was appointed as a lecturer at the School of Industrial Technology, USM (1994–1995) before working at UKM. Prof. Zakaria is an expert in utilizing bio-resources lignocellulosic materials especially from plant biomass in converting them into various high-value-added products. Her research expertise focuses on the area of pulp and paper technology, wood plastic composites, bio-resin, cellulose derivatives, and regenerated cellulose with the combination of nano-functional materials (organic and inorganic materials).

Ruey Shan Chen, PhD
Materials Science Program, School of Applied Physics, Universiti Kebangsaan Malaysia, Selangor, Malaysia

Ruey Shan Chen, PhD, is a Senior Lecturer at the Department of Applied Physics, Faculty of Science and Technology, Universiti Kebangsaan Malaysia (National University of Malaysia). She is a member of the Malaysia Board of Technologists (MBOT) and has authored several books, book chapters, and over 70 articles in professional journals. Her primary research focuses on polymer matrix biocomposites and nanocomposites, with a specific emphasis on enhancing the strength and performance of recycled polymers and biopolymers through the incorporation of agricultural fibers and nanoparticles. She also investigates the foaming process to customize their density and properties, aiming to develop potential applications in fields such as automotive, building materials, and insulation materials.

Md. Shahariar Chowdhury, PhD
Faculty of Environmental Management, Prince of Songkhla University, Thailand

Md. Shahariar Chowdhury, PhD, has a full-time position as a Lecturer in the Faculty of Environmental Management at Prince of Songkhla University, Thailand. Concurrently, he also extends his expertise as a part-time research fellow at the Faculty of Health and Life Sciences at INTI International University in Malaysia. In 2017, he was the recipient of a full scholarship from the Thailand Education Hub (THE-AC). His contributions to the scholarly community have been significant, including roles as a reviewer for ISI-indexed journals and a guest editor at an ISI journal. In terms of research output, Dr. Chowdhury's portfolio is robust, with more than 60 research articles, two review articles, and two chapters in academic books to his credit. His areas of research interest are multifaceted, covering nanomaterials, semiconductor materials, solar cells, thin films, PV recycling, e-waste recycling, and the recycling of materials more broadly.

Contents

Contributors

R. Abd-Shukor
Department of Applied Physics, Universiti Kebangsaan Malaysia, Bangi, Selangor, Malaysia

Mardina Abdullah
Space Science Center (ANGKASA), Institute of Climate Change, Universiti Kebangsaan Malaysia, UKM Bangi, Selangor, Malaysia; Department of Electrical, Electronic, and Systems Engineering, Faculty of Engineering and Built Environment Universiti Kebangsaan Malaysia, Bangi, Malaysia

Nur Khairun Atiyah Sagee Ahmad
Department of Applied Physics, Faculty of Science and Technology, Universiti Kebangsaan Malaysia, UKM Bangi, Selangor, Malaysia

Subki Ahmad
Environmental Health and Industrial Safety Program, Faculty of Health Sciences, Universiti Kebangsaan Malaysia, Jalan Raja Muda Abdul Aziz, Kuala Lumpur, Malaysia

Anin Sofya Mohd. Akhiri
Department of Applied Physics, Faculty of Science and Technology, Universiti Kebangsaan Malaysia, UKM Bangi, Selangor, Malaysia

Naif H. Al-Hardan
Department of Applied Physics, Faculty of Science and Technology, Universiti Kebangsaan Malaysia, UKM Bangi, Selangor, Malaysia

Roel Apas
Davao del Sur State College, Matti, Digos City, Philippines

A. Atiqah
Institute of Microengineering and Nanoelectronics, Universiti Kebangsaan Malaysia, UKM Bangi, Selangor, Malaysia

Rozidawati Awang
Department of Applied Physics, Faculty of Science and Technology, Universiti Kebangsaan Malaysia, UKM Bangi, Selangor Darul Ehsan, Malaysia

Azrina Abd Aziz
Faculty of Civil Engineering Technology, Universiti Malaysia Pahang, Gambang, Malaysia

Della Grace Bacaltos
Davao del Sur State College, Matti, Digos City, Philippines

Khairunnisa Bahaman
Environmental Health and Industrial Safety Program, Faculty of Health Sciences, Universiti Kebangsaan Malaysia, Jalan Raja Muda Abdul Aziz, Kuala Lumpur, Malaysia

M. A. Bakar
Institute of Microengineering and Nanoelectronics, Universiti Kebangsaan Malaysia, UKM Bangi, Selangor, Malaysia

Alberto Bandiola
Provincial Environment and Natural Resources Office, Digos City, Philippines

Basundhara Bhattarai
Institute for Study and Development Worldwide, Sydney, Australia

Arup Kumar Biswas
Faculty of Environmental Management, Prince of Songkhla University, Hat Yai, Songkhla, Thailand; Hydrocarbon Unit, Energy and Mineral Resources Division, Government of the People's Republic of Bangladesh, Dhaka, Bangladesh

Suhaila M. Buhari
Department of Physics, Faculty of Science, Universiti Teknologi Malaysia, UTM, Skudai, Johor, Malaysia

Jeff Camkin
Institute for Study and Development Worldwide, Sydney, Australia

Siverlyn Camposano
Davao del Sur State College, Matti, Digos City, Philippines

S. Domi Caroline
Department of Electronics and Communication Engineering, Kings Engineering College, Irungattukottai, Chennai, Tamil Nadu, India

Md. Shahariar Chowdhury
Research Center on Industrial Ecology in Energy Business, Faculty of Environmental Management, Prince of Songkhla University, Hat Yai, Songkhla, Thailand; Faculty of Health and Life Sciences, INTI International University, Persiaran Perdana BBN, Putra Nilai, Nilai, Negeri Sembilan, Malaysia

Megh Raj Dangal
School of Arts, Kathmandu University, Kathmandu, Nepal; School of Engineering, King Mongkut's Institute of Technology, Ladkrabang, Thailand

Nazim Uddin Emon
Department of Pharmacy, Faculty of Biological Science, International Islamic University Chittagong, Chittagong, Bangladesh; Department of Public Health, School of Science and Technology, Bangladesh Open University, Gazipur, Bangladesh

S. Domi Evangeline
Department of Computer Science and Engineering, Kings Engineering College, Irungattukottai, Chennai, Tamil Nadu, India

Mohd. Firdaus-Raih
Department of Applied Physics, Faculty of Science and Technology, Universiti Kebangsaan Malaysia, UKM Bangi, Selangor, Malaysia

Cendana Fitrahanjani
Biostatistics and Demography Department, Faculty of Public Health, Universitas Airlangga, Surabaya, Indonesia

Augie Fuentes
Davao del Sur State College, Matti, Digos City, Philippines

Saroj Gyawali
Sustainable Study and Research Institute, Kathmandu, Nepal

Ainul Hafiza Abdul Hair
Bioresources and Biorefinery Laboratory, Department of Applied Physics, Faculty of Science and Technology, Universiti Kebangsaan Malaysia, UKM Bangi, Selangor, Malaysia; Center of Foundation Studies, Univesiti Teknologi MARA, Cawangan Selangor, Kampus Dengkil, Dengkil, Selangor, Malaysia

Ermie Abdull Hamid
Environmental Health and Industrial Safety Program, Faculty of Health Sciences,
Universiti Kebangsaan Malaysia, Jalan Raja Muda Abdul Aziz, Kuala Lumpur, Malaysia

Muhammad Azmi Abdul Hamid
Department of Applied Physics, Faculty of Science and Technology, Universiti Kebangsaan Malaysia,
UKM Bangi, Selangor, Malaysia

Nurul Shazana Abdul Hamid
Department of Applied Physics, Faculty of Science and Technology, Universiti Kebangsaan Malaysia,
UKM Bangi, Selangor, Malaysia
Space Science Center (ANGKASA), Institute of Climate Change, Universiti Kebangsaan Malaysia,
UKM Bangi, Selangor, Malaysia

Md. Fuad Hasan
Mineral Processing Center (MPC), Institute of Mining, Mineralogy, and Metallurgy (IMMM),
Bangladesh Council of Scientific and Industrial Research (BCSIR), Joypurhat, Bangladesh

A. B. P. Ilhamsyah
College of Engineering, Universiti Tenaga Nasional, Jalan Ikram-Uniten, Kajang, Selangor, Malaysia;
Department of Applied Physics, Universiti Kebangsaan Malaysia, Bangi, Selangor, Malaysia

Md. Rafiqul Islam
Energy and Mineral Resources Division, Ministry of Power, Energy, and Mineral Resources,
Government of the People's Republic of Bangladesh, Dhaka, Bangladesh

Minhajul Islam
Department of Pharmacy, Faculty of Biological Science, International Islamic University Chittagong,
Chittagong, Bangladesh; Department of Pharmaceutical Science, North South University, Dhaka,
Bangladesh

Anuar Ithnin
Environmental Health and Industrial Safety Program, Faculty of Health Sciences,
Universiti Kebangsaan Malaysia, Jalan Raja Muda Abdul Aziz, Kuala Lumpur, Malaysia

Sharifah Nabihah Syed Jaafar
Department of Applied Physics, Faculty of Science and Technology, Universiti Kebangsaan Malaysia,
UKM Bangi, Selangor, Malaysia

A. Jalar
Department of Applied Physics, Faculty of Science and Technology, Universiti Kebangsaan Malaysia,
UKM Bangi, Selangor, Malaysia; Institute of Microengineering and Nanoelectronics, Universiti
Kebangsaan Malaysia, UKM Bangi, Selangor, Malaysia

Masnita Mat Jusoh
Center of Foundation Studies, Universiti Teknologi MARA Cawangan Selangor, Kampus Dengkil,
Dengkil, Selangor, Malaysia

Ain Zafirah Binti Kamaruddin
Department of Applied Physics, Faculty of Science and Technology, Universiti Kebangsaan Malaysia,
UKM Bangi, Selangor, Malaysia

Lim Kar Keng
School of Liberal Studies, Universiti Kebangsaan Malaysia, UKM Bangi, Selangor, Malaysia

Noor Hashimah Hashim Lim
Department of Urban and Regional Planning, Faculty of Built Environment, Universiti Malaya,
Kuala Lumpur, Malaysia; Research Center on Industrial Ecology in Energy Business,
Faculty of Environmental Management, Prince of Songkhla University, Songkhla, Thailand

Rodceal Malinao
Davao del Sur State College, Matti, Digos City, Philippines

Archielito Mallari
Provincial Environment and Natural Resources Office, Digos City, Philippines

Abdullah Al Mamun
Doctorate Degree Program in Public Health, Faculty of Public Health, Universitas Airlangga, Campus C, Surabaya, Indonesia; Center for Multidisciplinary and Contemporary Research, Rangpur, Bangladesh

Nor Hamizah Miswan
Department of Mathematical Sciences, Faculty of Science and Technology, Universiti Kebangsaan Malaysia, Bangi, Selangor, Malaysia

Jafar Sharafeldeen Albagir Mohammad
Faculty of Environmental Management, Prince of Songkhla University, Hat Yai, Songkhla, Thailand; Oil Exploration and Production Authority, Ministry of Energy and Petroleum, Republic of the Sudan

Roy Molina
Davao del Sur State College, Matti, Digos City, Philippines

Minhaj Uddin Monir
Department of Petroleum and Mining Engineering, Jashore University of Science and Technology, Jashore, Bangladesh; Energy Conversion Laboratory, Department of Petroleum and Mining Engineering, Jashore University of Science and Technology, Jashore, Bangladesh

Madihah Mujaini
College of Engineering, Universiti Tenaga Nasional, Jalan Ikram-Uniten, Kajang, Selangor, Malaysia; Department of Applied Physics, Universiti Kebangsaan Malaysia, Bangi, Selangor, Malaysia

Silmi Muna
Biostatistics and Demography Department, Faculty of Public Health, Universitas Airlangga, Surabaya, Indonesia

Asmaa Soheil Najm
Iraqi Oil Exploration Company, Ministry of Oil, Iraq

Hemant Ojha
Institute for Study and Development Worldwide, Sydney, Australia

A. A. Omar
Institute of Microengineering and Nanoelectronics, Universiti Kebangsaan Malaysia, UKM Bangi, Selangor, Malaysia

Fatin Saiha Omar
Department of Applied Physics, Faculty of Science and Technology, Universiti Kebangsaan Malaysia, UKM Bangi, Selangor Darul Ehsan, Malaysia

M. A. R. Othman
Department of Applied Physics, Faculty of Science and Technology, Universiti Kebangsaan Malaysia, Bangi, Selangor, Malaysia

Khamphe Phoungthong
Faculty of Environmental Management, Prince of Songkhla University, Hat Yai, Songkhla, Thailand

Tofan Agung Eka Prasetya
Health Department, Faculty of Vocational Studies, Universitas Airlangga, Campus B, Surabaya, East Java, Indonesia

Jovanie Putong
Davao del Sur State College, Matti, Digos City, Philippines

Doris Huai Xia Quay
Department of Applied Physics, Faculty of Science and Technology, Universiti Kebangsaan Malaysia, UKM Bangi, Selangor, Malaysia

M. A. A. Rahim
Department of Applied Physics, Faculty of Science and Technology, Universiti Kebangsaan Malaysia, Bangi, Selangor, Malaysia; Department of Physics, Faculty of Science, Universiti Putra Malaysia, Seri Kembangan, Selangor, Malaysia

Rozaidi Rasid
Department of Applied Physics, Faculty of Science and Technology, Universiti Kebangsaan Malaysia, Bangi, Selangor, Malaysia

V. Mahesh Kumar Reddy
Chaitanya Bharathi Institute of Technology, Proddatur, Andhra Pradesh, India

Muhammad Izz Rosli
Department of Applied Physics, Faculty of Science and Technology, Universiti Kebangsaan Malaysia, UKM Bangi, Selangor Darul Ehsan, Malaysia

Mohd. Adly Rahmat Saat
Department of Applied Physics, Faculty of Science and Technology, Universiti Kebangsaan Malaysia, Bangi, Selangor, Malaysia

Anthony Sales
Department of Science and Technology XI, Davao City, Philippines

Idahwati Sarudin
School of Physics, Universiti Sains Malaysia, USM, Pulau Pinang, Malaysia

Dilli Ram Sedai
Program of Sustainable Energy Management, Faculty of Environmental Management, Prince of Songkhla University, Songkhla, Thailand

Chan Chee Seng
Department of Artificial Intelligence, Faculty of Computer Science and Information Technology, University of Malaya, Kuala Lumpur, Malaysia

Nurul Raihan Mohd. Suib
Center of Foundation Studies, Universiti Teknologi MARA Cawangan Selangor, Kampus Dengkil, Dengkil, Selangor, Malaysia

R. Suresh
Department of Electronics and Communication Engineering, Kings Engineering College, Irungattukottai, Chennai, Tamil Nadu, India

Jarsey Krim Taub
Davao del Sur State College, Matti, Digos City, Philippines

Kuaanan Techato
Program of Sustainable Energy Management, Research Center on Industrial Ecology in Energy Business, Faculty of Environmental Management, Prince of Songkhla University, Hat Yai, Songkhla, Thailand; Faculty of Health and Life Sciences, INTI International University, Persiaran Perdana BBN, Putra Nilai, Nilai, Negeri Sembilan, Malaysia

M. Therasa
Department of Computer Science and Engineering, Panimalar Engineering College, Varadharajapuram, Poonamallee, Chennai, Tamil Nadu, India

Makatar Wae-Hayee
Department of Mechanical and Mechatronics Engineering, Faculty of Engineering, Prince of Songkhla University, Hat Yai, Songkhla, Thailand

Ratnaningtyas Wahyu Kusuma Wardani
Health Department, Faculty of Vocational Studies, Universitas Airlangga, Campus B, Surabaya, East Java, Indonesia

Chumpol Yuangyai
Department of Urban and Regional Planning, Faculty of Built Environment, Universiti Malaya, Kuala Lumpur, Malaysia

Nuttaya Yuangyai
Program of Sustainable Energy Management, Faculty of Environmental Management, Prince of Songkhla University, Songkhla, Thailand; Research Center on Industrial Ecology in Energy Business, Faculty of Environmental Management, Prince of Songkhla University, Songkhla, Thailand

Siti Najihah Zaini
Department of Applied Physics, Faculty of Science and Technology, Universiti Kebangsaan Malaysia, UKM Bangi, Selangor Darul Ehsan, Malaysia

Sarani Zakaria
Bioresources and Biorefinery Laboratory, Department of Applied Physics, Faculty of Science and Technology, Universiti Kebangsaan Malaysia, UKM Bangi, Selangor, Malaysia

Raihan Uz Zaman
Griffith Institute for Drug Discovery, Griffith University, Nathan QLD, Australia; Faculty of Traditional Thai Medicine, Prince of Songkhla University, Hat Yai, Songkhla, Thailand; Department of Pharmacy, Faculty of Biological Science, International Islamic University Chittagong, Chittagong, Bangladesh

Rashaduz Zaman
Department of Pharmacy, Faculty of Biological Science, International Islamic University Chittagong, Chittagong, Bangladesh; Department of Renewable Resources, University of Alberta, Edmonton, Canada

Preface

It is with great pleasure that we present to you *Engineering Design and Technical Applications of Physical Science: An Integrated Approach for Sustainable Development*. This book is the result of a collective effort, stemming from the Physical and Integrated Sciences Conference (PISCON) 2022, organized by the Department of Applied Physics, Faculty of Science and Technology, Universiti Kebangsaan Malaysia. The conference provided a platform for researchers, experts, and enthusiasts to gather together and share their knowledge and insights on the pressing issues of sustainable development.

In this era of rapidly advancing technology and global challenges, it has become crucial to adopt an integrated approach that bridges the gap between scientific principles and their practical applications. The chapters compiled in this book reflect this need for synergy, covering a diverse range of topics within four main parts: Sustainable Composite Materials Technology, Emerging Materials for Sustainable Energy, Case Studies, Environment, Health, and Sustainable Development.

Part I–Sustainable Composite Materials Technology explores innovative methods and advancements in composite materials, addressing their role in sustainable development. From surface modifications of nanosheets to the mechanical properties of recycled blends, this section delves into cutting-edge research on composite materials and their applications.

Part II–Emerging Materials for Sustainable Energy focuses on the development of materials that contribute to sustainable energy solutions. Chapters on single-photon generation, battery-type electrodes, and thin film photovoltaic devices provide valuable insights into the potential of materials science in advancing sustainable energy technologies.

Part III–Case Studies presents real-world examples and in-depth analyses of various sustainable development initiatives. From geostatistical analysis for coal gasification to the impact of climate change on livestock production, these case studies shed light on the challenges and opportunities involved in implementing sustainable practices.

Lastly, Part IV–Environment, Health, and Sustainable Development examines the crucial interconnections between the environment, human health, and sustainable development. Chapters exploring indoor air quality, pharmacological investigations, and intelligible models for healthcare underscore the importance of holistic approaches in achieving sustainable development goals.

The contributions within these chapters are the result of the dedication and expertise of the conference's speakers and participants. Their valuable research, diverse perspectives, and shared experiences have created a comprehensive resource that will undoubtedly inspire researchers, engineers, policymakers, and students to further explore the field of sustainable development.

We extend our sincere gratitude to the authors for their outstanding contributions and to the conference organizers for their vision and commitment to fostering collaboration and knowledge exchange. We also express our appreciation to the reviewers and editors who meticulously reviewed and refined the chapters, ensuring the highest quality of content.

As editors, it is our hope that this book serves as a catalyst for further advancements in engineering design and technical applications of physical science for sustainable development. By embracing an integrated approach and leveraging the power of science and technology, we can collectively strive toward a more sustainable and prosperous future.

—Editors

PART I

Sustainable Composite Materials Technology

CHAPTER 1

Single-Photon Generation Assisted by a Nonlinear Material with High Purity and Yield

M. A. A. RAHIM[1,2] and M. A. R. OTHMAN[1]

[1]*Department of Applied Physics, Faculty of Science and Technology, Universiti Kebangsaan Malaysia, Bangi, Selangor, Malaysia*

[2]*Department of Physics, Faculty of Science, Universiti Putra Malaysia, Seri Kembangan, Selangor, Malaysia*

ABSTRACT

Single-photon sources that emit only one photon at a time with high yield are important for security and efficiency in quantum communication and computation, respectively. One method to develop the desired single-photon source is through entangled single-photon pairs from spontaneous parametric down-conversion (SPDC) crystals where detection of idler photons "heralds" the presence of signal photons. If the SPDC crystal is pumped with weak intensity laser, high purity heralded single photons can be produced at the cost of low yield. Increasing the yield by using stronger laser intensity inevitably decreases the purity due to the presence of multiple photons. Here, we propose a conceptual method to overcome this trade-off between single-photon purity and single-photon yield. Our setup consists of a weakly pumped SPDC crystal seeded with strong coherent states (CS) at the idler mode. Clicks on a click detector at the idler mode conditionally prepare a superposition state of zero-, one-, and two-photon Fock states (FS) at the signal mode. By employing repeated processes of infinitesimal displacement (ID) and two-photon elimination (PE), we can obtain a final

Engineering Design and Technical Applications of Physical Science: An Integrated Approach for Sustainable Development. Chin Hua Chia, Sarani Zakaria, Ruey Shan Chen, & Md. Shahariar Chowdhury (Eds.)

state of one-photon FS. Based on the heralding probability (single-photon yield) and single-photon fidelity (single-photon purity), we found that our method transforms the initial state with high heralding probability but low single-photon fidelity (near minimum) to a final state with high heralding probability and high single-photon fidelity (maximum). Furthermore, we discussed the conditions required for the experimental implementations of the processes. Thus, our work paves the way to producing heralded single photons with high purity and yield in SPDC sources.

1.1 INTRODUCTION

Single-photon-based quantum applications (Eisaman et al., 2011; Becher et al., 2022) require single-photon sources that emit only one photon at a time (no multi-photons) and with high yield (O'Brien, Furusawa, & Vučković, 2009). This is because the absence of multi-photons is important for security in quantum communication (Sangouard & Zbinden, 2012; Basso Basset et al., 2021) and efficiency in quantum computation (Varnava, Browne, & Rudolph, 2008; Nagao, Horikiri, & Sasaki, 2019). Moreover, a high single-photon yield ensures a high rate of information sharing and processing (Slussarenko & Pryde, 2019). Therefore, various methods are being pursued to achieve the desired single-photon sources, such as in the class of single-particle emitters: single atom or ion (Becher et al., 2022), solid-state emitters: quantum dot or carbon nanotube (Luo et al., 2019), and non-linear processes: spontaneous four-wave mixing or spontaneous parametric down-conversion (SPDC) (Dell'Anno, De Siena, & Illuminati, 2006; Couteau, 2018). SPDC sources emit photons in pairs simultaneously at idler and signal mode, in which detection of idler photons "heralds" the presence of signal photons. Although the photon pairs are generated probabilistically, they are entangled, which makes them important in quantum teleportation, metrology, and fundamental tests of quantum mechanics (Couteau, 2018). Furthermore, SPDC sources are preferable in a wide range of applications (Meyer-Scott, Silberhorn, & Migdall, 2020) because of their robustness, cost-effectiveness, operability at room temperature, and integrability in optical circuits (Christ & Silberhorn, 2012). However, in a conventional setup where a click detector is used for heralding at the idler mode, heralded photons at the signal mode suffer a trade-off between single-photon purity and single-photon yield (purity-yield trade-off), due to inevitable multi-photon emission in the SPDC process (Tang et al., 2021). As shown in Figure 1.1(a), an SPDC crystal pumped with an extremely weak intensity laser ensures high single-photon purity, since multi-photons are negligible. However, this leads to an extremely low single-photon yield, which is close to 0% (Christ & Silberhorn, 2012). In this 'click detector' method, attempting to increase the yield by increasing the intensity of the pump will degrade the purity, since the click detector fails to distinguish between single-photon and multi-photon events.

The problem of purity-yield trade-off can be overcome by replacing the click detector with a photon-number-resolving (PNR) detector (Christ & Silberhorn, 2012). Since the PNR detector can be set to herald only single photons, higher single-photon yield can be achieved by increasing the pump intensity, without affecting the high single-photon purity. Nevertheless, the maximum yield achievable in this 'PNR detector' method is only 25%, constrained by the thermal distribution of the photon pairs in the SPDC process (Christ & Silberhorn, 2012). Fortunately, by further combining or 'multiplexing' 17 SPDC sources (Christ & Silberhorn, 2012; Migdall, Branning, & Castelletto, 2002), a single-photon yield of up to >99% can be accomplished. In general, this 'multiplexed SPDC' method involves rerouting heralded photons to a single output mode, in which the photons are generated from a set of multiplexed units, either in space (multiple SPDC processes) or in time (multiple release times from a storage loop) (Bodog et al., 2020). Although the 'multiplexed SPDC' method offers a theoretically promising single-photon yield, it is practically difficult to implement, since it requires a set of advanced devices that are not currently available and are still in development, such as a sensitive PNR detector and a lossless photon router (Meyer-Scott, Silberhorn, & Migdall, 2020; Ahumada et al., 2019; Nehra et al., 2020). Furthermore, optimum setups for the 'multiplexed SPDC' method are still actively investigated (Bodog et al., 2016; Kaneda & Kwiat, 2019; Bodog et al., 2020). Thus, achieving high single-photon purity and single-photon yield in SPDC sources remains a challenging task.

Here, we propose an alternative method to overcome the purity-yield trade-off in the 'click detector' method, by using displacement operation and photon elimination. Firstly, the displacement operation is based on the displacement operator $\hat{D}(\beta) = e^{\beta\hat{a}^\dagger - \beta^*\hat{a}}$, where applying to n-photon Fock states (FS) $|n\rangle$ generates a displaced n-photon FS (de Oliveira et al., 1990). Recent practical applications of displacement operation are in entanglement demonstration (Bruno et al., 2013; Lvovsky et al., 2013), quantum teleportation (Podoshvedov, 2008) and complete Bell state measurement (Sergey A. Podoshvedov, Kim, & Lee, 2008). Secondly, photon elimination is based on eliminating single or multiple-photon FS from a superposition state. Recently, n-photon FS-filtered coherent states (CS) resulting from eliminating a single n-photon FS from a CS $|\alpha\rangle$ have been studied (Meher

& Sivakumar, 2018). This type of elimination has applications such as in non-classical state generation (Meher & Sivakumar, 2018), quantum key distribution (Meher & Sivakumar, 2018) and optical interferometry (Meher & Sivakumar, 2020). Other possible ways of photon elimination are through quantum scissors (Pegg, Phillips, & Barnett, 1998) or destructive interference (Lu & Ou, 2001). Our proposed method employs infinitesimal displacement (ID), where the amplitude of the displacement is infinitesimal, $d\beta \rightarrow 0$, and two-photon elimination (PE), where two-photon FS $|2\rangle$ is eliminated. A schematic diagram of our proposed method is depicted in Figure 1.1(b), where a heralded state in the signal mode of a seeded SPDC setup is manipulated with a single PE, followed by multiple ID-then-PE, to achieve high purity and yield of single photons.

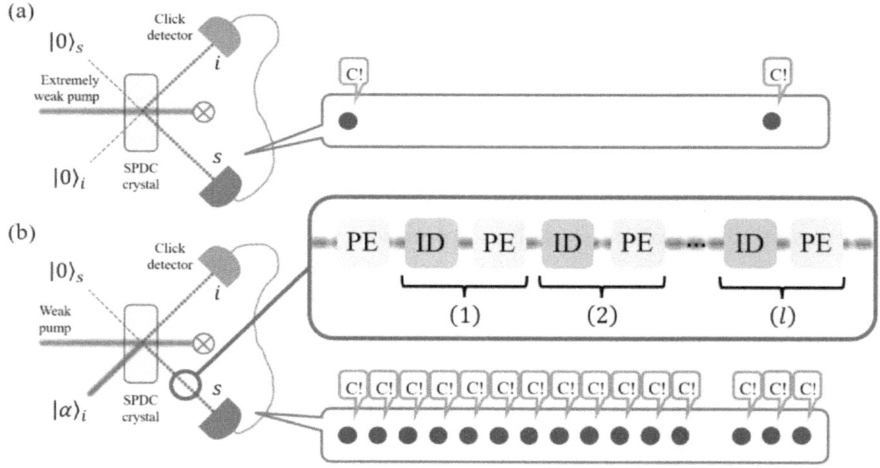

FIGURE 1.1 Schematic diagrams of (a) the 'click detector' method and (b) the proposed method to achieve high purity and yield of single photons. (C) represent clicks from the click detector at the idler mode and red circles represent single photons detected at the signal mode.

1.2 PROPOSED METHOD

1.2.1 IDLER MODE MANIPULATION: SUPERPOSITION STATE PREPARATION

Our method starts with the preparation of a superposition state from an SPDC setup. As shown in Figure 1.1(b), consider an SPDC crystal pumped with a weak pulsed pump and seeded with a pulsed CS in the idler mode (Resch,

Lundeen, & Steinberg, 2002). Hence, the output state $|\Psi\rangle$ at the signal and idler modes can be represented by

$$\begin{aligned}|\Psi\rangle &\approx \left(1+g\left(\hat{a}_s^\dagger\hat{a}_i^\dagger-\hat{a}_s\hat{a}_i\right)+\frac{g^2}{2}\left(\hat{a}_s^\dagger\hat{a}_i^\dagger-\hat{a}_s\hat{a}_i\right)^2\right)|0\rangle_s|\alpha\rangle_i \\ &=|0\rangle_s|\alpha\rangle_i-\frac{g^2}{2}|0\rangle_s\,\hat{a}_i\hat{a}_i^\dagger|\alpha\rangle_i+g|1\rangle_s\,\hat{a}_i^\dagger|\alpha\rangle_i+\frac{g^2}{\sqrt{2}}|2\rangle_s\,\hat{a}_i^{\dagger 2}|\alpha\rangle_i\end{aligned} \tag{1}$$

where g is proportional to the pump strength and the second-order nonlinear susceptibility of the SPDC crystal (Kalamidas, Gerry, & Benmoussa, 2008). Since the pump is weak, we have assumed here $g^v \approx 0$ when $v \geq 3$ (Kalamidas, Gerry, & Benmoussa, 2008). The subscripts s and i represent signal and idler mode, respectively. For extremely strong CS, $|\alpha|^2 \to \infty$, we may approximate all the three states $\hat{a}_i\hat{a}_i^\dagger|\alpha\rangle_i$, $\hat{a}_i^\dagger|\alpha\rangle_i$ and $\hat{a}_i^{\dagger 2}|\alpha\rangle_i$ with $|\alpha\rangle_i$, since the fidelity f between the three states with CS is very high (near unity), as shown in Figure 1.2. Hence, the output state in Eq. (1) is well approximated by;

$$|\Psi\rangle=\left(1-\frac{g^2}{2}\right)|0\rangle_s|\alpha\rangle_i+g|1\rangle_s|\alpha\rangle_i+\frac{g^2}{\sqrt{2}}|2\rangle_s|\alpha\rangle_i. \tag{2}$$

For an ideal click detector at the idler mode, its positive operator value measure for no-click and click are given by $\hat{\Pi}_{i,\text{"no-click"}}=|0\rangle_{i\,i}\langle 0|$ and $\hat{\Pi}_{i,\text{"click"}}=1-|0\rangle_{i\,i}\langle 0|$, respectively (Silberhorn, 2007). By normalizing the state in Eq. (2) with normalization constant $\mathcal{N}$, we derive the heralding probability *HP*, which is the probability of a successful heralding event per pulse (Christ & Silberhorn, 2012), to be;

$$HP=\langle\Psi|\hat{\Pi}_{i,\text{"click"}}|\Psi\rangle\approx 1. \tag{3}$$

From Eq. (3), the heralding probability is approximately unity because, for each pulse, the click detector at the idler mode most probably detects photons from the extremely strong CS.

As shown in Figure 1.1(b), the click detector at the idler mode will herald the state at the signal mode. Hence, by using the formula $|\Psi'\rangle\langle\Psi'|=\dfrac{Tr_i\left(\hat{\Pi}_{i,\text{"click"}}|\Psi\rangle\langle\Psi|\right)}{\langle\Psi|\hat{\Pi}_{i,\text{"click"}}|\Psi\rangle}$ (Christ & Silberhorn, 2012), the heralded state at the signal mode $|\Psi'\rangle\langle\Psi'|$ can be reduced to a superposition state of

$$|\Psi'\rangle=\left(1-\frac{g^2}{2}\right)|0\rangle_s+g|1\rangle_s+\frac{g^2}{\sqrt{2}}|2\rangle_s. \tag{4}$$

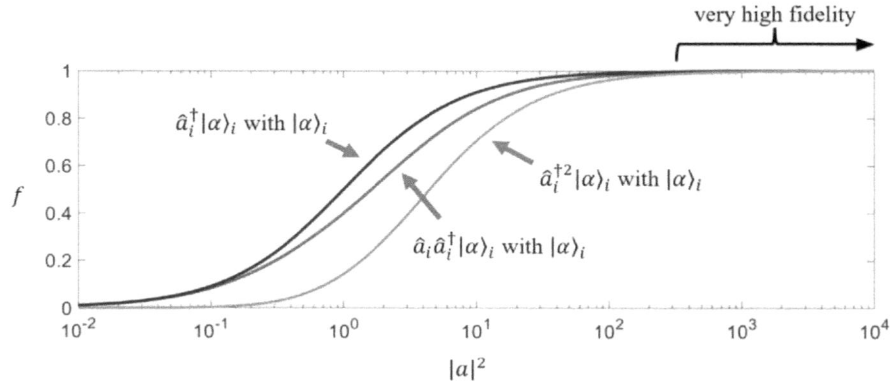

FIGURE 1.2 Fidelity between $\hat{a}_i\hat{a}_i^\dagger|\alpha\rangle_i$, $\hat{a}_i^\dagger|\alpha\rangle_i$ and $\hat{a}_i^{\dagger 2}|\alpha\rangle_i$ with $|\alpha\rangle_i$. Very high fidelity was obtained for large $|\alpha|^2$.

The superposition state in Eq. (4) means that for each click from the click detector at the idler mode, the heralded state at the signal mode is superposed between the states $|0\rangle_s$, $|1\rangle_s$, and $|2\rangle_s$.

Next, to quantify the single-photon purity of the heralded state at the signal mode, we derive the fidelity between the normalized state in Eq. (4) with one-photon FS

$$f = |\langle 1|\Psi'\rangle|^2 = N^2 g^2. \tag{5}$$

Eq. (5) suggests that the fidelity is very small since the state in Eq. (4) has a very small probability of $|1\rangle_s$. Hence, from both Eq. (3) and (5), it is observed that the heralded state has a very low single-photon purity, although the heralding probability is very high. The heralding probability only represents the single-photon yield of an SPDC source if the fidelity is unity. Therefore, the main goal afterward is to manipulate the state in Eq. (4) to obtain a fidelity of one.

1.2.2 SIGNAL MODE MANIPULATION: ZERO- AND TWO-PHOTON FS ELIMINATION

We begin our manipulation in signal mode by first applying PE to the superposition state in Eq. (4) so that the unwanted $|2\rangle_s$ can be eliminated. Hence, the PE transforms the state in Eq. (4) to a qubit state (Pegg, Phillips, & Barnett, 1998).

$$|\Psi'\rangle = \left(1-\frac{g^2}{2}\right)|0\rangle_s + g|1\rangle_s . \tag{6}$$

The transformation only eliminates $|2\rangle_s$ without altering the coefficients of $|0\rangle_s$ and $|1\rangle_s$. It is important to note that $|2\rangle_s$ is not eliminated from the approximation made earlier ($g^v \approx 0$ when $v \geq 3$) but through an optical process. Now, consider the displacement operator $\hat{D}(\beta)$ with $\theta_\beta = 0$. For infinitesimal amplitude of the displacement where $|d\beta|^w \approx 0$ when $w \geq 2$, application of ID $\hat{D}(d\beta)$ on the state in Eq. (6) results in (Alex & Mlynek, 2002; Resch, Lundeen, & Steinberg, 2002; Bimbard et al., 2010; Zhang & Zhang, 2018).

$$\begin{aligned}
\hat{D}(d\beta)|\Psi'\rangle &= |\Psi'\rangle_1 = \left(1-\frac{g^2}{2}\right)|d\beta\rangle_s + g|d\beta:1\rangle_s \\
&= \left(1-\frac{g^2}{2}\right)\left(|0\rangle_s + |d\beta||1\rangle_s\right) + g\left(-|d\beta||0\rangle_s + |1\rangle_s + \sqrt{2}|d\beta||2\rangle_s\right) \\
&= \left(1-\frac{g^2}{2} - g|d\beta|\right)|0\rangle_s + \left(\left(1-\frac{g^2}{2}\right)|d\beta| + g\right)|1\rangle_s + \sqrt{2}g|d\beta||2\rangle_s
\end{aligned} \tag{7}$$

where subscript 1 in $|\Psi'\rangle_1$ means $|\Psi'\rangle$ is displaced 1 time and $|d\beta{:}1\rangle_s$ is the displaced one-photon FS (de Oliveira et al., 1990; Lvovsky & Babichev, 2002). Based on the PE and ID that result in Eq. (6) and Eq. (7), respectively, applying the state in Eq. (6) with multiple numbers of ID-then-PE as shown in Figure 1.1(b) results in:

$$|\Psi'\rangle_l = \left(1-\frac{g^2}{2} - lg|d\beta|\right)|0\rangle_s + \left(l\left(1-\frac{g^2}{2}\right)|d\beta| + g\right)|1\rangle_s . \tag{8}$$

The manipulation causes the coefficient of the qubit state from only dependent on g in Eq. (6) to dependent on g, $|d\beta|$, and l in Eq. (8). The resulting state in Eq. (8) has no $|2\rangle_s$ because each time after ID, PE will eliminate the $|2\rangle_s$ from the superposition state.

Finally, by imposing the coefficient for $|0\rangle_s$ in Eq. (8) to be 0, $1-\frac{g^2}{2} - lg|d\beta| = 0$, we obtain the relation at the limit of large l

$$|d\beta| = \frac{1}{l}\left(\frac{1}{g} - \frac{g}{2}\right). \tag{9}$$

Large l is necessary because the infinitesimal $|d\beta|$ is inversely proportional to l. Eq. (9) means that if the parameters $|d\beta|$, l, and g are related based on Eq. (9), the $|0\rangle_s$ in the state of Eq. (8) will be eliminated, leaving only $|1\rangle_s$. Imposing the condition $0|0\rangle_s$ is analogous to the case of destructive two-photon interference $0|2\rangle$ to obtain anti-bunching light (Koashi, Matsuoka, & Hirano, 1996; Lu & Ou, 2001; Pittman, Franson, & Jacobs, 2007). Furthermore, eliminating $|0\rangle$ from manipulation of its coefficient is common in quantum optics, such as in odd Schrodinger's cat state (Dodonov & Malkin, 1974; Lee & Kim, 2019), where all even number FS (including $|0\rangle$) in a superposed CS are eliminated, or in truncated odd Schrodinger's cat state (Miranowicz et al., 2014), where for initial qubit state of $|0\rangle + |1\rangle$, $|0\rangle$ is eliminated, leaving only $|1\rangle$.

As shown in Figure 1.3, we have plotted Eq. (9) for several values of $g \geq 0.30$. It is shown that for $|d\beta| < 0.1$, higher g requires a smaller minimal number of l so that $|0\rangle_s$ can be eliminated from the state in Eq. (8). Finally since the heralded state at the signal mode is $|\Psi'\rangle = |1\rangle_s$ after elimination of $|0\rangle_s$, the single-photon fidelity of the state is then $f = 1$, which is our main goal here. Thus, since $f = 1$ and $HP \approx 1$ from Eq. (3), we have finally obtained theoretically heralded single photons with high single-photon purity and single-photon yield.

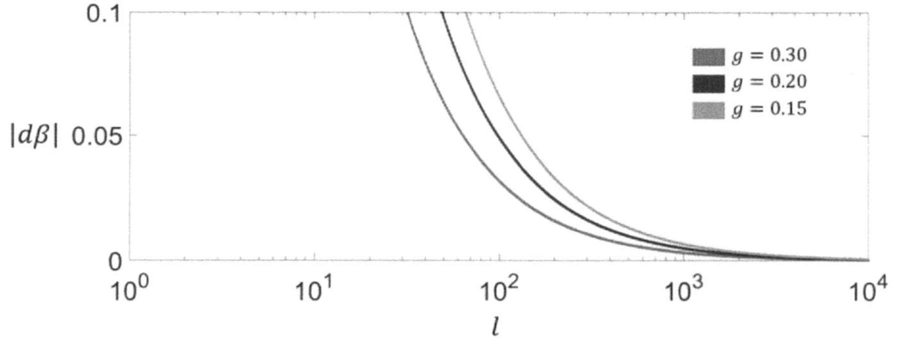

FIGURE 1.3 Plot of $|d\beta|$ against l with several values of g. A larger l is required for smaller $|d\beta|$ and g.

1.3 IMPLEMENTATION

We now discuss the possible experimental implementations of ID and PE processes for our method. Firstly, we found out that both ID and PE processes, which happen in signal mode, must not introduce additional heralding or conditional measurements. This is to ensure the high heralding probability obtained before the processes due to the seeded strong CS at idler mode is conserved. Furthermore, the PE process, must not alter the coefficient of zero- and one-photon FS. This is to ensure that only two-photon elimination occurs and does not increase (decrease) the coefficient of zero- (one-) photon FS.

Based on previous works, the ID process has been experimentally demonstrated using a highly reflective beam splitter with strong CS and one-photon FS as inputs to generate displaced one-photon FS (Lvovsky & Babichev, 2002). Fortunately, this experimental method follows the conditions mentioned above. On the other hand, the experimental method for the PE process was proposed based on a filter (Meher & Sivakumar, 2018), scissors (Pegg, Phillips, & Barnett, 1998), or interference (Lu & Ou, 2001). However, both the filter and scissor methods require a heralding process, hence they are not suitable for our method. For interference, the method requires no heralding process and does not alter the coefficient of zero- and one-photon FS. Unfortunately, the method is only applicable to the initial state of weak CS. Thus, at the moment, there is no known experimental method that is suitable for our PE process.

1.4 CONCLUSION

In conclusion, we have proposed an alternative method to overcome the purity-yield trade-off in SPDC setup, independent of advanced devices from the 'PNR detector' method and 'multiplexed SPDC' method. By manipulating a superposition state at the signal mode from a seeded SPDC crystal using PE and multiple ID-then-PE, we theoretically obtained heralded single photons with high single-photon purity and single-photon yield. Apparently, the roles of ID and PE in the manipulation are to eliminate $|0\rangle_s$ and $|2\rangle_s$ from the superposition state, respectively. Furthermore, to obtain the high purity and yield single photons, the parameters of pump strength g, ID strength $|d\beta|$, and repetition number l of ID-then-PE must be related to each other. For implementation, we have outlined some conditions required for the ID and PE processes, where only experimental demonstration of the ID process passes the conditions. Hence, future works include finding a suitable implementation for the PE process. Finally, our method paves the way for an alternative single-photon source that is useful for quantum optical applications.

ACKNOWLEDGMENT

The authors acknowledge the Fundamental Research Grant Scheme (FRGS), grant number FRGS/1/2020/STG07/UKM/02/2 funded by the Ministry of Higher Education Malaysia.

KEYWORDS

- **infinitesimal displacement**
- **quantum communication**
- **quantum computation**
- **single-photon source**
- **spontaneous parametric down-conversion**

REFERENCES

Ahumada, M., Orellana, P. A., Domínguez-Adame, F., & Malyshev, A. V. (2019). Tunable single-photon quantum router. *Physical Review A, 99*(3), 033827. https://doi.org/10.1103/PhysRevA.99.033827.

Basso Basset, F., Valeri, M., Roccia, E., Muredda, V., Poderini, D., Neuwirth, J., Spagnolo, N., Rota, M. B., Carvacho, G., Sciarrino, F., & Trotta, R. (2021). Quantum key distribution with entangled photons generated on demand by a quantum dot. *Science Advances, 7*(12), 1–8. https://doi.org/10.1126/sciadv.abe6379.

Becher, C., Höfling, S., Liu, J., Michler, P., Pernice, W., & Toninelli, C. (2022). Special topic on non-classical light emitters and single-photon detectors. *Applied Physics Letters, 120*(1), 010401. https://doi.org/10.1063/5.0078886.

Bimbard, E., Jain, N., MacRae, A., & Lvovsky, A. I. (2010). Quantum-optical state engineering up to the two-photon level. *Nature Photonics, 4*(4), 243–247. https://doi.org/10.1038/nphoton.2010.6.

Bodog, F., Adam, P., Mechler, M., Santa, I., & Koniorczyk, M. (2016). Optimization of periodic single-photon sources based on combined multiplexing. *Physical Review A, 94*(3), 033853. https://doi.org/10.1103/PhysRevA.94.033853.

Bodog, F., Mechler, M., Koniorczyk, M., & Adam, P. (2020). Optimization of multiplexed single-photon sources operated with photon-number-resolving detectors. *Physical Review A, 102*(1), 013513. https://doi.org/10.1103/PhysRevA.102.013513.

Bruno, N., Martin, A., Sekatski, P., Sangouard, N., Thew, R. T., & Gisin, N. (2013). Displacement of entanglement back and forth between the micro and macro domains. *Nature Physics, 9*(9), 545–548. https://doi.org/10.1038/nphys2681.

Christ, A., & Silberhorn, C. (2012). Limits on the deterministic creation of pure single-photon states using parametric down-conversion. *Physical Review A, 85*(2), 023829. https://doi.org/10.1103/PhysRevA.85.023829.
Couteau, C. (2018). Spontaneous parametric down-conversion. *Contemporary Physics, 59*(3), 291–304. https://doi.org/10.1080/00107514.2018.1488463.
Dell'Anno, F., De Siena, S., & Illuminati, F. (2006). Multiphoton quantum optics and quantum state engineering. *Physics Reports, 428*(2–3), 53–168. https://doi.org/10.1016/j.physrep.2006.01.004.
Dodonov, V. V., & Malkin, I. A. (1974). Even and odd coherent states and excitations of a singular oscillator. *Physica, 72*(4), 597–615.
Eisaman, M. D., Fan, J., Migdall, A., & Polyakov, S. V. (2011). Invited review article: Single-photon sources and detectors. *Review of Scientific Instruments, 82*(7), 071101. https://doi.org/10.1063/1.3610677.
Kalamidas, D., Gerry, C. C., & Benmoussa, A. (2008). Proposal for generating a two-photon added coherent state via down-conversion with a single crystal. *Physics Letters A, 372*(11), 1937–1940. https://doi.org/10.1016/j.physleta.2007.10.089.
Kaneda, F., & Kwiat, P. G. (2019). High-efficiency single-photon generation via large-scale active time multiplexing. *Science Advances, 5*(10), 1–7. https://doi.org/10.1126/sciadv.aaw8586.
Koashi, M., Matsuoka, M., & Hirano, T. (1996). Photon antibunching by destructive two-photon interference. *Physical Review A, 53*(5), 3621–3624. https://doi.org/10.1103/PhysRevA.53.3621.
Lee, S. W., & Kim, J. (2019). Single-photon characteristics of superposed weak coherent states. *Physical Review A, 99*(1), 013847. https://doi.org/10.1103/PhysRevA.99.013847.
Lu, Y. J., & Ou, Z. Y. (2001). Observation of nonclassical photon statistics due to quantum interference. *Physical Review Letters, 88*(2), 023601. https://doi.org/10.1103/PhysRevLett.88.023601.
Luo, Y., He, X., Kim, Y., Blackburn, J. L., Doorn, S. K., Htoon, H., & Strauf, S. (2019). Carbon nanotube color centers in plasmonic nanocavities: A path to photon indistinguishability at telecom bands. *Nano Letters, 19,* 9037–9044. https://doi.org/10.1021/acs.nanolett.9b04069.
Lvovsky, A. I., & Babichev, S. A. (2002). Synthesis and tomographic characterization of the displaced Fock state of light. *Physical Review A, 66*(1), 011801. https://doi.org/10.1103/PhysRevA.66.011801.
Lvovsky, A. I., & Mlynek, J. (2002). Quantum-optical catalysis: Generating nonclassical states of light by means of linear optics. *Physical Review Letters, 88*(25), 250401. https://doi.org/10.1103/PhysRevLett.88.250401.
Lvovsky, A. I., Ghobadi, R., Chandra, A., Prasad, A. S., & Simon, C. (2013). Observation of micro-macro entanglement of light. *Nature Physics, 9*(9), 541–544. https://doi.org/10.1038/nphys2682.
Meher, N., & Sivakumar, S. (2018). Number state filtered coherent states. *Quantum Information Processing, 17*(9), 233. https://doi.org/10.1007/s11128-018-1995-6.
Meher, N., & Sivakumar, S. (2020). Enhancing phase sensitivity with number state filtered coherent states. *Quantum Information Processing, 19*(2), 51. https://doi.org/10.1007/s11128-019-2553-6.
Meyer-Scott, E., Silberhorn, C., & Migdall, A. (2020). Single-photon sources: Approaching the ideal through multiplexing. *Review of Scientific Instruments, 91*(4), 041101. https://doi.org/10.1063/5.0003320.
Migdall, A. L., Branning, D., & Castelletto, S. (2002). Tailoring single-photon and multiphoton probabilities of a single-photon on-demand source. *Physical Review A, 66*(5), 053805. https://doi.org/10.1103/PhysRevA.66.053805.

Miranowicz, A., Paprzycka, M., Pathak, A., & Nori, F. (2014). Phase-space interference of states optically truncated by quantum scissors: Generation of distinct superpositions of qudit coherent states by displacement of vacuum. *Physical Review A, 89*(3), 033812. https://doi.org/10.1103/PhysRevA.89.033812.

Nagao, K., Horikiri, T., & Sasaki, T. (2019). Blind quantum computation with a heralded single-photon source. *Physical Review A, 99*(4), 042324. https://doi.org/10.1103/PhysRevA.99.042324.

Nehra, R., Chang, C. H., Yu, Q., Beling, A., & Pfister, O. (2020). Photon-number-resolving segmented detectors based on single-photon avalanche-photodiodes. *Optics Express, 28*(3), 3660. https://doi.org/10.1364/OE.380416.

O'Brien, J. L., Furusawa, A., & Vučković, J. (2009). Photonic quantum technologies. *Nature Photonics, 3*(12), 687–695. https://doi.org/10.1038/nphoton.2009.229.

Oliveira, F. A. M. de, Kim, M. S., Knight, P. L., & Buek, V. (1990). Properties of displaced number states. *Physical Review A, 41*(5), 2645–2652. https://doi.org/10.1103/PhysRevA.41.2645.

Pegg, D. T., Phillips, L. S., & Barnett, S. M. (1998). Optical state truncation by projection synthesis. *Physical Review Letters, 81*(8), 1604–1606. https://doi.org/10.1103/PhysRevLett.81.1604.

Pittman, T. B., Franson, J. D., & Jacobs, B. C. (2007). Investigation of a single-photon source based on quantum interference. *New Journal of Physics, 9*(6), 195. https://doi.org/10.1088/1367-2630/9/6/195.

Podoshvedov, S. A. (2008). Quantum teleportation through an entangled state composed of displaced vacuum and single-photon states. *Journal of Experimental and Theoretical Physics, 106*(3), 435–441. https://doi.org/10.1134/S1063776108030035.

Podoshvedov, S. A., Kim, J., & Lee, J. (2008). Generation of a displaced qubit and entangled displaced photon state via conditional measurement and their properties. *Optics Communications, 281*(14), 3748–3754. https://doi.org/10.1016/j.optcom.2008.03.073.

Resch, K. J., Lundeen, J. S., & Steinberg, A. M. (2002). Quantum state preparation and conditional coherence. *Physical Review Letters, 88*(11), 113601. https://doi.org/10.1103/PhysRevLett.88.113601.

Sangouard, N., & Zbinden, H. (2012). What are single photons good for? *Journal of Modern Optics, 59*(17), 1458–1464. https://doi.org/10.1080/09500340.2012.687500.

Silberhorn, C. (2007). Detecting quantum light. *Contemporary Physics, 48*(3), 143–156. https://doi.org/10.1080/00107510701662538.

Slussarenko, S., & Pryde, G. J. (2019). Photonic quantum information processing: A concise review. *Applied Physics Reviews, 6*(4), 041303. https://doi.org/10.1063/1.5115814.

Tang, J., Tang, L., Wu, H., Wu, Y., Sun, H., Zhang, H., Li, T., Lu, Y., Xiao, M., & Xia, K. (2021). Towards on-demand heralded single-photon sources via photon blockade. *Physical Review Applied, 15*(6), 064020. https://doi.org/10.1103/PhysRevApplied.15.064020.

Varnava, M., Browne, D. E., & Rudolph, T. (2008). How good must single photon sources and detectors be for efficient linear optical quantum computation? *Physical Review Letters, 100*(6), 060502. https://doi.org/10.1103/PhysRevLett.100.060502.

Zhang, S., & Zhang, X. (2018). Photon catalysis acting as noiseless linear amplification and its application in coherence enhancement. *Physical Review A, 97*(4), 043830. https://doi.org/10.1103/PhysRevA.97.043830.

CHAPTER 2

Surface Modification of Cobalt-Manganese Sulfide Nanosheets by Cobalt Hydroxide as Battery-Type Electrode for Supercapattery Application

MUHAMMAD IZZ ROSLI, SITI NAJIHAH ZAINI, FATIN SAIHA OMAR, and ROZIDAWATI AWANG

Department of Applied Physics, Faculty of Science and Technology, Universiti Kebangsaan Malaysia, UKM Bangi, Selangor Darul Ehsan, Malaysia

ABSTRACT

Nowadays, the speedy need for carbon neutrality has pressed the demand to produce and develop sustainable energy sources. Since supercapacitors (SCAP) manifest high-power/energy densities, ultrafast charge/discharge rates, and super long-life cycles, they have drawn much attention from scientists to date. In this work, a free-binding layered double hydroxide compound (CoMnS-$Co(OH)_2$) was synthesized through a facile hydrothermal method. The presence of metal hydroxide as a second layer serves as an additional platform for redox activities to take place. X-ray diffraction (XRD) proved that the CoMnS-Co (OH) compound has a crystalline structure. Moreover, the morphology and structural studies reveal that the crystal compound formed a flower-like petal and nanolumps nanoarchitecture. For electrochemical studies, the CoMnS-$Co(OH)_2$) electrode produced a maximum of 777.5 C g^{-1} (with a rate capability of 64.31%), higher than that of the CoMnS electrode (330 C g^{-1} with a rate capability of 68.23%) at 1 A/g. In short, the presence of the metal hydroxide layer greatly ameliorates

Engineering Design and Technical Applications of Physical Science: An Integrated Approach for Sustainable Development. Chin Hua Chia, Sarani Zakaria, Ruey Shan Chen, & Md. Shahariar Chowdhury (Eds.)

the electrochemical adsorption surface area and high specific capacity of the electrodes. However, further investigation is needed to elucidate the effect of Co $(OH)_2$ on the electrical conductivity of the CoMnS electrode.

2.1 INTRODUCTION

The global crisis of carbon emissions has been seen as a potential threat to human civilizations. The increasing severe weather, changes in food supply, geographical alterations, etc., have had a significant impact on nations worldwide. Yet, the continuous employment of conventional fossil fuel combustion as a primary energy source might lead to catastrophic consequences (Anggoro, Udaibah, Prasetyaningrum, & Zakaria, 2022). There are a few suggestions and options to name alternative energy sources for daily usage, namely solar energy systems, wind turbine energy, hydropower, biogas plants, etc. Nevertheless, these systems require structured and efficient energy storage for sustainable continuity without seasonal interruption (Şahin, Blaabjerg, & Sangwongwanich, 2022). Thus, to date, various efforts have been made to urge and uplift the empowerment of highly proficient energy ability and eco-friendly alternative energy sources (renewable energy) to help decarbonize. Among those, electrochemical energy storage (EES) such as supercapacitors and supercapatteries (SCAP) are seen as potential nominees to fulfill the needs (Lemian & Bode, 2022).

Supercapacitors can be categorized into two forms: electrostatic double-layer capacitors (EDLC) and pseudocapacitors. Nevertheless, these types of supercapacitors suffer from low energy density (Chen, 2017). Herein, SCAP offers a better choice in terms of both energy and power densities. They operate by the combination of faradaic reactions and electrostatic movement due to the utilization of different types of electrodes within the system (capacitive and battery-type electrodes). With the hybrid solution in this method, SCAPs are expected to have both attributes from supercapacitors and batteries in terms of high storage capacity, fast charging/discharging rate, and prolonged cyclic stability. In order to produce a vast quantity of SCAP devices, designing advanced electrode materials is critical. The performance of these devices is influenced by the number of active sites, porosity, morphology, crystallinity, and rate of interfacial electron transportation of their electrodes (Omar et al., 2018; Ramly, Omar, Chanlek, Aspanut, & Goh, 2022). Therefore, metal sulfides can meet the mentioned demand since they have been extensively studied for their well-known strong redox nature and high electrical conductivity.

Metal sulfides are admirable, especially as faradaic electrodes, as they possess oxidation states that can encourage high faradaic reactions, and hence, high specific energy of SCAP. However, single metal sulfides have low-rate capability attributed to their considerably low electrical conductivity. The formation of binary composites of metal sulfides is expected to offer higher specific capacity than single-metal sulfides due to the coexistence of different elements (and thus, oxidation states). Cobalt hydroxide (Co $(OH)_2$) has high theoretical capacitance, is cost-effective, eco-friendly, and abundantly available. It has been widely investigated for SCAP development (Ramly, Omar, Chanlek, Aspanut, & Goh, 2022; Rovetta, Browne, Harvey, Godwin, Coleman, & Lyons, 2017). Nevertheless, there is a lack of critical appraisal within this research to date. In this work, the discussion on the structure, morphologies, and performances of the free-binding CoMnS-Co $(OH)_2$ compound via a facile hydrothermal method was further analyzed. This research focuses on the effect of the growth compound on the electrode to increase the specific capacitance, energy density, and power density of SCAP devices for further development into EES applications.

2.2 METHOD

2.2.1 MATERIAL

Manganese (II) chloride dihydrate ($Cl_2H_4MnO_2$), cobalt chloride hexahydrate ($CoCl_2$), thiourea (CH_4N_2S), and potassium hydroxide (KOH) were purchased at Sigma-Aldrich, Malaysia. Deionized (DI) water was used throughout the experiment.

2.2.2 PREPARATION OF COMNS

The mixture of $Cl_2H_4MnO_2$, $CoCl_2$, and CH_4N_2S was stirred for 30 minutes to reach homogeneity. It was then poured into a Teflon autoclave reactor together with a nickel (Ni) foam electrode sheet (1.5 × 1.5 cm) before hydrothermal treatment took place (140 °C, 6 hours). After that, the Ni foam electrode was rinsed using DI water and dried overnight (50 °C) in the oven. It was then cooled down to ambient temperature and labeled as CoMnS electrode.

2.2.3 PREPARATION OF COMNS-CO $(OH)_2$

The CoMnS electrode was then immersed in a cobalt salt solution (1 mM). The mixture of CoMnS electrode and salt solution was then heated in an oven at 100 °C for 6 hours. After that, the electrode was rinsed using DI water and dried overnight. It was then cooled down to ambient temperature and labeled as a free-binding CoMnS-Co $(OH)_2$ electrode. The concentration of the Co salt solutions was varied (2 mM, 3 mM, 4 mM) for optimization.

2.2.4 CHARACTERIZATION OF THE SAMPLES

The crystalline phases of the electrodes were analyzed using X-ray diffraction (XRD; 0°–100°; AXS D8 Advance, Bruker) with copper Kα radiation (λ = 1.5406Å) at a scan rate of. $0.02s^{-1}$ Morphological analysis of the electrodes was performed using a field emission scanning electron microscope (FESEM, SUPRA PV 55). For electrochemical testing, a standard three-electrode cell configuration was used for measurements. The CoMnS-Co $(OH)_2$ electrode, Ag/AgCl electrode, and platinum wire were employed as working, reference, and counter electrodes, respectively. The measurements were carried out in 1 M KOH at room temperature. Cyclic voltammetry (CV) and galvanostatic charge-discharge (GCD) were conducted using an Autolab PGSTAT128N, Metrohm.

2.3 RESULTS AND DISCUSSION

The XRD diffractogram for CoMnS and CoMnS-Co $(OH)_2$ electrodes is depicted in Figure 2.1. Referring to the pattern, there are several obvious sharp peaks that indicate the presence of the crystal structure of materials. The two highest peaks matched in both compounds, indicating the presence of Ni foam. These peaks are located at diffraction angles of 44.5° and 51.8° corresponding to the (111) and (200) planes of Ni, respectively (JCPDS no. 04–0850) (Xin et al., 2020). There are four diffraction peaks closely related to both compounds' patterns at 2θ values of 29.70°, 30.78°, 47.03°, and 72.91° which correspond to the planes (311), (222), (511), and (731) displaying the presence of a Co_9S_8 cube phase, respectively (JCPDS no. 02–1459). The α-MnS crystal phase peaks appeared at 2θ values of 49.40° and 61.39° corresponding to planes (220) and (222), respectively (JCPDS no. 65–2919). The Mn element or MnS ion filled in the cobalt lattice is stated not to affect the structure of the cobalt crystal, and the lattice parameters are due to the atomic radii of Mn (117 pm) being almost the same as Co (116 pm) (Surender, Omar, Bashir, Pershaanaa, Ramesh, & Ramesh, 2021). Furthermore, the coordination of both phase cubes for these compounds: Co_9S_8 and MnS, becomes increasingly layered, and it is difficult to observe the peaks after the immersion of the CoMnS electrode with Co $(OH)_2$ salt solution. As a result, the presence of the β-Co $(OH)_2$ phase has successfully masked the presence of the CoMnS layer and interrupted the lattice coordination, altering the crystalline structure of CoMnS at once. The 2θ values of 18.71°, 32.35°, 3785°, 59.64°, and 61.61°, corresponding to the planes (001), (100), (101), (100), and (111), prove the presence of the β-Co $(OH)_2$ phase, respectively (JCPDS no. 30–0443) (Peng, Wei, Sun, Ma, Feng, Yang, & Lei, 2018).

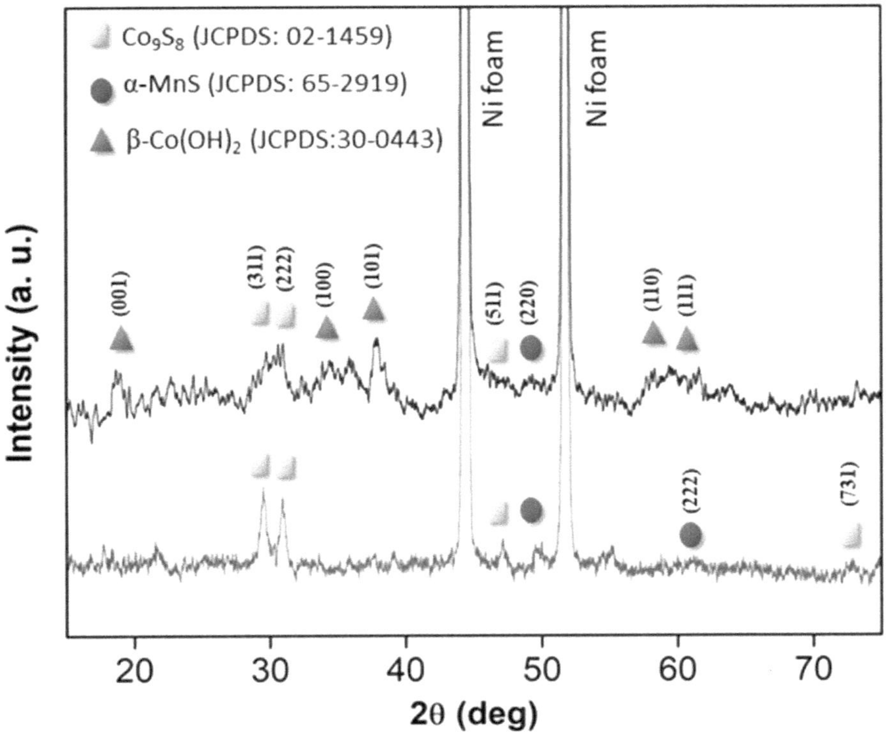

FIGURE 2.1 XRD spectra for CoMnS and CoMnS-Co $(OH)_2$.

Figure 2.2 displays three pieces of surface morphological images of CoMnS and CoMnS-$Co(OH)_2$ electrodes. By nature, Ni foam has a three-dimensional porous-like structure. The significant porosity values of the Ni foam electrode provide an advantage in exhibiting a high surface area. This will help with the transportation of electrolyte infiltrations and simultaneously reduce the ionic resistivity effect. Therefore, the conductivity rate of the Ni foam electrode will increase for better structural stability (Zheng, Liu, & Lee, 2020). The morphology of CoMnS grown on the Ni foam electrode shows a nanosheet structure that is arranged vertically close to each other, as depicted in Figure 2.2(a). This structure is advantageous in the electrochemical performance of electrodes as it is able to provide a high contact area between the electrode and electrolyte, enabling faster and more efficient redox activities (Wei, Zhou, & Wang, 2019). Figure 2.2(b) displays the morphology of CoMnS-Co $(OH)_2$ electrodes with denser nanoflakes, confirming that Co $(OH)_2$ has successfully grown on the surface of CoMnS.

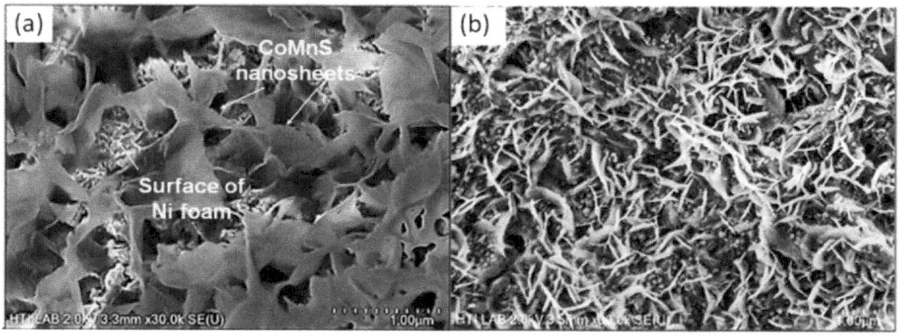

FIGURE 2.2 FESEM image of (a) CoMnS and (b) CoMnS-Co $(OH)_2$.

The electrochemical performance of CoMnS and CoMnS-Co $(OH)_2$ electrodes was analyzed using a three-electrode system. Figure 2.3 shows the redox peaks of cyclic voltammetry (CV) curves for CoMnS and CoMnS-$Co(OH)_2$ electrodes with various concentrations at $5mVs^{-1}$ different scan rates within a specific potential window. As a result, the presence of redox reactions for all various concentrations was positive. These redox reactions are related to the faradaic mechanisms of the active materials within the electrolytes, indicating that the compound behaves like a battery. The faradaic mechanism results from the interaction between Mn^{2+}/Mn^{3+}, $Co^{2+}/Co^{3+}/Co^{4+}$ in the first layer as well as the second layer of the electrode, and it is expressed in the following chemical equations:

$$CoS + OH^{-} \leftrightarrow CoSOH + e^{-} \tag{1}$$

$$CoSOH + OH^{-} \leftrightarrow CoSO + H_2O + e^{-} \tag{2}$$

$$MnS + OH^{-} \leftrightarrow MnSOH + e^{-} \tag{3}$$

$$MnSOH + OH^{-} \leftrightarrow MnSO + H_2O + e^{-} \tag{4}$$

While the equation for the second layer of Co $(OH)_2$ produces the redox reaction as below:

$$Co(OH)_2 + OH^{-} \leftrightarrow CoO(OH) + H_2O + e^{-} \tag{5}$$

$$CoO(OH) + OH^{-} \leftrightarrow CoO_2 + OH^{-} + e^{-} \tag{6}$$

Figure 2.3(a) obviously shows that the highest surface area of the CV curve occurs at the CoMnS/2mMCo $(OH)_2$ electrode. It has the highest oxidation peak (62.56 mAg^{-1}) and the highest reduction peak (55.03 mAg^{-1}) as compared to the other concentrations. Through the surface area of the CV curve, the specific capacitance of each electrode can be measured. However, although the increment of concentration has been made, unfortunately, the redox peaks did not reach beyond the optimized parameter (2 mM). With limited data presented in this work, we hypothesized that this can be attributed to the morphological effect of the material on the concentration of the cobalt salt. The particle size at the CoMnS/3mMCo $(OH)_2$ electrode might be larger than the CoMnS/2mMCo $(OH)_2$ electrode particle size. Theoretically, the higher the concentration, the higher the source for the nucleation process and growth rate of the particle to take place, thus, the lower the effective surface area (Numan et al., 2017; Rashid, Salman, Mohammed, & Mahdi, 2022). Meanwhile, Figure 2.3(b) illustrates the CV curve of the CoMnS-Co $(OH)_2$ electrode (2 mM) where the current density is measured at scan rates of 3, 5, 10 and 20 mVs^{-1}. The graph displays the formation of the anodic and cathodic peaks of the CV that shift upward further to the right and the left and also appear to increase as the applied scanning rate increases due to the polarization effect of the faradaic electrode (Pratheepa & Lawrence, 2020).

Figure 2.4 shows the GCD curve analysis on CoMnS and CoMnS-Co $(OH)_2$ (representing the optimized composite electrode based on CV data), and the experiments were conducted at current densities from 1 to 9 A/g. The nonlinear discharge curve in Figure 2.4 signifies that the charge and discharge process resemble a faradaic mechanism to indicate a battery-type electrode (Iqbal et al., 2020). Referring to Figure 2.4(a), the CoMnS-Co $(OH)_2$ electrode showed a longer discharge time (480 s) compared to the CoMnS electrode (300 s). Figure 2.4(b) displays the GCD curve for the CoMnS-Co

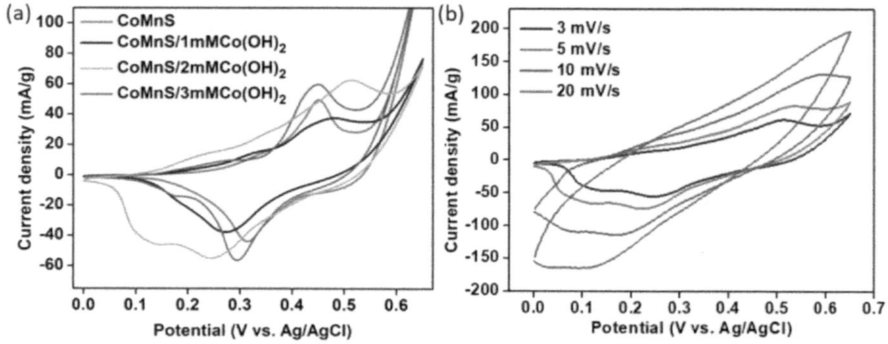

FIGURE 2.3 (a) CV curves of CoMnS and all the CoMnS-Co $(OH)_2$ electrodes at 5 mVs^{-1}; (b) CV curves of CoMnS-2mMCo$(OH)_2$ at various scan rates.

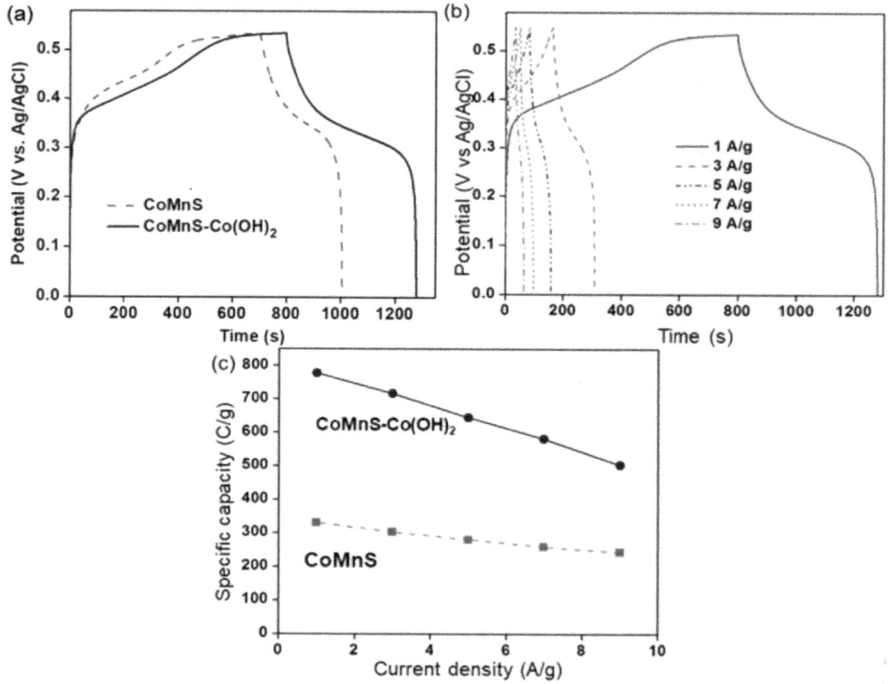

FIGURE 2.4 (a) GCD curves for CoMnS and CoMnS-Co $(OH)_2$ at 1 A/g; (b) GCD curve for CoMnS-Co $(OH)_2$ at 1–9 A/g; (c) Rate capability of CoMnS and CoMnS-Co $(OH)_2$

$(OH)_2$ electrode at various current densities. It can be observed that as the current density increases, the electrode discharge time decreases (Jose et al., 2016). Figure 2.4(c) plots the specific capacity versus current density for all electrodes calculated from the discharge curves shown in Figure 2.4(b). As a result, it can be seen that the specific capacity for CoMnS-$Co(OH)_2$ is higher than for CoMnS. The CoMnS electrode exhibited a Q_s value of 330.5 C/g and a minimum value of 225.5 C/g, while the CoMnS-Co $(OH)_2$ electrode exhibited higher Q_s of 777.5 C/g (max.) and 500.0 C/g (min). The rate capability can be demonstrated through the graph in Figure 2.4(c), where the graphs for both electrodes appear to decrease their specific capacity as the current density increases. This phenomenon of decreasing rate capability, which typically occurs in battery-type electrodes, is caused by the active region inside the electrode that is incapable of receiving the redox process at high current rates (Rodríguez-Pérez, Zhang, Leonard, & Ji, 2019). In addition, based on measurements, the percentage rate capability for the CoMnS-Co $(OH)_2$ electrode was 64.31%, and the CoMnS electrode was 68.23%. Even though the presence of Co $(OH)_2$ enhanced the specific capacity of the CoMnS electrode, the CoMnS-Co $(OH)_2$ electrode has provided a lower capability rate than the CoMnS electrode owing to the low conductivity of Co $(OH)_2$ (Shakir, Almutairi, Shar, & Nafady, 2020).

2.4 CONCLUSION

Herein, CoMnS-Co $(OH)_2$ nanocomposite was prepared by hydrothermal method. The XRD results confirmed the successful synthesis of CoMnS-Co $(OH)_2$ while the morphological study of FESEM manifested the depositions of the nanocomposite. Electrochemical analysis revealed that CoMnS-Co $(OH)_2$ bears improved energy storage performance compared to CoMnS. It produced a maximum of 777.5 C g^{-1} with a rate capability of 64.31%. Whereas the maximum specific capacitance of the CoMnS electrode is 330 C g^{-1} with a rate capability of 68.23% using a three-electrode cell system. The issue of low specific capacity of the CoMnS electrode was addressed by adding Co $(OH)_2$; however, the effect on the rate capability still needs further improvement.

ACKNOWLEDGMENT

The authors would like to thank Geran Galakan Penyelidik Muda (GGPM) for their continuous support in this research (GGPM-2020-019).

KEYWORDS

- **binder-free electrodes**
- **electrostatic double-layer capacitors**
- **metal sulfides**
- **nanocomposites**
- **supercapattery**
- **X-ray diffraction**

REFERENCES

Anggoro, D. D., Udaibah, W., Prasetyaningrum, A., & Zakaria, Z. Y. (2022). Feasibility of hydrogen production from cellulose and prediction of the product distribution: Thermodynamics analysis. *Sains Malays, 51*(3), 747–756.

Chen, G. Z. (2017). Supercapacitor and supercapattery as emerging electrochemical energy stores. *International Materials Reviews, 62*(3), 173–202.

Iqbal, J., Numan, A., Ansari, M. O., Jafer, R., Jagadish, P. R., Bashir, S., Hasan, P. M. Z., Bilgrami, A. L., Mohamad, S., Ramesh, K., & Ramesh, S. (2020). Cobalt oxide nanograins and silver nanoparticles decorated fibrous polyaniline nanocomposite as battery-type electrodes for high-performance supercapattery. *Polymers, 12*(12), 1–19.

Jose, R., Krishnan, S. G., Vidyadharan, B., Misnon, I. I., Harilal, M., Aziz, R. A., Ismail, J., & Yusoff, M. M. (2016). Supercapacitor electrodes deliver high energy and power densities. In *Materials Today: Proceedings* (Vol. 3, pp. S48–S56). Elsevier Ltd.

Lemian, D., & Bode, F. (2022). Battery-supercapacitor energy storage systems for electrical vehicles: A review. *Energies, 15(15), 5683*. MDPI. https://doi.org/10.3390/en15155683.

Numan, A., Duraisamy, N., Saiha Omar, F., Gopi, D., Ramesh, K., & Ramesh, S. (2017). Sonochemical synthesis of nanostructured nickel hydroxide as an electrode material for improved electrochemical energy storage application. *Progress in Natural Science: Materials International, 27*(4), 416–423.

Omar, F. S., Numan, A., Bashir, S., Duraisamy, N., Vikneswaran, R., Loo, Y. L., Ramesh, K., & Ramesh, S. (2018). Enhancing rate capability of amorphous nickel phosphate supercapattery electrode via composition with crystalline silver phosphate. *Electrochimica Acta, 273*, 216–228.

Peng, H., Wei, G., Sun, K., Ma, G., Feng, E., Yang, X., & Lei, Z. (2018). Integrated and heterostructured cobalt manganese sulfide nanoneedle arrays as advanced electrodes for high-performance supercapacitors. *New Journal of Chemistry, 42*(22), 18328–18334.

Pratheepa, M. I., & Lawrence, M. (2020). The eco-friendly approach in supercapacitor application: CuZnCdO nanosphere decorated in reduced graphene oxide nanosheets. *SN Applied Sciences 2,* 318.

Ramly, M. M., Omar, F. S., Chanlek, N., Aspanut, Z., & Goh, B. T. (2022). Control growth of high density and morphological uniformity of taper-free Ni_3Si_2 NWs for enhancement in supercapacitor. *Electrochimica Acta, 431*, 141076.

Rashid, I. M., Salman, S. D., Mohammed, A. K., & Mahdi, Y. S. (2022). Green synthesis of nickel oxide nanoparticles for adsorption of dyes. *Sains Malays, 51*(2), 533–546.

Rodríguez-Pérez, I. A., Zhang, L., Leonard, D. P., & Ji, X. (2019). Aqueous anion insertion into a hydrocarbon cathode via a water-in-salt electrolyte. *Electrochemistry Communications, 109*, 106599.

Rovetta, A. A. S., Browne, M. P., Harvey, A., Godwin, I. J., Coleman, J. N., & Lyons, M. E. G. (2017). Cobalt hydroxide nanoflakes and their application as supercapacitors and oxygen evolution catalysts. *Nanotechnology, 28*(37), 375401.

Şahin, M. E., & Blaabjerg, F., & Sangwongwanich, A. (2022). A comprehensive review of supercapacitor applications and developments. *Energies, 15(3), 674*. MDPI. https://doi.org/10.3390/en15030674.

Shakir, I., Almutairi, Z., Shar, S. S., & Nafady, A. (2020). Synthesis of $Co(OH)_2$/CNTs nanocomposite with superior rate capability and cyclic stability for energy storage applications. *Materials Research Express, 7*(12), 125501.

Surender, G., Omar, F. S., Bashir, S., Pershaanaa, M., Ramesh, S., & Ramesh, K. (2021). Growth of nanostructured cobalt sulfide-based nanocomposite as faradaic binder-free electrode for supercapattery. *Journal of Energy Storage, 39*, 102599.

Wei, S., Zhou, R., & Wang, G. (2019). Enhanced electrochemical performance of self-assembled nanoflowers of MoS_2 nanosheets as supercapacitor electrode materials. *ACS Omega, 4, 15780–15788.*

Xin, C., Ang, L., Musharavati, F., Jaber, F., Hui, L., Zalnezhad, E., Bae, S., Hui, K. S., & Hui, K. N. (2020). Supercapacitor performance of nickel-cobalt sulfide nanotubes decorated using Ni-Co layered double hydroxide nanosheets grown in situ on Ni foam. *Nanomaterials, 10*(3), 584.

Zhang, L., Cai, P., Wei, Z., Liu, T., Yu, J., Al-Ghamdi, A. A., & Wageh, S. (2021). Synthesis of reduced graphene oxide supported nickel-cobalt-layered double hydroxide nanosheets for supercapacitors. *Journal of Colloid and Interface Science, 588*, 637–645.

Zheng, W., Liu, M., & Lee, L. Y. S. (2020). Best practices in using foam-type electrodes for electrocatalytic performance benchmark. *ACS Energy Letters, 5*(10), 3260–3264.

CHAPTER 3

Formation of Cellulose Spheres from Cotton Linter

NUR KHAIRUN ATIYAH SAGEE AHMAD, MOHD. FIRDAUS-RAIH, DORIS HUAI XIA QUAY, and SHARIFAH NABIHAH SYED JAAFAR

Department of Applied Physics, Faculty of Science and Technology, Universiti Kebangsaan Malaysia, UKM Bangi, Selangor, Malaysia

ABSTRACT

Cellulose-based particles such as cellulose spheres (CS) offer unique characteristics such as a high specific surface area with many hydroxyl groups, excellent stability, as well as being non-toxic. Herein, we present a facile water-in-oil (W/O) emulsion method to assess the effects of using cotton linter (CL) on the fabrication of CS. The shape of the CS was analyzed using an optical microscope (OM), while the mean diameter and surface charge of the CS were measured by dynamic light scattering (DLS) and zeta potential (ZP), respectively. Based on the results, the 2:9 CS showed homogeneous and stable size formation.

3.1 INTRODUCTION

There is a growing interest in using alternative biopolymers to produce micro and nanoparticles. Among the existing polymers, the ubiquitous and inexpensive cellulose is exceptionally interesting due to its renewability, biocompatibility, good mechanical performance, and customizable surface chemistry (Joesph et al., 2020). This leads to attraction toward biomedical, chemistry, and physics applications.

Engineering Design and Technical Applications of Physical Science: An Integrated Approach for Sustainable Development. Chin Hua Chia, Sarani Zakaria, Ruey Shan Chen, & Md. Shahariar Chowdhury (Eds.)

As a result, several advanced preparation methods for cellulose-based particles have been presented in the literature such as microfluidics, spray-drying, 3D printing, and electrospray. For instance, Ke et al., (2017) utilized an electro-spraying technique to prepare cellulose acetate microspheres embedded with Ag-nanocubes. On the other hand, Zhang et al. (2020) synthesized porous cellulose acetate microspheres by combining microfluidic and flash-freezing methods. However, these advanced methods were found to be very sophisticated and complex as they required more attention regarding delicate engineering and material manipulation.

Therefore, the emulsion method has been introduced earlier as one of the alternative methods that mainly offer simplicity in obtaining cellulose particles with desired sizes. Despite being considered a low-cost method, the emulsion method, especially water-in-oil (W/O), provides advantages like high purity level, low-temperature reaction, and good composition control. In fact, the W/O emulsion method has also been practiced extensively to obtain spherical cellulose particles. Herein, we aimed to assess the formation of cellulose spheres (CS) through W/O emulsion by varying the ratio of water and oil phases.

3.2 METHOD

Cellulose cotton linter (CL) (degree of polymerization, DP: 660) (α-cellulose > 95%) was purchased from George Weil & Sons Ltd, Guildford, United Kingdom, while paraffin oil was obtained from R&M Chemicals. Sorbitan monooleate (S80) (density: 0.992 g/cm^3), liquid paraffin, hydrochloric acid (HCl) (36-38%), sodium hydroxide (NaOH, 96%), urea (ACS reagent), and acetone (99.9%) were obtained from Sigma-Aldrich.

The water phase, containing 100 g of NaOH/urea/H_2O (7/12/81), was prepared and pre-cooled to –12.5°C. Subsequently, CL was added, stirred, and then centrifuged. A 1% S80 solution was added to liquid paraffin and stirred for 30 min (oil phase). The cellulose solution was loaded into a plastic syringe and fitted with a stainless-steel needle before being placed on a syringe pump (Nfiber) horizontally. The cellulose solutions were controllably extruded continuously at a fixed constant rate of 2.00 mL/min and stirred. The ratio of W/O was varied at 2:7, 2:9, and 2:11. After 3 h, 100 mL of 10% hydrochloric acid solution was slowly added to the beaker until the pH reached 7. Later, two phases were obtained, the upper phase was oil and the lower was CS. The CS was filtered and rinsed with acetone three times. Later, the weight of the CS solution was varied to 1 and 3 wt.%.

Optical microscopy (OM) Axiolab A1.MAT was used to view the formation of CS. A drop of CS in an aqueous solution was placed between two glass slides. The shape formations of CS were determined by using digital image analyses with 40X magnification. The average size and zeta potential (ZP) of the CS were characterized by Malvern Zetasizer Nano ZS. All measurements for ZP were performed at 25°C using a capillary electrophoresis cell, and the samples were suspended in an aqueous solution at pH 7.

3.3 RESULTS AND DISCUSSION

The CS exhibits a perfectly spherical shape without the obvious occurrence of aggregation (Figure 3.1). The 2:9CS shows the best homogeneous formation of spheres and the occurrence of coalescence happened in 2:7CS. The 2:11CS shows the poorest particle size distribution because of the instability of the droplets. The instability of droplets in the emulsion system happened due to low interfacial tension. Sufficient interfacial tension is important to increase the distance between the cellulose solution and paraffin oil between them (Ho, Razzaghi, Ramachandran, & Mikkonen, 2022).

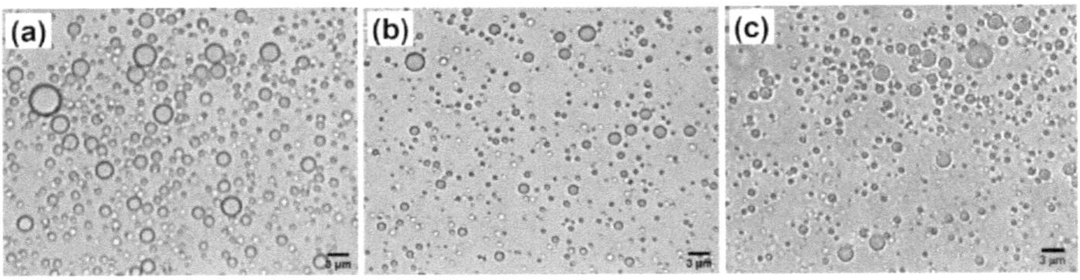

FIGURE 3.1 CS formation from (a) 2:7, (b) 2:9, and (c) 2:11 of W/O.

From the particle size analysis (Table 3.1), it was found that the mean size of 2:9CS was the smallest (1.83 μm), compared to 2:7CS (2.59 μm) and 2:11CS (3.83 μm). In terms of zeta potential, the CS samples had negative charge values because of their hydroxyl groups (Ramli, Aziz, Omar, Yusoff et al., 2022). The 2:7CS depicts the lowest value, followed by 2:9CS and

2:11CS. The 2:11CS seems to have low zeta potential due to flocculation. The van der Waals attractive forces acting on the particles can cause them to flocculate (become unstable) when the zeta potential is low (Zhang, Guo, Ren, & Ren, 2020).

TABLE 3.1 Size Average and Zeta Values of 2:7, 2:9 and 2:11CS

Samples	Z Average (d.μm)	Zeta Potential (mV)
2:7	2.59	–14.1
2:9	1.83	–12.8
2:11	3.83	–3.43

Later, 1% and 3% wt of CL were tested in a continuous phase ratio of 9 to observe the formation of spheres. As shown in Figure 3.2, the CS exhibits imperfect size formations. This indicates that 2% wt of CL was the ideal amount to form a stable droplet and thus create a perfect sphere. It is evident that 1:9CS experienced coalescence similar to 2:7CS. Additionally, the average sizes of 1:9CS and 3:9CS were small, measuring 0.63 μm and 0.59 μm, respectively (Table 3.2). Their zeta potential values were very low, –44.8 mV (1:9CS) and –25.7 mV (3:9CS). These values indicate that 1:9CS and 3:9CS were unstable systems.

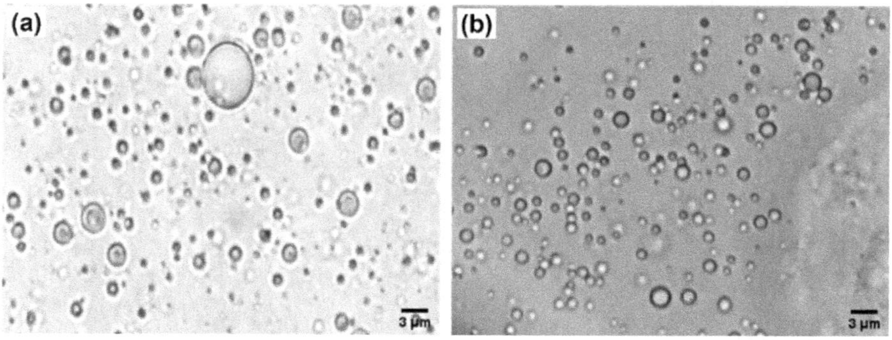

FIGURE 3.2 CS formed from (a) 1 wt.% and (b) 3 wt.% CL.

TABLE 3.2 Size Average and Zeta Values of 1:9 and 3:9CS

Samples	Z Average (d.μm)	Zeta Potential (mV)
1:9	0.63	–44.8
3:9	0.59	–25.7

3.4 CONCLUSION

In this paper, cellulose spheres have successfully formed through controlled injection of W/O emulsion. It was found that the ratio of cellulose solution to paraffin oil and cellulose weight had a significant influence on the size formation and stability of CS.

ACKNOWLEDGMENT

Thank you for the financial support from the research grant university (GUP-2021-041) of Universiti Kebangsaan Malaysia (UKM) and the Center for Research and Instrumentation, UKM, for the testing services.

KEYWORDS

- **cellulose spheres**
- **continuous phase**
- **emulsion**
- **spheres**
- **water-in-oil**
- **zeta potential**

REFERENCES

Ho, T. M., Razzaghi, A., Ramachandran, A., & Mikkonen, K. S. (2022). Emulsion characterization via microfluidic devices: A review on interfacial tension and stability to coalescence. *Advances in Colloid and Interface Science, 299*, 102541–102562.

Joseph, B., Sagarika, V. K., Sabu, C., Kalarikkal, N., & Thomas, S. (2020). Cellulose nanocomposites: Fabrication and biomedical applications. *Journal of Bioresources and Bioproducts, 5*(4), 223–237.

Ke, Y., Meng, G., Huang, Z., & others. (2017). Electrosprayed large-area membranes of Ag-nanocubes embedded in cellulose acetate microspheres as homogeneous SERS substrates. *Journal of Materials Chemistry C, 5*(6), 1402–1408.

Ramli, S. F., Aziz, H. A., Omar, F. M., Yusoff, M. S., Halim, H., Kamaruddin, M. A., Ariffin, K. S., & Hung, Y. T. (2022). Influence of particle size and zeta potential in treating highly colored old landfill leachate by tin tetrachloride and rubber seed. *International Journal of Environmental Research and Public Health, 19*(4), 3016–3029.

Saidane, D., Perrin, E., Cherhal, F., Guellec, F., & Capron, I. (2016). Some modification of cellulose nanocrystals for functional Pickering emulsions. *Philosophical Transactions of the Royal Society A: Mathematical, Physical and Engineering Sciences, 374*(2072), 20150139–20150150.

Zhang, M., Guo, W., Ren, M., & Ren, X. (2020). Fabrication of porous cellulose microspheres with controllable structures by microfluidic and flash freezing methods. *Materials Letters, 262*, 127193–127197.

CHAPTER 4

Mechanical Properties of Recycled PE-PET-PS-PP Blends Produced by High Shear Mixing

MOHD. ADLY RAHMAT SAAT and ROZAIDI RASID

Department of Applied Physics, Faculty of Science and Technology, Universiti Kebangsaan Malaysia, Bangi, Selangor, Malaysia

ABSTRACT

This work focused on the mechanical behavior of polymers by implementing the High Shear Mixing (HSM) method of polymer blending. In this study, recycled polyethylene (PE), recycled polyethylene terephthalate (PET), recycled polystyrene (PS), and recycled polypropylene (PP) were used as they are the most abundant polymers used at the current time. The research was carried out using a counter-rotating twin-screw extruder at a constant speed of 250 rotations per minute (rpm) and at 120°C with a weight mixing ratio of 1:1:1:1. According to the impact study, the HSM produced a material that possesses comparable impact strength with some of the other polymer blends that used conventional methods. According to the flexural results, the HSM materials show significant improvement compared to the other polymer blends.

4.1 INTRODUCTION

The most common process of blending immiscible polymers often involves the usage of chemical substances as the compatibilizing agent since the polymers have different types of functional groups and most of them are not

Engineering Design and Technical Applications of Physical Science: An Integrated Approach for Sustainable Development. Chin Hua Chia, Sarani Zakaria, Ruey Shan Chen, & Md. Shahariar Chowdhury (Eds.)

compatible physically blended (Flaifel, Sahrim, Hassan, Bahri, Tarawneh, & Lih-Jiun, 2013). Hence, the usage of compatibilizers is essential to create better polymer blends (Chen, Sahrim, Sinyee, Nazry, Hafizuddin, & Mou'ad, 2016). However, the usage of compatibilizers is mostly effective when only two types of polymers are involved. If more polymers are involved, it will bring much complexity to the process as more types of compatibilizers are needed, resulting in inconsistent properties (Jia-Hong, Yi-Jun, Chi-Fan, Chien-Lin, Chien-Teng, & Chih-Kuang, 2015).

The most favored way of dealing with these problems is by segregating the polymers according to their types. The segregation process can be the answer to the compatibility issue, but unfortunately, this method is inefficient as it is very labor-intensive (Sihama, Jawad, & Sajid, 2018). It can also be very costly as experts are needed to recognize the types of polymers involved. In today's world, since the pandemic has caused such unstable security in the economic sectors, industries are forced to cut all unnecessary costs (Fernandes, 2020).

Nowadays, more and more developed and developing countries export their waste to poorer countries for they to deal with the waste (Margareth, 2019; Jawad, Sihama, & Rawaa, 2015). This global waste trade is getting out of hand as it will cause the affected countries many problems, with more and more landfills emerging (Carey, Carty, Donlon, Howley, & Nealon, 2000). Because of this, a disturbance in the ecosystem can lead to future environmental issues since heat is needed to degrade those plastics, and this can cause dangerous gases and unwanted byproducts from the degradation process to be dispersed into the air, land, and water sources (Kibria, 2017). The recycling process of plastics might need a new unconventional way to overcome these problems and produce a better outcome, reducing the risk of unwanted side effects (Joung, Hyungsu, Myung, & Jae, 2004).

One of the unconventional methods that has yet to be fully explored is high shear mixing (HSM). This method is called the newplast-process. High shear mixing is a mixing method that adapts the homomicronization technique, which is the process of mixing polymer under ultra-high shear mixing where the mixing speeds reach up to 1500–3000 rotations per minute (rpm). Due to the process, the materials are subjected to ultra-high shearing force between the blades and the materials themselves, which will subsequently melt the material and blend it. This method blends all sorts of plastics in a customized mixing chamber consisting of an ultra-high shearing mixer that melts and blends the plastics by using shearing force and is done by subjecting the plastics to high speed and torque, enhancing the mixing and increasing the heat through the force of shearing (La Mantia, 1996; Wantinee, Richard, & Jayant, 2007). The friction caused by the blades against the plastics and between the plastics themselves will melt the material and mix it at high speed. The resulting blend consists of a homogeneous dough. Unfortunately, this method of polymer blending is somehow being halted, and there have been no new reports of this work recently. This method will not involve the use of any compatibilizers. One of the common machineries that can mimic high-shear mixing is a twin-screw extruder. The counter-rotating twin screws can generate enough shearing force inside the barrel to improve the blending. Thus, the process of segregation and the use of chemicals can be set aside to aid in the blending of the polymers involved.

The adaption of this technique using the counter-rotating twin-screw extruder is the first step in understanding the fundamentals of homogenization in terms of the mechanical characteristics of the blends. However, the homogenization process is done in a chamber under a specific time period, and the process is not continuous. On the other hand, mixing using a twin-screw extruder is a continuous process in which the residence time of the process depends on the speed of the screws (Chung, 2000). Therefore, a way to overcome this is by using a thin slip at the die of the extruder to increase the residence time and prolong the mixing period. This will help the materials blend longer under the high shearing force.

Commonly, the mechanical properties of secondary plastic materials obtained by recycling plastic containers are poor due to the incompatibility between primary polymers and the degradation of components during heterogeneous reprocessing (La Mantia, 2004). To make the secondary plastics usable, they must at least meet certain minimum mechanical properties required for daily life essentials. It is expected that HSM will yield lower mechanical properties compared to virgin materials because there are no compatibilizers involved in the mixing process. However, the advantage of HSM lies in its ability to enhance the mixing of incompatible polymers by exerting extra shearing on the materials, resulting in comparable results to conventional methods of polymer blending. Therefore, this current work reports preliminary results of the mechanical behavior of polymers.

4.2 MATERIALS AND METHODS

The recycled polymers used in this research were polypropylene (PP), polyethylene (PE), polyethylene terephthalate (PET), and polystyrene (PS), all of which are commercially available in various forms and shapes. The polymers were obtained from bottled water, as the bottle is PET, the bottle cap is PE, the PS is from old compact disk casings, and the PP is from a used plastic chair. The plastics were divided according to their types, cleaned to remove unwanted substances, and air-dried. Subsequently, the plastics were crushed by the Ming Lee Strong Crusher type ML-SC 1.5 KW crusher machine into smaller plastic flakes. The flakes were then dried in the oven at 80°C for 12 hours and stored in zip-lock bags according to their type (Rahmat, Rozaidi, Azman, Maria, & Emee, 2019).

The second step was the high shear mixing process, carried out using a twin-screw counter-rotating extruder (Thermo Scientific™ HAAKE™ Rheomex CTW 100 OS Twin-Screw Extruder). The extruder's temperature was set at 120°C, and the rotation speed was set at 250 rpm to induce shearing of mixing, as the maximum rotation speed of the machinery is 250 rpm and the common mixing speed used was about 80–120 rpm. The die of the extruder was a slit type with dimensions of 1 mm thickness and 50 mm wide. The materials were weighed with a 1:1:1:1 weight ratio, manually mixed, and stirred by hand to achieve a homogeneous mixture of the flakes in a container. The materials were then poured into the feeder of the twin-screw counter-rotating extruder. The materials were extruded into sheets with dimensions according to the die's slit (Joung, Hyungsu, Myung, & Jae, 2004). The materials were then cooled down to room temperature and crushed using the crusher to produce smaller flakes. The raw materials of HSM were dried in an oven at 65°C for 24 hours (Chen, Salleh, Ghani, Sahrim, & Gan, 2015).

The next step was carried out to investigate the mechanical properties of the HSM. The raw materials obtained from the extruder were subjected to a thermal compression process to prepare the sample in stick form (Zainab, 2019). The raw materials were poured into a 3 mm thick popsicle stick mold and compressed by the hot press machine (LABTECH Engineering Company LTD). The pressure was set at 6.9 MPa, the processing temperature was 180°C, the preheating time was 2 minutes, the venting time was 6 minutes, the pressing time was 6 minutes, and the cooling time was 5 minutes (Jia, Royan, Yong, Hafizuddin, & Sahrim, 2014).

The samples for the Izod impact test were notched at a 45° angle using a (machine name) and then tested using a (machine name) Izod impact test machine to evaluate the mechanical properties. The speed of the pendulum was 3.49 m/s with a weight of 0.61 kg. For the flexural test, the samples were tested using a universal tensile test machine (Testometric M500-50CT) to evaluate the mechanical properties. The test was performed at a constant strain rate of 150 mm/minute.

4.3 RESULTS AND DISCUSSION

Figure 4.1 shows the various polymer blends compared to HSM blends. From the chart, it shows that HSM impact strength is 27 J/m, which is quite good compared to the others. The lowest one is the PP/PET mix, which has only 48% of the impact strength of HSM. PP/PS blend has the highest impact strength compared to the others; this is mainly because the mix of PP, which is a very impact-resistant material, into the PS matrix will subsequently increase the impact strength of the blend. The blend has 40% more impact strength compared to HSM. On the other hand, the PS/HDPE blend has almost the same impact strength as the HSM, in which they only vary by 9%. Compared to the blend of PS/LDPE, the HSM is 58% more impact-resistant in comparison to these two. For the PP/PET blend, the HSM impact strength is almost double that of the PP/PET. The HSM mixing method did consist of a compatibilizer; hence, when compared to a blend consisting of a compatibilizer, such as PP/PET/SEBS-g-MA, the HSM puts out 33% more impact strength.

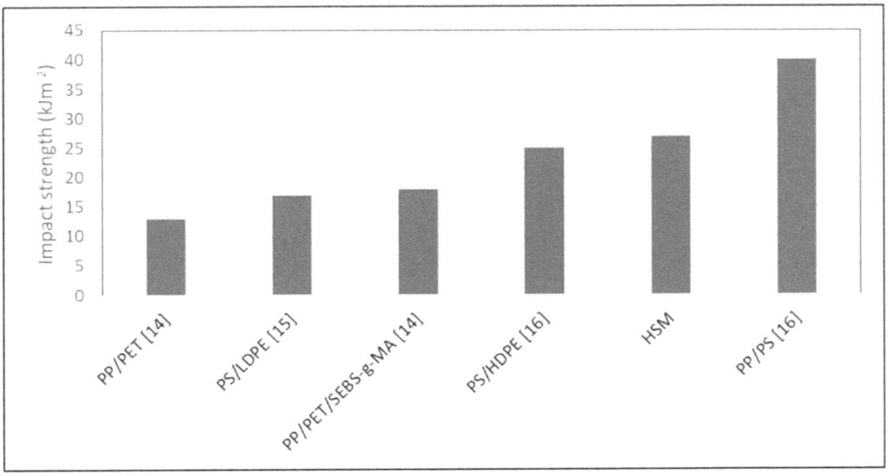

FIGURE 4.1 Impact strength of HSM compared to other blends.

From the results and comparison, the use of PET and PP for HSM might have compensated for the low-impact strength of PE, and the high shearing inside the barrel of the twin-screw extruder might also have an impact on the quality of the blending of all four polymers involved. Because of the low temperature of the HSM process, the higher melting point polymers such as

PET might not all be melted thoroughly and might have been dispersed inside the matrix of PP, PE, and PS, acting as fillers to strengthen the blending.

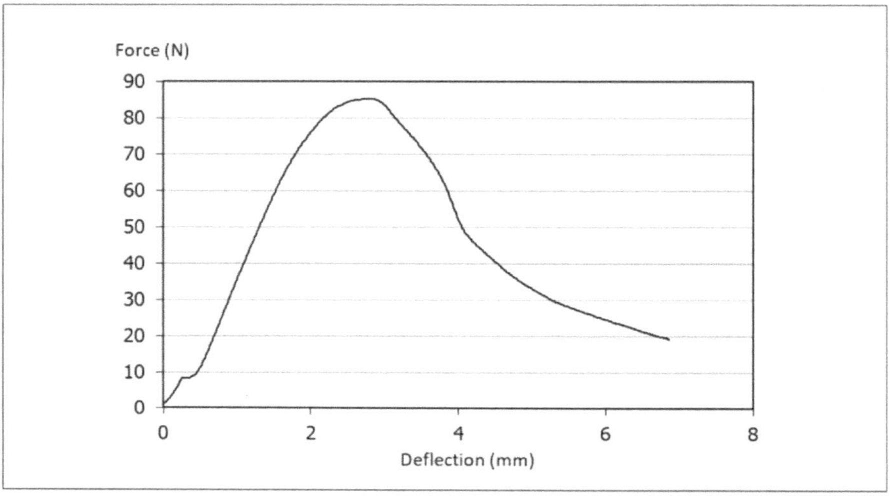

FIGURE 4.2 Flexural stress-strain curve of HSM.

Figure 4.2, shows that the HSM does not have a typical flexural curve. The most noticeable difference is a slight scission early on. This phenomenon might be related to the process of rearrangement of polymer chains upon stress. During this process, the agglomerated polymer chains will start to align when force is applied, resulting in a stronger polymer chain structure. Hence, after this process is done, greater force will be needed to bend the polymer, explaining the sharp increase in force after the slight flat of the curve (Luciana & Ana, 2018).

Figure 4.3 shows that the HSM method does give better flexural properties compared to other polymer-polymer composites. As shown, the HSM peak strength is 70.3 MPa, which is 66.6% higher than the PE/PVC (15 wt.%) and 51.7% higher than PE/PP (15 wt.%). Both samples are processed using friction stir processing. In comparison to a virgin polymer, HSM flexural strength is about 10.5% higher than virgin PP. Compared to PS/PMMA (20 wt.%), HSM possesses 7% higher strength, 14.7% stronger than PP/HDPE (20 wt.%), and 46% stronger than PS/PP (20 wt.%). All three of these polymer blends are blended using conventional low-speed high-temperature mixing.

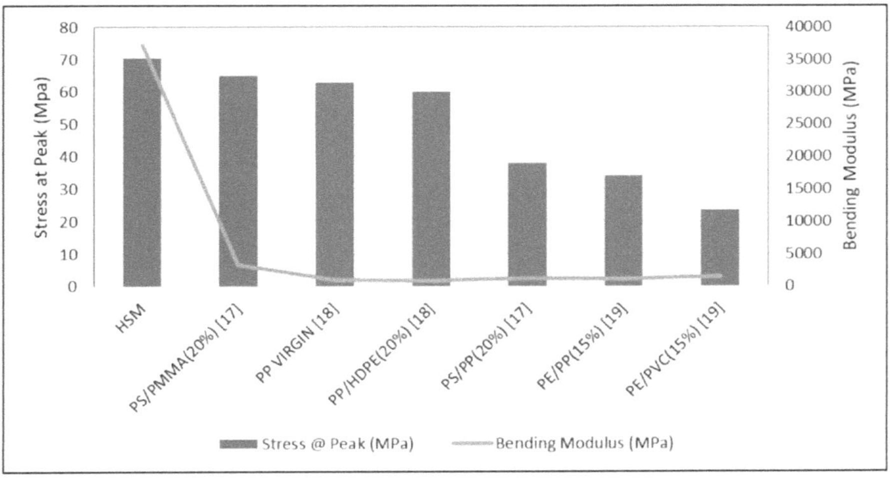

FIGURE 4.3 Comparison of stress at yield and bending modulus.

Flexural modulus is the value of how much a material can withstand bending under a load. For HSM, the value of its modulus is 37116 MPa, which is 91% higher than the closest comparable value, which is the PS/PMMA (20 wt.%). On the other hand, the other blends' flexural modulus is less than 5% each compared to HSM.

4.4 CONCLUSION

The process of blending polymers by not using additives through the HSM method produces a polymer blend that is comparable to other types of polymer blends. HSM also exceeds some of the blends processed by using the conventional slow mixing method. HSM also shows better flexural strength in comparison to certain other types of polymer blends using different processing methods, and HSM possesses a very high flexural modulus compared to others.

ACKNOWLEDGMENT

The authors gratefully acknowledge the financial support from Assoc. Professor Dr. Rozaidi Rasid for sponsoring this research and the Ministry of Higher Education of Malaysia for the MyPhD scholarship.

KEYWORDS

- **high-speed mixing**
- **mechanical blending**
- **mechanical properties**
- **polyethylene terephthalate**
- **polymer blending**
- **polymer recycling**

REFERENCES

Carey, P., Carty, G., Donlon, B., Howley, D., & Nealon, T. (2000). Landfill manuals landfill site design. Environmental Protection Agency, Wexford, p. 11.

Chen, R. S., Sahrim, A., Sinyee, G., Nazry, S. M., Hafizuddin, M., & Mou'ad, A. T. (2016). Thermochimica. *Thermochimica Acta, 640*, 52–61.

Chen, R. S., Salleh, M. Ghani, M., Sahrim, A., & Gan, S. (2015). *BioResources, 10*(4), 6872–6885. DOI: 10.15376/biores.

Chung, C. I. (2000). Extrusion of Polymers Theory & Practice. Hanser Publishers, Munich • Hanser Publications, Cincinnati.

Fernandes, N. (2020). IESE Business School Working Paper, No. WP-1240-E.

Flaifel, M. H., Sahrim, A., Hassan, A., Bahri, S., Tarawneh, M. A., & Lih-Jiun, Y. (2013). Composites: Part B. *Composites Part B, 52*, 334–339.

Jawad, K. O., Sihama, S., & Rawaa, A. A. A. (2015). Engineering & Technology, *33*(6), 1450–1461.

Jia, Y. T., Royan, N., Yong, C. N., Hafizuddin, M., & Sahrim, A. (2014). *Advances in Materials Science and Engineering, 2014*, Article ID 938961.

Jia-Hong, L., Yi-Jun, P., Chi-Fan, L., Chien-Lin, H., Chien-Teng, H., & Chih-Kuang, C. (2015). Materials. *Materials, 8*, 8850–8859.

Joung, G. R., Hyungsu, K., Myung, H. K., & Jae, W. L. (2004). Korea-Australia Rheology Journal, *16*(3), 147–152.

Kibria, G. (2017). Plastic waste, plastic pollution – A threat to all nations. DOI: 10.13140/RG.2.2.11169.51048.

Luciana, M. G. A., & Ana, R. M. (2018). *Polímeros, 28*(1), 84–91.

Mantia, F. P. L. (1996). Recycling of PVC and Mixed Plastics. Francesco Paolo La Mantia ChemTec Publishing, 1996 – Nature, Toronto, 122–127.

Mantia, F. P. L. (2004). Polymer Mechanical Recycling: Downcycling or Upcycling? *Progress in Rubber, Plastics, and Recycling Technology, 20*(1), 11–24.

Margareth, S. (2019). NTS Insight, No. IN19-02.

Rahmat, M. A., Rozaidi, R., Azman, J., Maria, A. B., & Emee, M. S. (2019). Polymery, *64*(10), 676–679.

Sihama, I. S., Jawad, K. O., & Sajid, A. A. (2018). Journal of University of Babylon Engineering Sciences. *Journal of University of Babylon Engineering Sciences, 26*(5), 339–346.
Wantinee, V., Richard, L. L., & Jayant, J. (2007). Society of Plastics Engineers [SPE] annual technical conference proceedings Cincinnati.
Zainab, R. (2019). ASTM International. Tensile Testing, ASTM D790, ISO 178.

CHAPTER 5

Surface Modification of Composite Fabric Structures for Wound Dressing: New Trends and Achievements

ROZIDAWATI AWANG,[1] ANIN SOFYA MOHD. AKHIRI,[1]
AINUL HAFIZA ABDUL HAIR,[2,3] and SARANI ZAKARIA[2]

[1]*Department of Applied Physics, Faculty of Science and Technology, Universiti Kebangsaan Malaysia, UKM Bangi, Selangor, Malaysia*

[2]*Bioresources and Biorefinery Laboratory, Department of Applied Physics, Faculty of Science and Technology, Universiti Kebangsaan Malaysia, UKM Bangi, Selangor, Malaysia*

[3]*Center of Foundation Studies, Univesiti Teknologi MARA, Cawangan Selangor, Kampus Dengkil, Dengkil, Selangor, Malaysia*

ABSTRACT

Plasma treatment is currently being investigated for its technology that can be used as a tool for surface modification of materials. In this work, the surface modification of woven cotton gauze wound dressing was done using a plasma treatment generated by a home-built radio-frequency plasma-enhanced chemical vapor deposition (RF-PECVD) system. The RF power is fixed at 50 W, and the nitrogen gas flow rate was varied at 20, 40, 60, 80, and 100 sccm for 20 minutes of treatment. The surface morphology and chemical bonding properties were investigated using field emission scanning electron microscopy (FESEM) and Fourier-transform infrared spectroscopy (FTIR), respectively. As a result, the FESEM image showed the modification of surface morphology of plasma-treated cotton gauze. Meanwhile, the FTIR analysis exhibited an intensification of the existing functional group

Engineering Design and Technical Applications of Physical Science: An Integrated Approach for Sustainable Development. Chin Hua Chia, Sarani Zakaria, Ruey Shan Chen, & Md. Shahariar Chowdhury (Eds.)

for all treated samples. The results of the study can then be a reference for surface modification of cotton gauze wound dressing that can be used by the community in the future.

5.1 INTRODUCTION

Gauze wound dressings are used to clean wounds, protect them from infection, and are believed to stimulate wound healing. This is due to their large surface area, mechanical properties, and high ability to absorb wound discharge. However, the use of gauze wound dressing in treating chronic wounds with excessive wound discharge is irrelevant because it will adhere to the wound upon drying, thus making it difficult to remove during dressing changes. These properties have also caused the gauze wound dressing's inability to control the wound's moisture, chemical, and biological environments (Dhivya et al., 2015). This circumstance requires the modification of the gauze wound dressing in order to enhance its potential as a wound dressing. In general, several techniques are available to improve the properties and functionality of gauze wound dressing, including direct coating techniques via deposition methods, surface modification methods using chemicals or ionized gases such as plasma, and hybrid approaches (John & Anandjiwala, 2009).

Plasma treatment technology has gained more attention for medical applications in recent years. It appears to be a promising and effective tool for improving surface properties without affecting the bulk properties of materials such as textiles, polymers, and biomedical products. Without using water or any chemical solvents, the plasma treatment technique yields a dry, environmentally friendly process (Eswaramoorthy & McKenzie, 2017) that causes no harm to the treated material. During the plasma treatment, excited atoms, free radicals, electrons, ions, and molecules interact physically and chemically with the material's surface to modify its properties (Karahan & Özdoğan, 2008). The properties of the modified surfaces depend on the exposure time, gas flow rate, pressure, and energy used to generate the plasma (Awang et al., 2018), (Aziz & Awang, 2017), (Awang et al., 2016). In addition, the working gases used to create plasma, such as oxygen, nitrogen, argon, or ammonia, affect the chemical bond properties of the treated surfaces. Different gases will expose different plasma compositions on the material's surface, resulting in different modifications to the functional group on the material's surface (Jelil, 2015).

In this study, a home-built radio-frequency plasma-enhanced chemical vapor deposition (RF-PECVD) system was used to conduct plasma treatment on cotton gauze wound dressing surfaces at various gas flow rates. The modification of cotton gauze wound dressing surfaces was characterized using field emission scanning electron microscopy (FESEM) and Fourier-transform infrared spectroscopy (FTIR). The objective is to investigate the suitability of plasma treatment methods in improving the properties of wound dressing fabrics by studying the influence of nitrogen gas flow rate for plasma treatment on the morphology and chemical bond properties of wound dressing fabrics.

5.2 METHOD

5.2.1 MATERIAL PREPARATION

The plain pre-cut cotton gauze wound dressing with the dimensions of 25 cm × 15 cm was placed on the substrate holder in the chamber of the RF-PECVD system as shown in Figure 5.1.

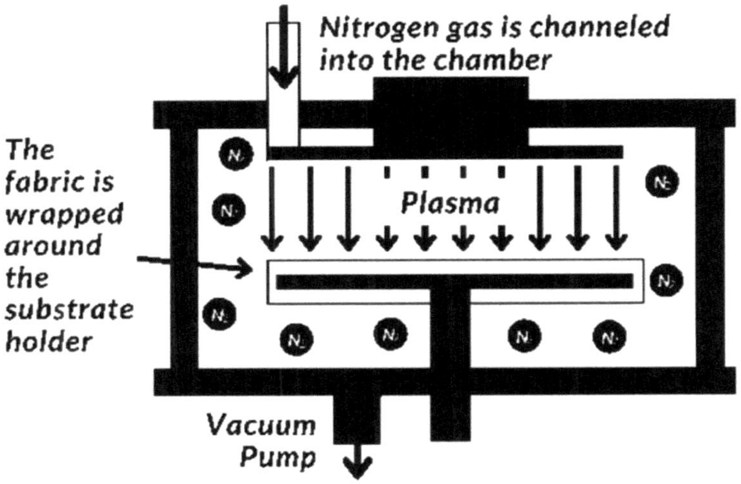

FIGURE 5.1 Schematic diagram of RF-PECVD chamber.

5.2.2 NITROGEN PLASMA TREATMENT

The plasma chamber was tightly closed by tightening the screws on the chamber cover once the cotton gauze was placed over the substrate holder.

A rotary pump was used to pump out atmospheric gases such as oxygen, carbon dioxide, and hydrogen from the plasma chamber and the pressure was reduced to 8×10^{-2} mbar. After the atmospheric gases were removed from the plasma chamber, nitrogen gas (N_2) flowed into the plasma chamber. Then, the plasma treatment was carried out with a fixed RF power of 50 W and pressure at 6.4×10^{-1} mbar for 20 minutes. The plasma treatment was varied at different N_2 gas flow rates of 20, 40, 60, 80, and 100 sccm. The dissociation of nitrogen gas by RF power produced a pink-colored plasma, as shown in Figure 5.2.

FIGURE 5.2 RF-PECVD chamber with nitrogen plasma.

5.2.3 CHARACTERIZATIONS OF SAMPLES

The untreated (UCG) and treated cotton gauze (TCG) were characterized for morphology and chemical bonding by field emission scanning electron microscopy (FESEM) and Fourier-transform infrared spectroscopy (FTIR), respectively.

5.3 RESULTS AND DISCUSSION

5.3.1 SURFACE MORPHOLOGICAL ANALYSIS

Figure 5.3 shows a picture of the cotton gauze wound dressing with treated and untreated areas. Visually, there were no significant differences in the treated area between the UCG and TCG.

No physical damage was observed in the treated area, indicating that this technique does not modify or change the bulk properties of the cotton gauze wound dressing.

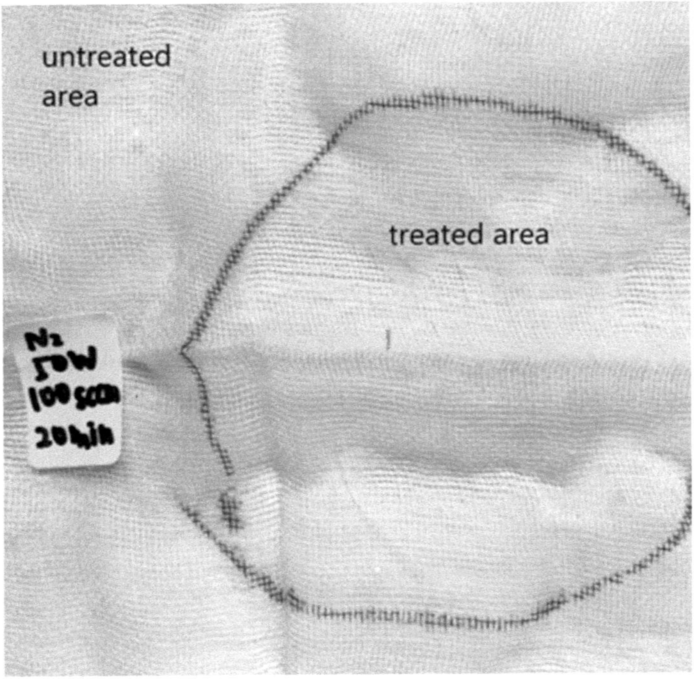

FIGURE 5.3 The treated and untreated areas of cotton gauze.

Figure 5.4 shows the FESEM micrograph images of UCG and TCG at different nitrogen gas flow rates of 20, 40, 60, 80, and 100 sccm. Morphologically, at a 500× magnification scale, there was a significant change in the surface morphology of the samples after being treated with nitrogen gas plasma. The UCG surfaces had a smoother surface than the TCG. Different micro-cracks and groove sizes on TCG were formed and unevenly dispersed on the plasma-treated surfaces. These micro-cracks were produced due to the plasma etching effect during the plasma treatment process (Nithya et al., 2012). The etched area continues to become more prominent on the TCG's surface as the gas flow rate increases to 60 sccm. This was because the higher number of plasma particles bombarded the cotton gauze's surface at a higher gas flow rate, enhancing physical etching. However, the development of crack structures on the TCG's surface was reduced as the gas flow rate was increased above 80 sccm. As the nitrogen gas flow rate was increased above 60.0 sccm, there was an increase in the number of gas molecules in the chamber while the power supplied was maintained at 50 W (Aadim et al., 2020). This will limit the generation of plasma in the chamber. As a result, the insufficient RF power further reduces the effect of plasma etching on the fabric and reduces the formation of micro-cracks on the TCG's surface, as shown in Figures 5.4(e) and 4(f).

5.3.2 CHEMICAL BONDING ANALYSIS

Figure 5.5 shows the FTIR spectra for all samples with the presence of nitrogen-based functional groups of C-N and N-H at the peak of 1028 cm^{-1} and 3300 cm^{-1}, respectively, and C-O, C=C, O-H, C-H peaks at 1159 cm^{-1}, 1313 cm^{-1}, 1426 cm^{-1}, 2900 cm^{-1}, respectively. Samples with plasma treatment clearly show an increase in transmission intensity at each of those functional groups. This suggests that plasma treatment modified the surface chemical composition of the TCG by introducing nitrogen elements and intensifying the existing functional groups. The presence of N-H bonds is the focus of the study to ensure the fabric's suitability for incorporating specific functionality to the surface, which in this case will help to increase cell adhesion ability (Luna et al., 2011) and fabric wettability (Ghimire et al., 2017). These characteristics are very important in the wound-healing process and help to stimulate wound healing. Based on the figure, the formation of N-H bonds on TCG is maximum at a gas flow rate of 60 sccm. Therefore, the optimum gas flow rate for the surface modification of cotton gauze surfaces in this work is 60 sccm.

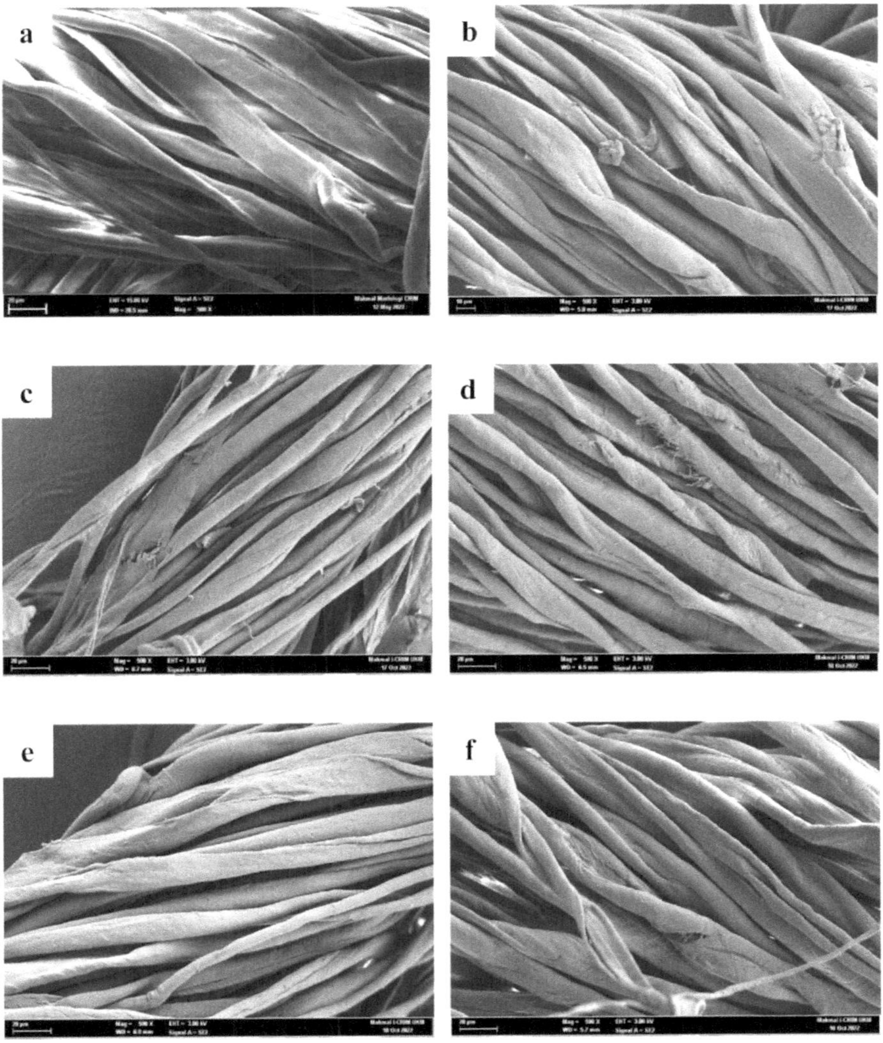

FIGURE 5.4 FESEM image of (a) UCG, and TCG at (b) 20 sccm, (c) 40 sccm, (d) 60 sccm, (e) 80 sccm, and (f) 100 sccm.

5.4 CONCLUSION

The surface of the cotton gauze wound dressing was successfully altered by the nitrogen plasma treatment generated from the home-built RF-PECVD system without affecting its bulk properties. This was confirmed by the

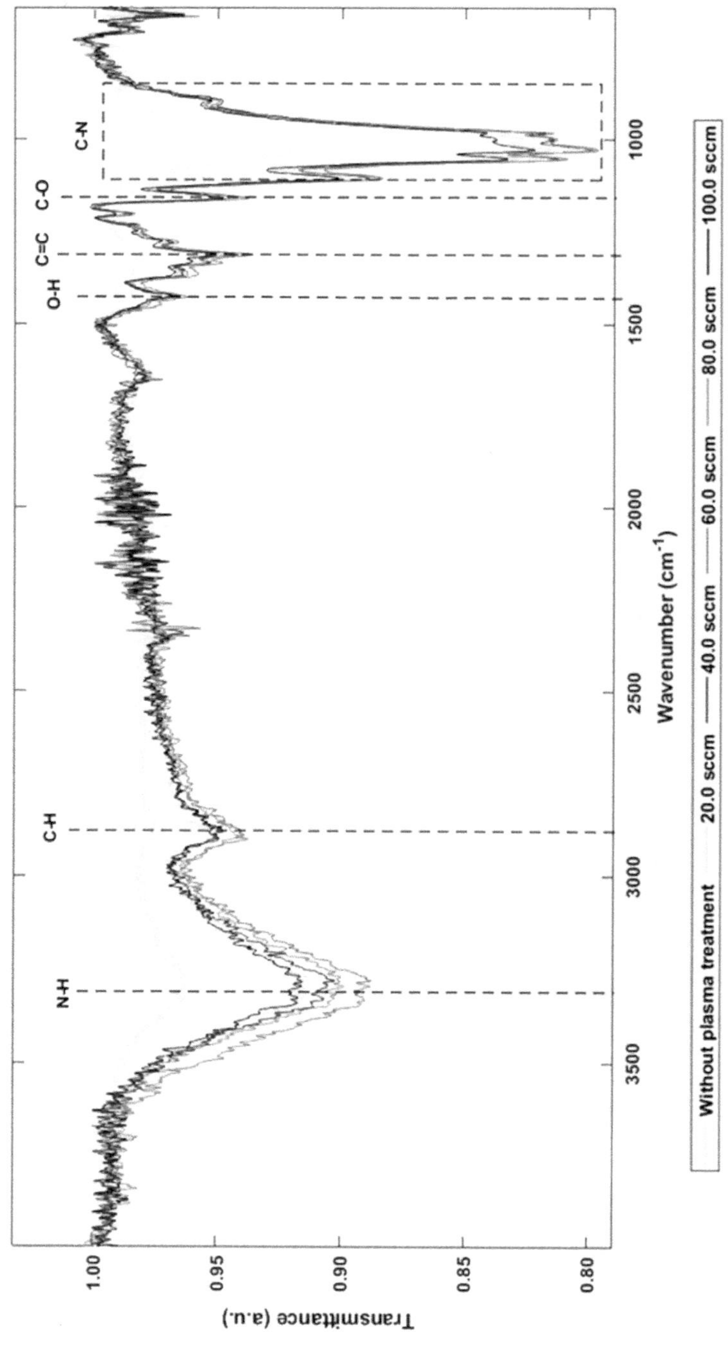

FIGURE 5.5 FTIR spectrum of UCG and TCG at 20, 40, 60, 80, and 100 sccm.

morphology and chemical bonding properties study on the cotton gauze wound dressing. The TCG treated at a gas flow rate of 60 sccm exhibited the most micro-crack structure with the highest abundance of N-H. Plasma treatments can be used as a viable approach for modifying the cotton surfaces by introducing an etching effect on the TCG surface and enhancing the existing functional groups. The nitrogen-based functional groups can provide specific assistance for further post-treatment to stimulate wound healing.

ACKNOWLEDGMENT

We would like to thank Universiti Kebangsaan Malaysia for the technical support and Kementerian Pengajian Tinggi (KPT) for the grant of TGRS/1/2019/UKM/02/1/2.

KEYWORDS

- **field emission scanning electron microscopy**
- **Fourier-transform infrared spectroscopy**
- **gauze wsound dressing**
- **nitrogen plasma treatment**
- **PECVD System**
- **surface modification**

REFERENCES

Aadim, K. A., Mazhir, S. N., Abdalameer, N. K., & Ali, A. H. (2020). Influence of gas flow rate on plasma parameters produced by a plasma jet and its spectroscopic diagnosis using the OES technique. *IOP Conference Series: Materials Science and Engineering, 987*(1), 1–10. https://doi.org/10.1088/1757-899X/987/1/012020.

Awang, R., Azman, N. I., & Purhanudin, N. (2016). Influence of deposition time on the bonding and morphology of amorphous carbon nitride thin films. *Materials Science Forum, 846*(100), 657–662. https://doi.org/10.4028/www.scientific.net/MSF.846.657.

Awang, R., Purhanudin, N., & Salman, N. S. (2018). Effect of radio frequency power on A-CN X film properties and its performance as humidity sensors. *Sains Malaysiana, 47*(11), 2863–2867. https://doi.org/10.17576/jsm-2018-4711-29.

Aziz, S. A. A., & Awang, R. (2017). Kesan kuasa frekuensi radio terhadap komposisi ikatan kimia filem nipis A-CNxsebagai pengesan kelembapan. *Sains Malaysiana, 46*(10), 1951–1958. https://doi.org/10.17576/jsm-2017-4610-34.

Dhivya, S., Padma, V. V., & Santhini, E. (2015). Wound dressings–A review. *Biomedical, 5*(4), 24–28. https://doi.org/10.7603/s40681-015-0022-9.

Eswaramoorthy, N., & McKenzie, D. R. (2017). Plasma treatments of dressings for wound healing: A review. *Biophysical Reviews, 9*(6), 895–917. https://doi.org/10.1007/s12551-017-0327-x.

Ghimire, B., Subedi, D. P., & Khanal, R. (2017). Improvement of wettability and absorbency of textile using atmospheric pressure dielectric barrier discharge. *AIP Advances, 7*(8), 85213–85221. https://doi.org/10.1063/1.4993084.

Jelil, R. A. (2015). A review of low-temperature plasma treatment of textile materials. *Journal of Materials Science, 50*(18), 5913–5943. https://link.springer.com/article/10.1007/s10853-015-9152-4#citeas (accessed on 5th June 2024).

John, M. J., & Anandjiwala, R. D. (2009). Surface modification and preparation techniques for textile materials. In Q. Wei (Ed.), *Surface Modification of Textiles* (pp. 1–25). Woodhead Publishing. DOI: 10.1533/9781845696689.1.

Karahan, H. A., & Özdoğan, E. (2008). Improvements of surface functionality of cotton fibers by atmospheric plasma treatment. *Fibers and Polymers, 9*(1), 21–26. https://doi.org/10.1007/s12221-008-0004-6.

Luna, S. M., Silva, S. S., Gomes, M. E., Mano, J. F., & Reis, R. L. (2011). Cell adhesion and proliferation onto chitosan-based membranes treated by plasma surface modification. *Journal of Biomaterials Applications, 26*(1), 101–116. https://doi.org/10.1177/0885328210362924.

Nithya, E., Radhai, R., Rajendran, R., Jayakumar, S., & Vaideki, K. (2012). Enhancement of the antimicrobial property of cotton fabric using plasma and enzyme pre-treatments. *Carbohydrate Polymers, 88*(3), 986–991. https://doi.org/10.1016/j.carbpol.2012.01.049.

CHAPTER 6

Potentiometric pH-Sensing Characteristic of Hydrothermally Grown Undoped Zinc Oxide and Aluminum-Doped Zinc Oxide Nanostructures

AIN ZAFIRAH BINTI KAMARUDDIN,[1] LIM KAR KENG,[2] MUHAMMAD AZMI ABDUL HAMID,[1] and NAIF H. AL-HARDAN[1]

[1]*Department of Applied Physics, Faculty of Science and Technology, Universiti Kebangsaan Malaysia, UKM Bangi, Selangor, Malaysia*

[2]*School of Liberal Studies, Universiti Kebangsaan Malaysia, UKM Bangi, Selangor, Malaysia*

ABSTRACT

Metal oxide semiconductors were studied extensively as membranes for pH sensing and have been the subject of numerous studies to enhance their sensitivity and stability. Here we report on the effect of doping zinc oxide (ZnO) with aluminum (Al) to enhance the sensitivity of the pH sensor. Undoped and doped ZnO with 1 at% Al were prepared on glass coated with fluorine-doped tin oxide (FTO) via the hydrothermal method. The prepared samples were characterized to explore the impact of aluminum content on the structural and morphological properties by X-ray diffraction analysis (XRD) and field-effect scanning electron microscope (FESEM). The XRD pattern reveals a highly polycrystalline structure for both samples. Furthermore, the FESEM reveals the growth of an array of nanorods perpendicular to the surface of FTO. The samples were prepared as electrodes for pH sensing. The highly sensitive and stable performance of the pH electrodes was noticed in the range of (pH 4–pH 12). The sensitivity of the undoped

Engineering Design and Technical Applications of Physical Science: An Integrated Approach for Sustainable Development. Chin Hua Chia, Sarani Zakaria, Ruey Shan Chen, & Md. Shahariar Chowdhury (Eds.)

ZnO and the Al-doped ZnO was 26.03 mV/pH with a linearity of 0.96 and 30.99 mV/pH with a linearity of 0.94, respectively. The enhancement of the sensitivity was due to the doping process that increases the binding sites on the nanorods' surface where the protonation or deprotonation process occurs.

6.1 INTRODUCTION

Zinc oxide is an excellent n-type semiconductor with a wide and direct energy bandgap (3.37 eV) at room temperature. ZnO has been intensively researched due to its incredible chemical, electronic, optical, mechanical, and magnetic properties (Belkhaoui et al., 2019; Ghazai et al., 2016). Recently, several research studies have reported developments utilizing metal oxide semiconductor materials with potential as pH sensor elements. Compared to other metal oxides, ZnO has been selected as a favorable material in various applications such as optoelectronic devices, solar cells, and the development of biosensors (Al-Khalqi et al., 2022; Hashim et al., 2017). However, the conductivity of pure ZnO thin film is low and can be enhanced by doping ZnO with different elements (Ahmed, 2018). According to Agarwal et al., (2020), Al-Hardan et al., (2021), and Al Farsi et al., (2021), aluminum has been the most widely utilized dopant element due to its cost-effectiveness and ability to improve the electrical properties by increasing the conductivity of ZnO. Additionally, as mentioned by (Mishra et al., 2022), the process of doping ZnO nanorods can be achieved through various methods such as chemical bath deposition (CBD), spray pyrolysis, magnetron sputtering, sol-gel, and the hydrothermal method. Among all these methods, the hydrothermal process is preferred and versatile as it is simple, cost-effective, environmentally friendly, and provides repeatable results (Kumar et al., 2019; Pawar & Deshmukh, 2018; Zan et al., 2015).

In this paper, we studied the effect of doping zinc oxide with aluminum (0 and 1 at% of Al) by the hydrothermal process. The undoped ZnO and doped ZnO:Al nanorods were characterized morphologically by field-effect scanning electron microscopy (FESEM) and structurally by X-ray diffraction (XRD). The sensitivity and linearity of the pH-sensing sensor will be discussed in detail.

6.2 METHOD

In general, conductive glass FTO was cleaned with acetone, methanol, and ethanol, followed by cleaning with deionized water through an ultrasonic process. The hydrothermal process was used to grow undoped ZnO and Al-doped ZnO nanorods. The precursor solution was made to grow ZnO by combining 0.025 M zinc nitrate hexahydrate ($Zn(NO_3)_2$•$6H_2O$) and 0.025 M hexamethylenetetramine (HMTA). Similar procedures were followed, with the addition of aluminum nitrate nanohydrate ($Al(NO_3)_2$•$9H_2O$) for growing Al-doped ZnO into the precursor solution. The atomic ratio of Al dopant was designated as 0 at% (undoped) and 1 at%, respectively. The growth process was performed in a conventional oven at 95 °C for 6 hours. Then, the samples were cooled to room temperature and washed with deionized water several times. The samples were dried for 20 minutes at 100°C in a conventional oven. Then, the samples were prepared as electrodes following the procedure mentioned in the reference (Al-Hardan et al., 2016). The fabricated pH sensors were immersed in reverse osmosis (RO) water for 12 hours before the sensing membranes were tested.

6.3 RESULTS AND DISCUSSION

The surface morphology of undoped and Al-doped ZnO nanorods grown on conductive glass FTO by hydrothermal method is shown in Figure 6.1. Figure 6.1(a) reveals the FE-SEM images of 0 at% of Al (undoped), and Figure 6.1(b) shows FE-SEM images of 1 at% of Al. Both figures indicate that the ZnO nanorods are vertically aligned with hexagonally shaped nanorods. The EDX spectra are shown in Figures 6.1(c) and 6.1(d). Based on the EDX spectra, well-defined peaks of Zn, O, and Al were noticed. There were no other impurity-related peaks found in the EDX spectra.

The X-ray diffraction (XRD) peaks of the produced undoped and Al-doped ZnO NRs are depicted in Figure 6.2. All produced materials were analyzed using the Bruker D8 Advance XRD equipment in the diffraction angle range of 20–80° using Cu K_α radiation ($\lambda = 1.54056$ Å). The diffraction peaks indicate the presence of a wurtzite-structured hexagonal phase of ZnO. These nanorods are extremely pure as no secondary phase could be found. The intensity of the (002) peak was found to be particularly high in contrast to the other peaks. The other diffraction peaks that have been observed, such as (004), do not imply that some ZnO nanorods were produced in other directions, but rather that c-axis ZnO nanorods were randomly oriented in such a way that these diffraction planes happen to fulfill Bragg's law. The peak position at 2θ values shows at 34.6° and 72.7° corresponding to the planes (002) and (004). Based on the diffraction peaks, it was perfectly matched with the standard Joint Committee on Powder Diffraction (JCPDS) card No. 35–1451.

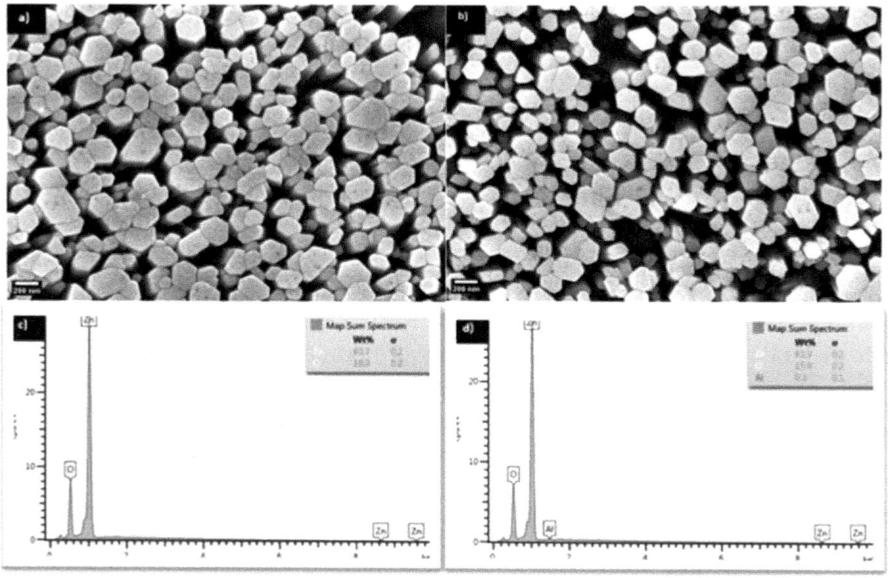

FIGURE 6.1 The FE-SEM hexagonal nanorods images of (a) 0 at% (undoped ZnO), (b) 1 at% (doped ZnO:Al), and the EDX spectra of (c) 0 at% of Al, (d) 1 at% (doped ZnO:Al).

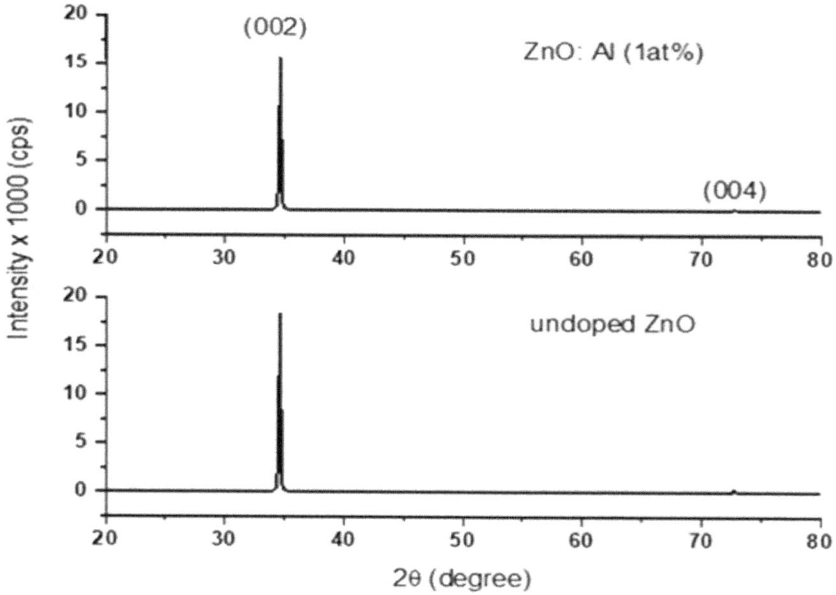

FIGURE 6.2 X-ray diffraction patterns of 0 at% (undoped ZnO) and 1 at% (doped ZnO:Al).

Figure 6.3 depicts the potentiometry response of the prepared ZnO and ZnO films with aluminum doping as a function of pH buffer values in the range 4 to 12, which covers both the acidic and alkaline environments. The binding sites on the film surface were protonated or deprotonated during the chemical reaction between the sensing film and the pH buffer solution, which caused the sensor to produce various response voltages in solutions with various pH values (Yang et al., 2022). From the figure, the sensitivity

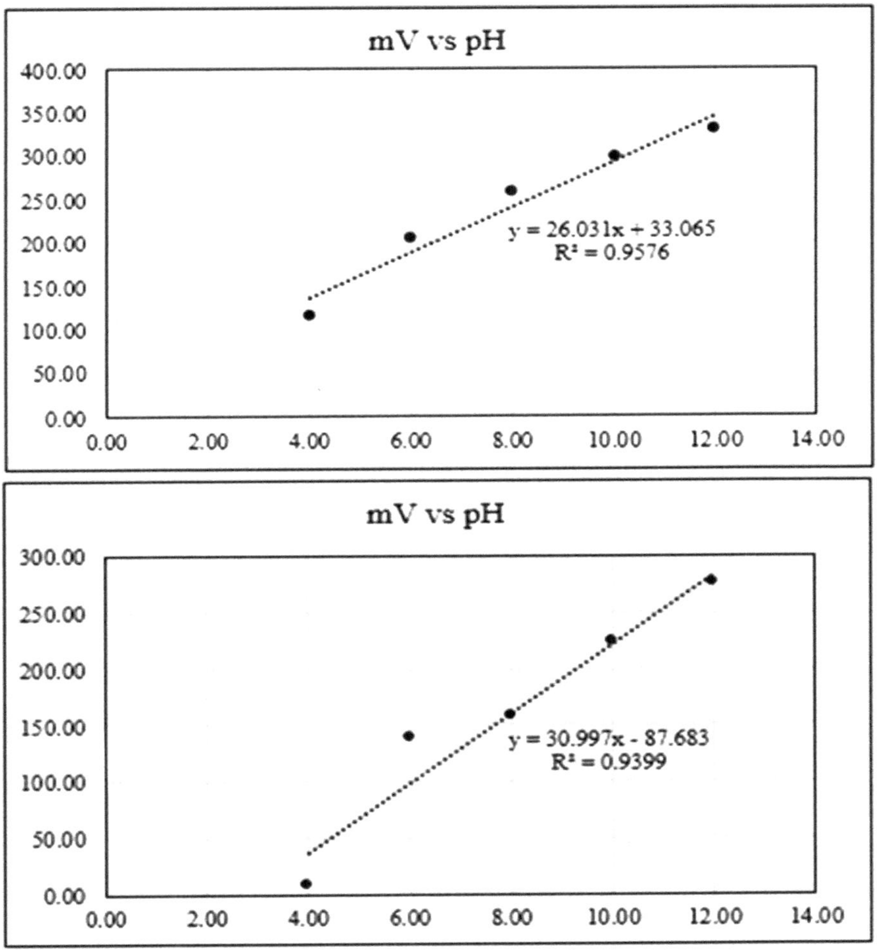

FIGURE 6.3 The response voltage with pH value in the range pH 4 to pH 12 of (a) 0 at% (undoped ZnO) and (b) 1 at% (doped ZnO: Al).

and linearity of (a) undoped ZnO and (b) ZnO doped with 1 at% of Al are depicted based on pH sensor response voltage (mV) versus different pH values. The sensitivity for the undoped ZnO was 26.03 mV/pH with a linearity of 0.96, while the sensitivity for 1 at % Al was 30.99 mV/pH with a linearity of 0.94. These results reveal that the prepared pH sensor exhibits good and excellent linearity. However, the results also show that ZnO doped with Al demonstrates better sensitivity and linearity compared to the undoped ZnO. Furthermore, this finding is consistent with (Wang et al., 2013), where the sensitivity value increases from 0 at % Al to 1 at % Al, with the value of 0 at % being 35.23 mV/pH and the value of 1 at % being 49.79 mV/pH. Therefore, it proves that by adding aluminum to ZnO, it is able to have a flexible sensor and increase the voltage response of the pH sensor (Tsai et al., 2019). The improvement of the Al doping of ZnO might be due to the enhancement of the site density that improves the reaction site by increasing charge carriers (Alkahlout et al., 2014).

6.4 CONCLUSION

The undoped ZnO and doped ZnO:Al were synthesized on glass coated with fluorine-doped tin oxide (FTO) using a simple hydrothermal method at 95°C. Based on the FESEM and XRD results, the structure was revealed to be polycrystalline with hexagonal-shaped nanorods. Additionally, the pH test for the doped ZnO:Al showed higher sensitivity at 30.99 mV/pH and a linearity of 0.94. Therefore, the addition of aluminum to zinc oxide increased the sensor's sensitivity. Furthermore, the pH sensor exhibited a linearity of 0.94, indicating good accuracy. Overall, the fabrication of pH sensing ZnO and ZnO:Al shows great potential and can be advantageous for disposable biosensors.

ACKNOWLEDGMENT

This work was supported by the Universiti Kebangsaan Malaysia (UKM) through a short-term grant GGPM-2020-039 (UKM). The authors are also thankful to the Centre for Research and Instrumentation Management (CRIM) at UKM for providing the FESEM, EDX, and XRD measurements.

KEYWORDS

- **aluminum-doped-zinc oxide nanostructures**
- **field-effect scanning electron microscope**
- **fluorine-doped tin oxide**
- **hydrothermal Method**
- **pH potentiometric**
- **X-ray diffraction analysis**

REFERENCES

Agarwal, M. B., Malaidurai, M., Sharma, A., & Thangavel, R. (2020). Effect of Al doping on hydrothermal growth and physical properties of doped ZnO nanoarrays for optoelectronic applications. *Materials Today: Proceedings, 21*, 1781–1786.

Ahmed, S. M. (2018). Characterization of Al-doped ZnO nanorods grown by a chemical bath deposition method. *Innovaciencia Facultad de Ciencias Exactas, Físicas y Naturales, 6*, 1–9.

Al Farsi, B., Al Marzouqi, F., Al-Maashani, M., Souier, M. T., Tay Zar Myint, M., & Al-Abri, M. Z. (2021). Rapid microwave-assisted fabrication of Al-doped zinc oxide nanorods on a glass substrate for photocatalytic degradation of phenol under visible light irradiation. *Materials Science and Engineering B: Advanced Functional Solid-State Materials, 264*, 114977.

Al-Hardan, N. H., Abdul Hamid, M. A., Ahmed, N. M., Jalar, A., Shamsudin, R., Othman, N. K., Keng, L. K., Chiu, W., & Al-Rawi, H. N. (2016). High-sensitivity pH sensor based on porous silicon (PSi) extended gate field-effect transistor. *Sensors, 16*(1), 1–12.

Al-Hardan, N. H., Abdul Hamid, M. A., Firdaus-Raih, M., & Kar Keng, L. (2021). Aluminum–modified ZnO nanoparticles synthesized through co-precipitation. *Jurnal. Teknologi., 2*, 1–6.

Alkahlout, A., Al Dahoudi, N., Grobelsek, I., Jilavi, M., & de Oliveira, P. W. (2014). Synthesis and characterization of aluminum-doped zinc oxide nanostructures via hydrothermal route. *Journal of Materials, January, 2014*, 1–8.

Al-Khalqi, E. M., Abdul Hamid, M. A., Al-Hardan, N. H., Keng, L. K., & Jalar, A. (2022). Magnesium-doped ZnO nanorod electrolyte–insulator–semiconductor (EIS) sensor for detecting organic solvents. *IEEE Sensors Journal, 22*, 11783–11790.

Belkhaoui, C., Mzabi, N., Smaoui, H., & Daniel, P. (2019). Enhancing the structural, optical, and electrical properties of ZnO nanopowders through (Al + Mn) doping. *Results in Physics, 12*, 1686–1696.

Ghazai, A. J., Salman, E. A., & Jabbar, Z. A. (2016). Effect of aluminum doping on zinc oxide thin film properties synthesis by spin coating method. *American Scientific Research Journal for Engineering, Technology, and Sciences (ASRJETS), 26*, 202–211.

Hashim, U., Fathil, M. F. M., Arshad, M. K. M., Gopinath, S. C. B., & Uda, M. N. A. (2017). Characterization of zinc oxide thin film for pH detector. *AIP Conference Proceedings, 1808*, 020020(1)–020020(5).

Kumar, A. A., Kumar, S. K., Aniley, A. A. N., Fernandez, R. E., & Bhansali, S. (2019). Hydrothermal growth of zinc oxide (ZnO) nanorods (NRs) on screen printed IDEs for pH measurement application. *Journal of The Electrochemical Society, 166*, B3264–B3270.

Mishra, P. N., Mishra, P. K., & Pathak, D. (2022). The influence of Al doping on the optical characteristics of ZnO nanopowders obtained by the low-cost sol-gel method. *ChemRxiv, 4*, 1136–1146.

Pawar, S., & Deshmukh, H. (2018). Hydrothermal method for synthesis of different nanostructure metal oxide thin film. *International Journal of Innovative Knowledge Concepts, 6*, 126–129.

Tsai, Y. T., Chang, S. J., Ji, L. W., Hsiao, Y. J., & Tang, I. T. (2019). Fast detection and flexible microfluidic pH sensors based on Al-doped ZnO nanosheets with a novel morphology. *ACS Omega, 4*, 19847–19855.

Yang, P. H., Chang, Y. S., & Chan, C. T. (2022). ZnO and AZO film potentiometric pH sensors based on flexible printed circuit board. *Chemosensors, 10*, 293.

Zan, L., Wei, Q., & Xiaohong, W. (2015). Controllable hydrothermal synthesis of Al-doped ZnO with different microstructures, growth mechanisms, and gas sensing properties. *RSC Advances, 5*, 56325–56332.

CHAPTER 7

Influence of Ferrimagnetic Cr_2S_3 Addition on the AC Susceptibility and Critical Current Density of (Bi,Pb)-Sr-Ca-Cu-O Superconductors

MASNITA MAT JUSOH,[1] NURUL RAIHAN MOHD. SUIB,[1]
A. B. P. ILHAMSYAH,[2,3] MADIHAH MUJAINI,[2,3] and R. ABD-SHUKOR[3]

[1]*Center of Foundation Studies, Universiti Teknologi MARA Cawangan Selangor, Kampus Dengkil, Dengkil, Selangor, Malaysia*

[2]*College of Engineering, Universiti Tenaga Nasional, Jalan Ikram-Uniten, Kajang, Selangor, Malaysia*

[3]*Department of Applied Physics, Universiti Kebangsaan Malaysia, Bangi, Selangor, Malaysia*

ABSTRACT

Superconductors are efficient electrical energy transport materials. The current carrying capacity can be increased by adding magnetic particles. In this work, we investigated the effect of ferrimagnetic Cr_2S_3 addition on the properties of $(Bi_{1.6}Pb_{0.4}) Sr_2Ca_2Cu_3O_{10}$ (Bi-2223). Superconductivity and magnetism are mutually exclusive phenomena, and their juxtaposition is an interesting topic. The samples were prepared via co-precipitation with starting composition $(Bi_{1.6}Pb_{0.4}) Sr_2Ca_2Cu_3O_{10} (Cr_2S_3)_x$. The structure, electrical resistance, transport critical current density, and AC susceptibility were determined. The $x = 0.05$ wt.% sample showed the highest superconducting transition temperature. The highest transport critical current density, J_{ct} at 30 K, was shown by the $x = 0.10$ wt.% sample (8.37 A cm^{-2}). J_{ct} for all added samples was significantly higher

Engineering Design and Technical Applications of Physical Science: An Integrated Approach for Sustainable Development. Chin Hua Chia, Sarani Zakaria, Ruey Shan Chen, & Md. Shahariar Chowdhury (Eds.)

than the non-added sample. The peak temperature of the imaginary part of the susceptibility, T_p, for all added samples, was much higher than the non-added sample, indicating that Cr_2S_3 enhanced flux pinning energy and intergrain coupling. A small amount of Cr_2S_3 (x = 0.05%) optimized the transition temperature, while x = 0.10–0.20 wt.% optimized the critical current density.

7.1 INTRODUCTION

The potential applications of superconductors in transforming the energy sector are one of the reasons why these materials are actively studied. The positive impact of superconductors in reducing energy losses has been widely reported. The (Bi, Pb)-Sr-Ca-Cu-O system is one of the cuprate superconductors with a relatively higher transition temperature, Tc (Hamadneh et al., 2006, Tanaka et al., 1990). The ideal structural formula for the (Bi,Pb)-Sr-Ca-Cu-O system is $Bi_2Sr_2Ca_{n+1}Cu_nO_{4+2n}$, where n = 1, 2, and 3. CuO_2 layers formed in the system are associated with the occurrence of superconductivity and can be used to describe the phases. Studies have shown that T_c increases with the number of CuO_2 layers. The $Bi_2Sr_2Ca_2Cu_3O_{10}$, also known as the Bi-2223 phase, exhibits the highest T_c (110 K) (Zelati et al., 2014). Tc can be increased to 117 K by Pb substitution with the composition of $(Bi_{1.6}Pb_{0.4})Sr_2Ca_2Cu_3O_{10}$ (Takada et al., 1990). The co-precipitation method has been used to fabricate higher-quality samples (Ismail et al., 2004).

The Bi-2223 phase superconductor has a high potential for technological application. However, weak intergrain links and weakened flux pinning are the main drawbacks of this system. This reduces the critical current density, J_c of the bulk sample. Several studies on adding nano- or micro-sized metal oxides have been conducted to improve the superconductivity properties of the Bi-2223 superconductor. Besides that, the inclusion of magnetic nanoparticles into the Bi-2223 superconductor is also expected to enhance the flux pinning capability. The addition of Ho (Abdeen et al., 2016), Dy_2O_3 (Zelati et al., 2014), MgO, TiO_2 (Siswayanti et al., 2018), FeO (Suazlina et al., 2017) and γ-Al_2O_3 (Aljurani, 2019) are a few examples of addition that showed different effects on Bi-2223 superconductor.

Instead of metal oxides which are common addition elements, (Bi,Pb)-Sr-Ca-Cu-O systems have been added with metal sulfide. This is interesting because oxygen and sulfur belong to the same group (VI) in the periodic table. Adding CuS nanoparticles to the Bi-2223 phase exhibited higher upper critical magnetic fields and Jc than the non-added samples (Loudhaief et al., 2020). Some works on metal sulfide addition have been done for

$(Bi_{1.6}Pb_{0.4}) Sr_2CaCu_2O_8$ (Bi-2212) superconductor. The appropriate amount of FeS addition increased the flakiness and flux pinning energy of the Bi-2212 superconductor (Masnita & Abd-Shukor, 2020). PbS added Bi-2212 improved the flux pinning strength, although the transition temperature was suppressed (Masnita et al., 2022). Adding ferrimagnetic Cr_2S_3 improved the layered-like microstructure but decreased the critical current density and transition temperature (Farah-Elia et al., 2019).

In Tl-based superconductors, substituting PbS resulted in a higher onset transition temperature, $T_{c\text{-}onset}$, than PbO (Nur-Akasyah et al., 2021). High purity and quality samples were obtained for $(Tl_{1-x}Cr_x) Sr_2CaCu_2O_{7-\delta}$ when using Cr_2S_3 as starting materials compared to Cr_2O_3 (Nur-Akasyah et al., 2020). In the YBCO system, the ferrimagnetic Cr_2S_3 added sample showed the highest transition temperature and critical current density for $x = 0.05$ wt.% (Nur-Syazwani et al., 2018).

Cr_2S_3 is an interesting metal sulfide. Cr_2S_3 is a ferrimagnetic semiconductor with a Curie temperature of 120 K (Maignan et al., 2018; Van Bruggen et al., 1970). The melting point of Cr_2S_3 is 1350°C, which is higher than the formation temperature of the Bi-2223 phase (850°C). The transition temperature of Bi-2223 is approximately 117 K, slightly lower than the Curie temperature of Cr_2S_3. Hence, it is interesting to investigate the interaction between the Bi-2223 superconductor and the ferrimagnetic Cr_2S_3. This work aimed to investigate the effects of ferrimagnetic Cr_2S_3 on the Bi-2223 superconductor. The structural, electrical resistance, AC susceptibility, and critical current density were measured.

7.2 METHOD

The Bi-2223 samples were prepared with a nominal starting composition of $(Bi_{1.6}Pb_{0.4}) Sr_2Ca_2Cu_3O_{10}$ by the co-precipitation method. Bismuth acetate, lead acetate, strontium acetate, calcium acetate, and copper acetate were mixed with acetic acid to form a solution (A). Solution (B), which consisted of oxalic acid, was dissolved in a mixture of water and isopropanol with a ratio of 1:1.5 to achieve a concentration of 0.5 M. Solution (B) was then poured into solution (A) in an ice bath and stirred until a uniform blue suspension was obtained. The slurry was filtered after 10 minutes of reaction. The precipitate was dried at 80°C overnight. The dry powders were ground and pre-calcined for 12 hours at 730°C in the air to remove the remaining volatile materials. Then, the calcined powder was reground and heated at 845°C in the air for 24 hours to eliminate the remaining oxalates. The product was black powder, which was reground and mixed with Cr_2S_3 at 0.05, 0.1, 0.2, and 0.3 wt.% and pressed into pellets of approximately 12.5 mm in diameter and 2 mm in thickness. The pellets were heated at 850°C for 48 hours and slowly cooled to room temperature.

X-ray diffraction (XRD) method using a Bruker D8 Advance Diffractometer with a CuK_α source was used to analyze the crystal structure and phases. The volume fraction of the Bi-2223 and Bi-2212 phases was determined by the sum of the intensities of each phase. All samples underwent resistance measurement using the four-point probe technique with silver paste contacts. The CTI Cryogenics Model 22 closed-cycle helium cryostat and temperature controller Lake Shore Model 340 were used. The transport critical current density, J_{ct}, was measured from 40 K to 77 K using the 1 μV/cm criterion.

An AC susceptometer from Cryo Industry model number REF-1808-AS was used to measure the complex susceptibility. The frequency used was 295 Hz, and the magnetic field was $H = 5$ Oe. Bean's model (Bean, 1964) was used to calculate the critical current density at the peak temperature, T_p, of χ'' with the formula $J_c(T_p) = H/(lw)^{1/2}$ where H is the applied field, l and w are the dimensions of the cross-section of the bar-shaped sample.

7.3 RESULTS AND DISCUSSION

The XRD diffraction patterns of $(Bi_{1.6}Pb_{0.4})\ Sr_2Ca_2Cu_3O_{10}\ (Cr_2S_3)\ x$ for $x = 0$–0.30 wt.% are shown in Figures 7.1(a) and 7.1(b). The Bi-2223 phase can be indexed based on orthorhombic unit cell structure. The volume fraction of Bi-2223, V_{2223}, and Bi-2212, V_{2212}, were calculated using:

$$V_{2223} = \frac{\sum I_{2223}}{\sum I_{2223} + \sum I_{2212}} \times 100\%$$

$$V_{2212} = \frac{\sum I_{2212}}{\sum I_{2223} + \sum I_{2212}} \times 100\%$$

where; I_{2223} and I_{2212} are the sum of intensities of Bi-2223 and Bi-2212 phase, respectively. All samples showed a mixed phase with a nearly 50:50 ratio between the Bi-2212 and Bi-2223 phases. Peaks with H indicated the high T_c Bi-2223 phase and L denoted the low T_c Bi-2212 phase. Cr_2S_3 peak was observed at $2\theta = 52°$ in all added samples. The addition of Cr_2S_3 slightly improved the formation of the Bi-2223 phase.

The lattice parameters for $x = 0$ wt.% sample were $a = 5.383$ Å, $b = 5.420$ Å, and $c = 37.135$ Å. There was an increase in the lattice parameters a and b

as well as the volume of the unit cell when Cr_2S_3 was added to the Bi-2223 superconductor. This suggested that Cr could be substituted in the crystal lattice; however, this assertion must be supported by more direct measurements. Table 7.1 shows the lattice parameters and the percentage of volume fraction for 2223 and 2212 phase superconductors.

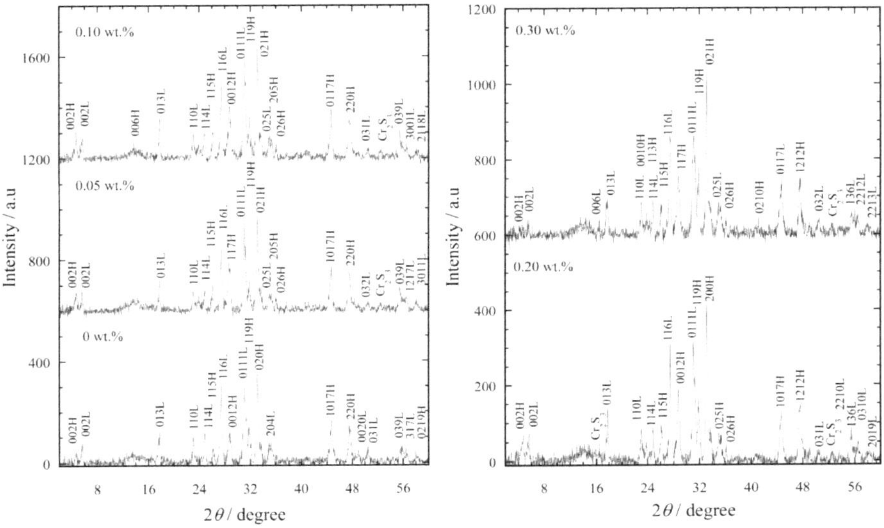

(a) (b)

FIGURE 7.1 XRD patterns of $(Bi_{1.6}Pb_{0.4})\ Sr_2Ca_2Cu_3O_{10}\ (Cr_2S_3)_x$ for (a) $x = 0–0.10$ wt.% and (b) $x = 0.20–0.30$ wt.%. Peaks denoted as 'H' indicate the Bi-2223 phase and 'L' indicate the Bi-2212 phase.

TABLE 7.1 Lattice Parameters, Unit Cell Volume, V_{2223}, and V_{2212} of $(Bi_{1.6}Pb_{0.4})\ Sr_2Ca_2Cu_3O_{10}$ $(Cr_2S_3)x$ for $x = 0$, 0.05, 0.10, 0.20, and 0.30 wt.%.

x/wt.%	*a*/Å	*b*/Å	*c*/Å	*V*/Å³	V_{2223}/%	V_{2212}/%
0.0	5.383	5.420	37.135	1083.4	46	54
0.05	5.385	5.436	37.162	1087.8	52	48
0.10	5.384	5.429	37.112	1084.8	51	49
0.20	5.406	5.436	37.148	1091.7	51	49
0.30	5.385	5.445	37.091	1087.6	49	51

Figure 7.2(a) and 7.2(b) show the electrical resistivity versus temperature curves of $(Bi_{1.6}Pb_{0.4})\ Sr_2Ca_2Cu_3O_{10}\ (Cr_2S_3)x$ for x = 0–0.30 wt.%. All samples

demonstrated the normal state metallic behavior above $T_{c\text{-}onset}$. The x = 0.05 wt.% sample showed the highest $T_{c\text{-}onset}$ and $T_{c\text{-zero}}$, which were 118 K and 104 K, respectively. All Cr_2S_3 added samples generally showed a higher $T_{c\text{-}onset}$ and zero transition temperature, $T_{c\text{-zero}}$, compared with the non-added sample. However, $T_{c\text{-}onset}$ and $T_{c\text{-zero}}$ were slightly decreased as the amount of Cr_2S_3 addition was increased (Figure 7.2). This indicated that only a small amount of Cr_2S_3 is needed to optimize the transition temperature. The transition width, ΔT_c for all samples was between 6 and 14 K.

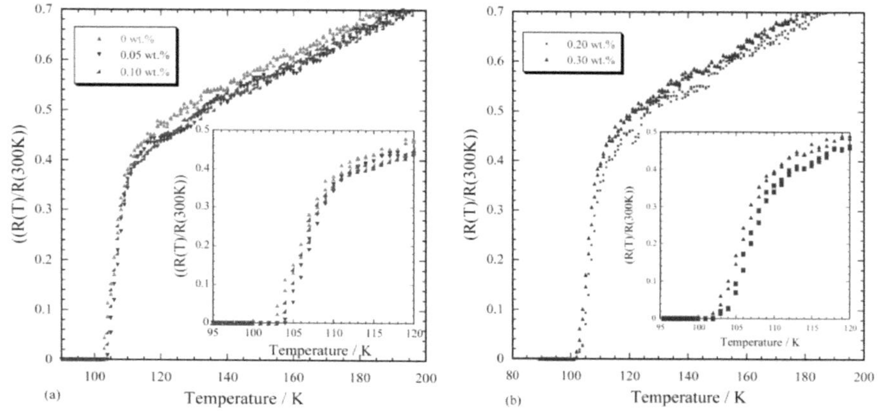

(a) (b)

FIGURE 7.2 Electrical resistance versus temperature of $(Bi_{1.6}Pb_{0.4})$ $Sr_2Ca_2Cu_3O_{10}$ $(Cr_2S_3)_x$ for (a) x = 0–0.10 and (b) x = 0.20–0.30.

The AC susceptibility versus temperature curves of $(Bi_{1.6}Pb_{0.4})$ $Sr_2Ca_2Cu_3O_{10}$ $(Cr_2S_3)_x$ for x = 0–0.30 wt.% are shown in Figures 7.3(a) and 7.3(b). The susceptibility transition temperature of the bulk sample, $T_{c\chi'}$ is indicated by a sudden decrease in the real part, χ' of the complex susceptibility, $\chi = \chi' + i\chi.'$ At this temperature, diamagnetic shielding behavior begins to appear, indicating the onset of bulk superconductivity. The peak at the imaginary part of susceptibility χ, " Tp represents AC losses. When Tp is shifted to lower temperatures, the superconductor's flux pinning energy and intergrain coupling are weakened.

$T_{c\chi'}$ for the non-added sample was 107 K. The $x = 0.3$ wt.% sample showed the highest $T_{c\chi'}$ (109 K). $T_{c\chi'}$ was enhanced when Cr_2S_3 was added to the Bi-2223 superconductor. $Tc\chi'$ showed a plateau for x = 0.05–0.20 wt.% (108 K) samples before increasing to 109 K for x = 0.3 wt.% sample. T_p shifted to a higher temperature when added with Cr_2S_3, indicating the enhancement of intergrain coupling and the flux pinning energy. The highest T_p was 91 K for $x = 0.05$ wt.% sample. At T_p, the intergrain currents reach the value that allows the AC field to pass through the intergrain matrix to the sample center. Hence, using the Bean Model Field, we can calculate the critical current density at T_p, $J_{pcs}(T_p)$ for all samples (Bean, 1964). $J_{cs}(T_p)$ for all samples were between 15 and 17 A cm^{-2}.

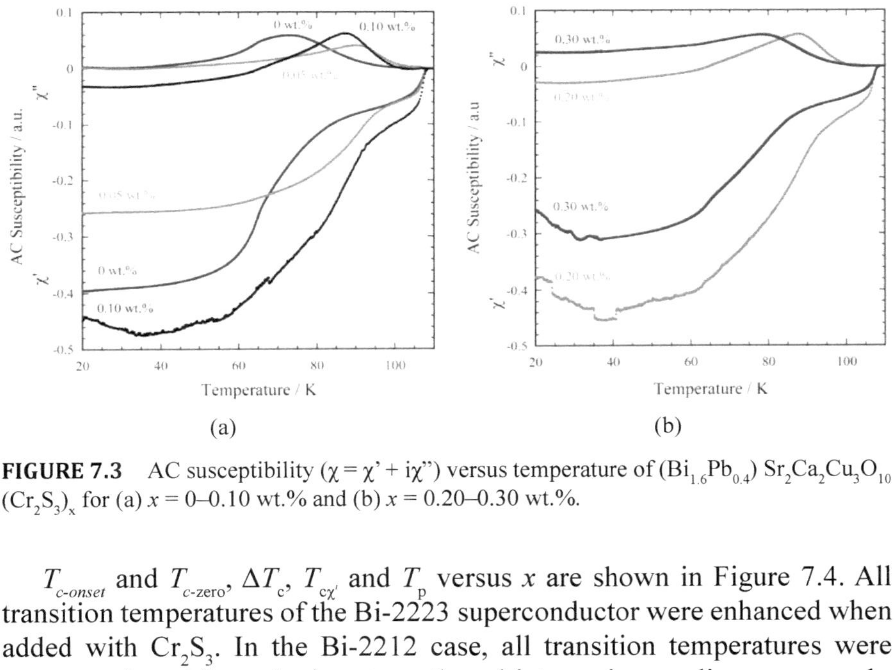

FIGURE 7.3 AC susceptibility ($\chi = \chi' + i\chi''$) versus temperature of $(Bi_{1.6}Pb_{0.4})\ Sr_2Ca_2Cu_3O_{10}$ $(Cr_2S_3)_x$ for (a) $x = 0$–0.10 wt.% and (b) $x = 0.20$–0.30 wt.%.

$T_{c\text{-}onset}$ and $T_{c\text{-}zero}$, ΔT_c, $T_{c\chi'}$ and T_p versus x are shown in Figure 7.4. All transition temperatures of the Bi-2223 superconductor were enhanced when added with Cr_2S_3. In the Bi-2212 case, all transition temperatures were suppressed, and flux pinning strength and intergrain coupling were weakened when added with Cr_2S_3 (Farah-Elia et al., 2019). It was proposed that adding Cr_2S_3 destroyed the superconductivity of Bi-2212 by the emergence of magnetic order at the end of the phase diagram's superconducting dome (Sarkar et al., 2020). However, our results were consistent with previous studies that magnetic addition increased the transition temperatures and critical current density of the Bi-2223 superconductor (Mohd Suib et al., 2019; Kong & Abd-Shukor, 2009). $T_{c\text{-}onset}$, $T_{c\text{-}zero}$, ΔT_c, $T_{c\chi'}$, T_p, $J_{cs}(T_p)$, J_{ct} at 30 and 77 K for $(Bi_{1.6}Pb_{0.4})\ Sr_2Ca_2Cu_3O_{10}\ (Cr_2S_3)_x$ for $x = 0$–0.30 wt.% are shown in Table 7.2.

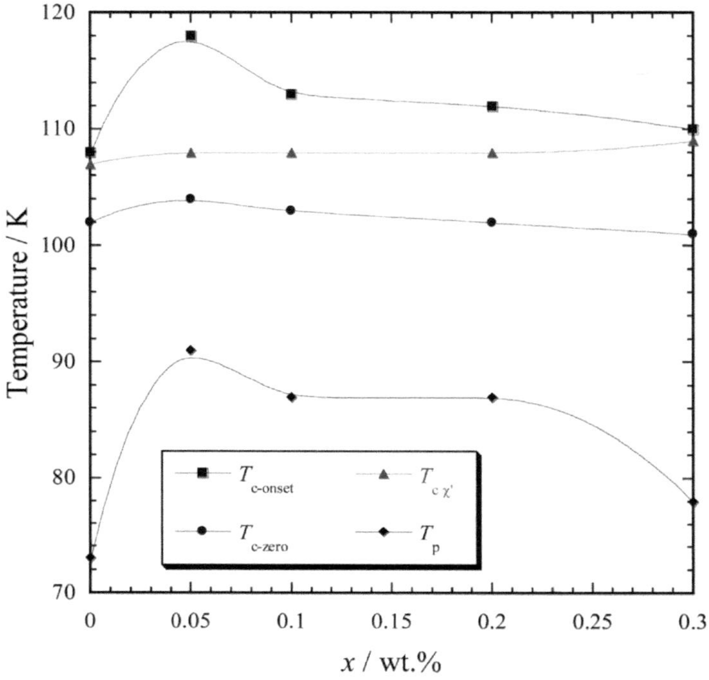

FIGURE 7.4 $T_{c\text{-onset}}$, $T_{c\text{-zero}}$, ΔT_c, $T_{c\chi'}$ and T_p versus x.

TABLE 7.2 $T_{c\text{-onset}}$, $T_{c\text{-zero}}$, ΔT_c, $T_{c\chi'}$, T_p, J_{cs} (T_p), J_{ct} at 30 and 77 K of $(Bi_{1.6}Pb_{0.4})\ Sr_2Ca_2Cu_3O_{10}\ (Cr_2S_3)_x$ for x = 0, 0.05, 0.10, 0.20, and 0.30 wt.%

x/wt.%	$T_{c\text{-onset}}$/K	$T_{c\text{-zero}}$/K	ΔT_c/K	$T_{c\chi'}$/K	T_p/K	J_{cs} (T_p)/Acm^{-2}	J_{ct}/Acm^{-2} (30 K)	J_{ct}/Acm^{-2} (77 K)
0.0	108	102	6	107	73	15	4.71	0.53
0.05	118	104	14	108	91	15	0.39	0.38
0.10	113	103	10	108	87	17	8.37	2.70
0.20	112	102	10	108	87	16	7.32	5.25
0.30	110	101	9	109	78	16	6.12	0.28

Figure 7.5 shows the temperature dependence of the transport critical current density, J_{ct}, of $(Bi_{1.6}Pb_{0.4})\ Sr_2Ca_2Cu_3O_{10}\ (Cr_2S_3)_x$ for x = 0–0.30 wt.%. J_{ct} was generally enhanced for all added samples except for the x = 0.05 wt.% sample. Despite having the highest $T_{c\text{-onset}}$ and $T_{c\text{-zero}}$, the x = 0.05 wt.% sample showed a lower transport critical current density. The highest J_{ct} at 30 K was for the x = 0.10 wt.% sample, which was 8.37 A cm^{-2}, while at 77

K it was 5.25 A cm^{-2} for the x = 0.20 wt.% sample. This demonstrates that the optimum amount of Cr_2S_3 addition to enhance the critical current density was x = 0.10–0.20 wt.%. This result also shows that Cr_2S_3 could act as an effective pinning center and improve grain connectivity.

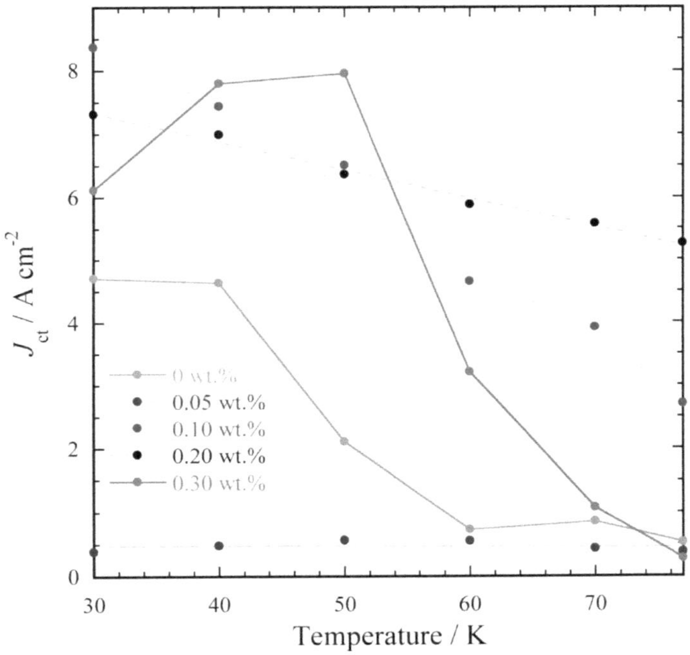

FIGURE 7.5 J_{ct} versus temperature of $(Bi_{1.6}Pb_{0.4}) Sr_2Ca_2Cu_3O_{10} (Cr_2S_3)_x$ for x = 0–0.30 wt.%.

7.4 CONCLUSION

In conclusion, the effect of ferrimagnetic Cr_2S_3 addition on $(Bi_{1.6}Pb_{0.4})Sr_2Ca_2Cu_3O_{10}$ superconductor has been investigated. The ferrimagnetic Cr_2S_3 addition significantly improved $T_{c\text{-onset}}$, $T_{c\text{-zero}}$, $T_{c\chi'}$ and T_p. T_p shifted to higher temperatures in all added samples indicating that Cr_2S_3 enhanced the flux pinning energy and intergranular coupling of the superconductor. Besides that, Cr_2S_3 addition increased the transport critical current density by more than two times, demonstrating that Cr_2S_3 can act as an effective flux pinning center and improve intergrain connectivity. Further works on Cr_2S_3 addition on other forms of Bi-2223 superconductor such as in thin film and tape can be useful in determining the application potential for this superconductor.

ACKNOWLEDGMENT

This research was supported by the Ministry of Higher Education of Malaysia (grant no. FRGS/1/2020/STG07/UKM/01/1) and the DUCS-Faculty Grant from Universiti Teknologi MARA (grant no. 600-UiTMSEL (PI.5/4) (123/2022).

KEYWORDS

- **chromium sulfide**
- **co-precipitation method**
- **critical current density**
- **X-ray diffraction**

REFERENCES

Abdeen, W., Marahba, S., Awad, R., Abou Aly, A. I., Ibrahim, I. H., & Matar, M. (2016). Electrical and mechanical properties of (Bi,Pb)-2223 substituted by holmium. *Journal of Advanced Ceramics, 5*, 54–69.

Aljurani, B. A. (2019). Influence of γ Al_2O_3 nanoparticles addition and pressure on the electrical, structural, and microstructure properties of Bi-2223 superconducting system. *Indian Journal of Science, 9*, 16570–16577.

Bean, C. P. (1964). Magnetization of high-field superconductors. *Reviews of Modern Physics, 36*, 31–39.

Farah-Elia, N. A. R., Ilhamsyah, A. B. P., & Abd-Shukor, R. (2019). Ferrimagnetic Cr_2S_3 effects on $(Bi_{1.6}Pb_{0.4})Sr_2CaCu_2O_8$ superconductor. *Journal of Materials Science: Materials in Electronics, 30*, 12031–12035.

Hamadneh, I., Halim, S. A., & Lee, C. K. (2006). Characterization of $Bi_{1.6}Pb_{0.4}Sr_2Ca_2Cu_3Oy$ ceramic superconductor prepared via coprecipitation method at different sintering times. *Journal of Materials Science, 41*, 5526–5530.

Ismail, M., Abd-Shukor, R., Hamadneh, I., & Halim, S. A. (2004). Transport current density of Ag-sheathed superconductor tapes using Bi-Sr-Ca-Cu-O powders prepared by the co-precipitation method. *Journal of Materials Science, 39*, 3517–3519.

Kong, W., & Abd-Shukor, R. (2009). Enhanced electrical transport properties of nano $NiFe_2O_4$-added $(Bi_{1.6}Pb_{0.4})$ $Sr_2Ca_2Cu_3O_{10}$ superconductor. *Journal of Superconductivity and Novel Magnetism, 23*, 257–263.

Loudhaief, N., Ben Salem, M., Labiadh, H., & Zouaoui, M. (2020). Electrical properties and fluctuation-induced conductivity studies of Bi-based superconductors added by CuS nanoparticles synthesized through the aqueous route. *Materials Chemistry and Physics, 242*, 122464.

Maignan, A., Guilmeau, E., Gascoin, F., Bréard, Y., & Hardy, V. (2012). Revisiting some chalcogenides for thermoelectricity. *Science and Technology of Advanced Materials, 13*, 053003.

Masnita, M. J., & Abd-Shukor, R. (2020). Iron sulfide effects on AC susceptibility and electrical properties of $Bi_{1.6}Pb_{0.4}Sr_2CaCu_2O_8$ superconductor. *Results in Physics, 17*, 103177.

Masnita, M. J., Awang, R., & Abd-Shukor, R. (2022). AC susceptibility and electrical properties of PbS added Bi. *Sains Malaysiana, 51*, 315–328.

Mohd Suib, N. R., Abu Bakar, I. P., Ibrahim, N. B., & Abd-Shukor, R. (2019). Critical current density and AC susceptibility of $Bi_{1.6}Pb_{0.4}Sr_2Ca_2Cu_3O_{10}$ superconductor with $Ni_{0.5}Zn_{0.5}Fe_2O_4$ nanomaterial. *Journal of Superconductivity and Novel Magnetism, 32*, 145–150.

Nur-Akasyah, J., Ilhamsyah, A. B. P., & Abd-Shukor, R. (2020). Elemental substitution at Tl-site of $T_1Sr_2CaCu_2O_7$ superconductor: A review. *Ceramics International, 46*, 18413–18418.

Nur-Akasyah, J., Nur-Diyana, M. A. F., & Abd-Shukor, R. (2021). Influence of lead (II) sulfide on the formation and superconductivity of (Tl1-xPbx)$Sr_2CaCu_2O_7$ (x = 0.1–0.8). *International Journal of Electrochemical Science, 16*, 2, 211015.

Nur-Syazwani, A. C. A., & Abd-Shukor, R. (2018). Effect of ferrimagnetic Cr_2S_3 on AC susceptibility and superconducting properties of $YBa_2Cu_3O_7$-δ. *Journal of Superconductivity and Novel Magnetism, 32*, 863–868.

Sarkar, T., Wei, D. S., Zhang, J., Poniatowski, N. R., Mandal, P. R., Kapitulnik, A., & Greene, R. L. (2020). Ferromagnetic order beyond the superconducting dome in a cuprate superconductor. *Science, 368*, 532–534.

Siswayanti, B., Yudanto, S. D., Imaduddin, A., Sebleku, P., Hendrik, Nugraha, H., Lusiana, Amal, I., Darsono, N., Amri, F., Syuhada, N., Syahfina, R., Herbirowo, S., & Pramono, A. W. (2018). The effect of MgO, CNT, and TiO_2 addition on transport properties and formation (Bi,Pb)-2223 prepared by solid-state method and recurrent sintering. *AIP Conference Proceedings, 1964*, 020039.

Suazlina, M. A., Azhan, H., Syamsyir, S. A., Azman, K., Razali, W. A. W., Yahya, N., Mohamed, R., & Yusainee, S. Y. S. (2017). Fabrication and characterization of $Bi_{1.6}Pb_{0.4}Sr_2Ca_2Cu_3$-xFexOy superconductor. *Solid State Phenomena, 268*, 311–314.

Takada, J., Kitaguchi, H., Egi, T., Oda, K., Miura, Y., Mazaki, H., Ikeda, Y., Hiroi, Z., Takano, M., & Tomii, Y. (1990). Superconductor with Tc = 117 K in the Bi-Pb-Sr-Ca-Cu-O system. *Physica C: Superconductivity, 170*, 249–253.

Van Bruggen, C. F., Vellinga, M. B., & Haas, C. (1970). Semiconducting and magnetic properties of rhombohedral Cr_2S_3. *Journal of Solid-State Chemistry, 2*, 303–308.

Zelati, A., Amirabadizadeh, A., Kompany, A., Salamati, H., & Sonier, J. (2014). Effects of Dy_2O_3 nanoparticle addition on structural and superconducting properties of BSCCO. *Indian Journal of Science and Technology, 7*, 123–124.

CHAPTER 8

The Influence of Varying Donor Concentration of N-CDS Buffer Layer on the Performance of Thin Film Photovoltaic Devices

ASMAA SOHEIL NAJM

Iraqi Oil Exploration Company, Ministry of Oil, Iraq

ABSTRACT

A CD's thin film buffer layer has been widely employed as a traditional n-type heterojunction partner in both existing and new thin film photovoltaic devices. Despite this, the main issue with CDs is controlling their resistivity without affecting their optical and structural characteristics. The main objective of this study was to utilize numerical modeling to examine how changing CD carrier concentrations as a buffer layer affects the performance of various absorber layers. Absorption layer evaluations assume a bandgap range of 0.9 to 1.5 eV and a conductive band offset of –0.5 to 0.5. Open-circuit voltage (Voc), short-circuit current (Jsc), fill factor (FF), and photoelectric conversion efficiency (η) were all numerically examined. According to the results, a buffer layer with a higher bandgap and higher carrier concentration is necessary to achieve a device with a higher conversion efficiency and a wider bandgap-CBO window. The modeling results, most notably, reveal that the electrical characteristics of the buffer layer are critical in determining the performance of the solar cells.

Engineering Design and Technical Applications of Physical Science: An Integrated Approach for Sustainable Development. Chin Hua Chia, Sarani Zakaria, Ruey Shan Chen, & Md. Shahariar Chowdhury (Eds.)

8.1 INTRODUCTION

Today, much of the world's power manufacturing comes from fossil sources. These sources present an enormous environmental pollution disadvantage (Zhuk, Kushwaha, Wong, & Masudy-panah, 2016). Thin film solar cells like copper indium gallium disulfide (CIGS) and cadmium telluride (CdTe) compete with silicon solar cells in the PV market. Recently, Cu (In, Ga) Se_2 (CIGS) solar cells have achieved a tremendous increase in efficiency to 22.9% (Green, Hishikawa, Dunlop, Levi, Hohl-Ebinger, Yoshita, & Ho-Baillie, 2018). The scarcity and the skyrocketing material cost of In and Ga in CIGS are major setbacks (Feurer et al., 2016). Although CdTe is stable with a conversion efficiency reaching 22.1%, the toxicity of selenium in CIGSSe and CdTe is a concern for their progress (Green, Hishikawa, Dunlop, Levi, Hohl-Ebinger, & Ho-Baillie, 2018). Kesterite-symmetrical non-toxic earth-abundant materials such as CZTS (Cu_2ZnSnS_4), CZTSe (Cu_2Zn-$SnSe_4$) (Ananthakumar, Ram Kumar, & Moorthy Babu, 2016; Adewoyin, Olopade, & Chendo, 2017; Boutebakh, Zeggar, Attaf, & Aida, 2017), CFTS (Cu_2FeSnS_4), CFTSe (Cu_2FeSnS_4), and their alloys are emerging as the most promising replacement for the chalcopyrite absorbers (CIGS, CIGSe) (Zhou, Hsu, Duan, Bob, Yang, Song et al., 2013). Thus, regarding thin-film solar cells, the selection of the proper buffer layer is a crucial factor. The role of the buffer layer is to absorb energy light from the spectrum's high-energy end. The layer should be thin and broad (bandgap 2.8 eV or above) to allow all available light to reach the absorbing layer via an interface (hetero-connection) (Zhou, Hsu, Duan, Bob, Yang, Song et al., 2013). The choice of the buffer layer is essential for thin film structures to obtain better features of solar cells (Najm et al., 2022). In addition to that, providing the device's structural stability and fixing the electrostatic conditions within the layer of absorber were considered the main two objectives for selecting the buffer layer (Najm et al., 2020). Different materials can be utilized as a thin film solar cell buffer layer, such as ZnO, ZnS, CDs, etc., (Holi, Al-Zahrani, Najm, Chelvanathan, & Amin, 2020). Nevertheless, these buffer layers are affected by a complicated reaction mechanism and light-soaking effects, posing a significant concern for cell reliability (Priya & Singh, 2021). Cadmium sulfide (CDs) is a significant II-VI compound semiconductor with remarkable transparency, direct bandgap transition (e.g., 2.42 eV), efficient electron affinity, and good n-type conductivity (Slonopas, Ryan, Foley, Sun, Sun, Globus, & Norris, 2016). Additionally, CDs enhance the lattice heterojunction interface match, increase the excess carrier lifetime, and optimize the device's band alignment (Voznyi et al., 2016).

Throughout this work, various scenarios that significantly affect power conversion efficiency in relation to the bandgap and carrier concentration in the buffer level are shown. It utilizes a generic CIGS solar cell-inspired absorber/buffer/window architecture. In addition, it will use the (X) term to describe the absorber layer in this paper. Modeling devices and numerical cell simulations are achieved to evaluate the effects of different donor concentrations on the efficiency of solar cells within absorption layers. In a two-dimensional performance contour, physical, optoelectronic, and energy gap, carrier concentration, and electron affinity, as electrical parameters, are varied and modifications are recorded and plotted for modifications in J_{SC}, V_{OC}, FF, and PCE (Ƞ).

8.2 METHODOLOGY

In order to better understand device characteristics, a modeling device was used to bridge the gap between empirical and theoretical research. Numerical simulations provide an excellent opportunity to analyze the operational principles of several PV systems before the practical cell is established. Consequently, our device simulations were carried out for solar cells p-X, n-CDs, which represent the typical device architecture of solar cells with thin film heterojunction. We used the device simulation tool SCAPS, a one-dimensional simulator of solar cell devices created at ELIS, University of Gent. SCAPS' modeling capabilities are primarily designed to simulate the characteristics of CIGS solar cells; however, it has been shown to be efficient in a variety of cell types. The device structure in our simulation is (Al/i-ZnO/n–CDs/p-Absorber/Mo/SLG), as described in Figure 8.1.

Each layer is described in detail in the following, together with significant theoretical concepts to provide a sensible reason for the selection of material and hetero-interface electrical characteristics, as shown in Table 8.1.

8.3 RESULTS AND DISCUSSION

8.3.1 ABSORBER LAYER WITH VARIABLE DONOR CONCENTRATION

The primary purpose of this numerical simulation study is to examine the impact of optical and electrical characterizations of buffer layers, specifically the bandgap and carrier density, on the performance of CIGS photovoltaic devices. In the simulation, the carrier density (N_D) from (1×10^{13} to 1×10^{17})

cm^{-3} has been used for the buffer layer, with a conduction band offset (–0.5 to 0.5) and a bandgap (0.9 to 1.5) eV for the absorber layer.

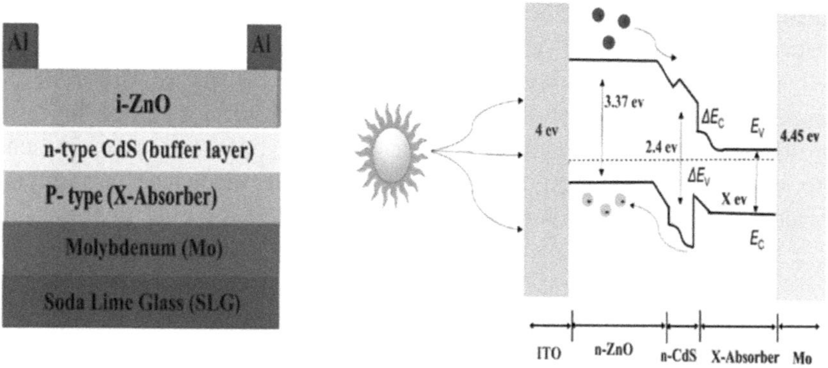

(a) (b)

FIGURE 8.1 The simulated device for CIGS solar cell.

TABLE 8.1 List of Physical and Electronic Properties (Najm et al., 2022; Hossain et al., 2019; Belghachi & Limam, 2017; Basak & Singh, 2021)

Parameters	ZnO	CDs	Absorber
Layer thickness (nm)	50	50	2500
The dielectric constant, ε_r	9	10	13.6
Electron mobility, μ_n ($cm^2/V.s$)	100	100	100
Hole mobility, μ_p ($cm^2/V.s$)	25	25	25
Acceptor concentration, NA (cm−3)	0	0	1×10^{16}
Donor concentration, ND (cm−3)	1×10^{18}	~	0
Bandgap, E_g (eV)	3.3	2.4	0.9–1.5
Electron affinity, χ (eV)	4.4	4.2	3.7–4.7
Effective density of states in Conduction band, NC (cm^{-3})	2.2×10^{18}	2.2×10^{18}	2.2×10^{18}
Effective density of states in Valence band, NV (cm^{-3})	1.8×10^{19}	1.8×10^{19}	1.8×10^{19}
Electron thermal velocity (cm s^{-1})	1×10^7	1×10^7	1×10^7
Hole thermal velocity (cm s^{-1})	1×10^7	1×10^7	1×10^7

Additionally, numerical results demonstrated that when the donor concentration (N_D) increased, the conduction band edge (E_C) of the buffer layer decreased in all simulated situations (Basak & Singh, 2021). As previously indicated, as the N_D increases, the effective density of states in the conduction band (N_C) decreases, which is associated with a low effective mass. A low effective mass leads to higher charge carrier mobility and low exciton separation energy (Abdulghani, Najm, Holi, Al-Zahrani, Al-Zahrani, & Moria, 2022). Here, an attempt is made to explain the observed pattern illustrated in the next section by analyzing the important photovoltaic performance factors that eventually determine the conversion efficiency of solar cells.

8.3.2 SHORT CIRCUIT CURRENT (J_{SC})

Short-circuit current (J_{SC}) in Figure 8.2 displays the simultaneous impact of buffer layer carrier concentration and bandgap. There is a direct correlation between increasing carrier concentration and increasing photo-generated electron lifetime, which results in an increased possibility of carrier collection

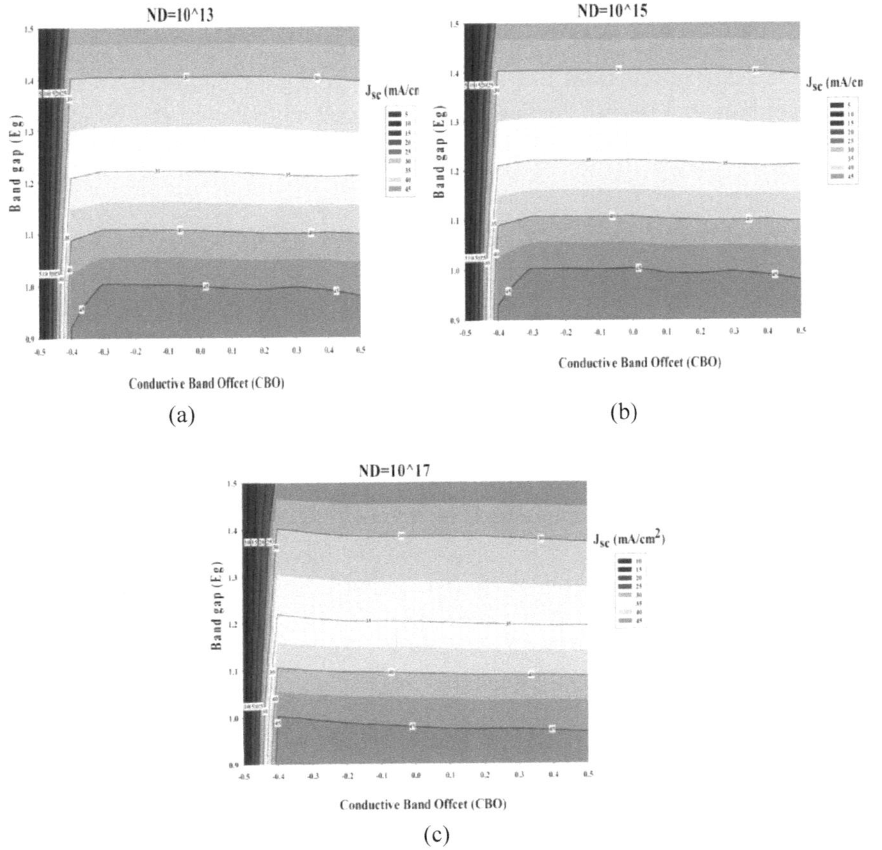

FIGURE 8.2 Photovoltaic performance parameters for J_{SC} at; (a) ND: 10^{13} cm^{-3} (b) ND: 10^{15} cm^{-3} (c) and ND: 10^{17} cm^{-3}.

and thus higher J_{SC} (Gunawan et al., 2019). It can be seen that in all cases, contours from the higher buffer layer to the lower donor concentration region (bottom) are almost similar.

8.3.3 OPEN CIRCUIT VOLTAGE (V_{OC})

It is known that the open circuit voltage (V_{OC}) is sensitive to changing the defect state density, in particular, to change the location of the defect state, and increases linearly with E_g due to the increased built-in electrical potential (Ostadebrahim & Dehghani, 2020). From our assumption, it is suggested that no defect can be effective in the absorber layer; thus, in this case, V_{OC} is dependent only on the buffer/absorber layer band alignment where the relative bandgap changes also indicate that the V_{OC} is directly affected, as shown in Figure 8.3. Based on the general theory of semiconductors, photo-generated charge carriers are reduced with a higher bandgap value, while V_{OC} is enhanced (Weiss, Bissig, Feurer, Carron, Buecheler, & Tiwari, 2019). Despite this, the results indicated that V_{OC} almost did not change.

8.3.4 FILL FACTOR (FF)

The fill factor is defined as the changing measurement of the "rectangularity" of the current-voltage curve for a solar cell. This value varies from 0.26 to 0.84 in our situation. FF depends mainly on two parameters, V_{OC} and the recombination method in the depletion region. As shown in Figure 8.4, a linear dependency of FF on carrier concentration has been noted, with the variation in buffer layer mobility having an extremely minimal impact. The FF onset of the carrier concentration from $10^{13}/cm^3$ up to $10^{17}/cm^3$ slightly increased.

8.3.5 THE CONVERSION EFFICIENCY (H)

The conversion efficiency (PCE) is primarily the J_{sc}, V_{oc}, and FF output. Since the buffer layer carrier concentration is up to $10^{17}/cm^3$, V_{OC} and FF parameters indicate a net increase. A comparable pattern for PCE will be achieved as shown in Figure 8.5. Moreover, carrier concentration plays an important role, as it is apparent from the results how the efficiency improved, and the wide range is observed. This is due to higher N_D induced higher photo-generated in the n-CdS/X junction towards the back contact, thus causing an enhanced electron current (Islam, Kumar, & Thakur, 2021). For example, according to Table 8.2, the bandgap of the CIGS solar cell is 1.14 eV matching with CBO as 0 eV, higher efficiency does not reach 20%, but when the N_D increases it becomes closer to 20%, Table 8.3. As well as in Perovskite solar cell with a bandgap equal to 1.5 eV and CBO (–0.27) eV, here also, it can clearly be seen the difference, with increasing N_D the efficiency improves to become 7.3%. As much as the electron affinity becomes higher, the highest results are obtained (Najm, Chelvanathan, Tiong, Ferdaous et al., 2021).

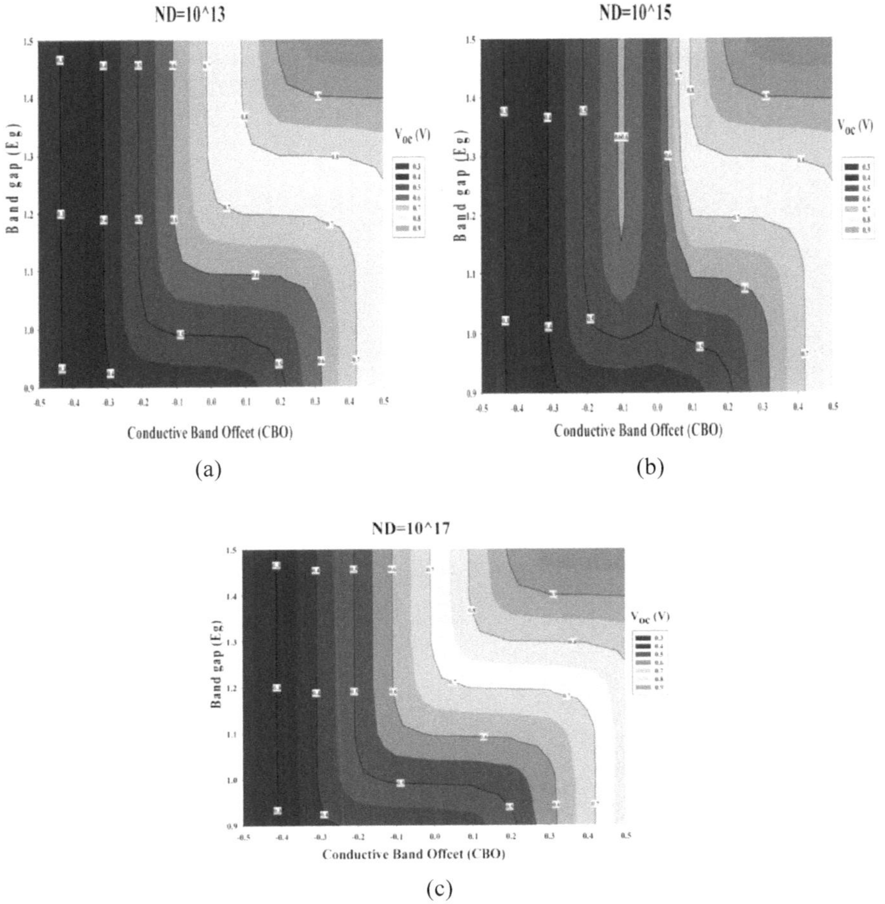

FIGURE 8.3 Photovoltaic performance parameters for V_{OC} at; (a) ND: 10^{13} cm^{-3} (b) ND: 10^{15} cm^{-3} (c) and ND: 10^{17} cm^{-3}.

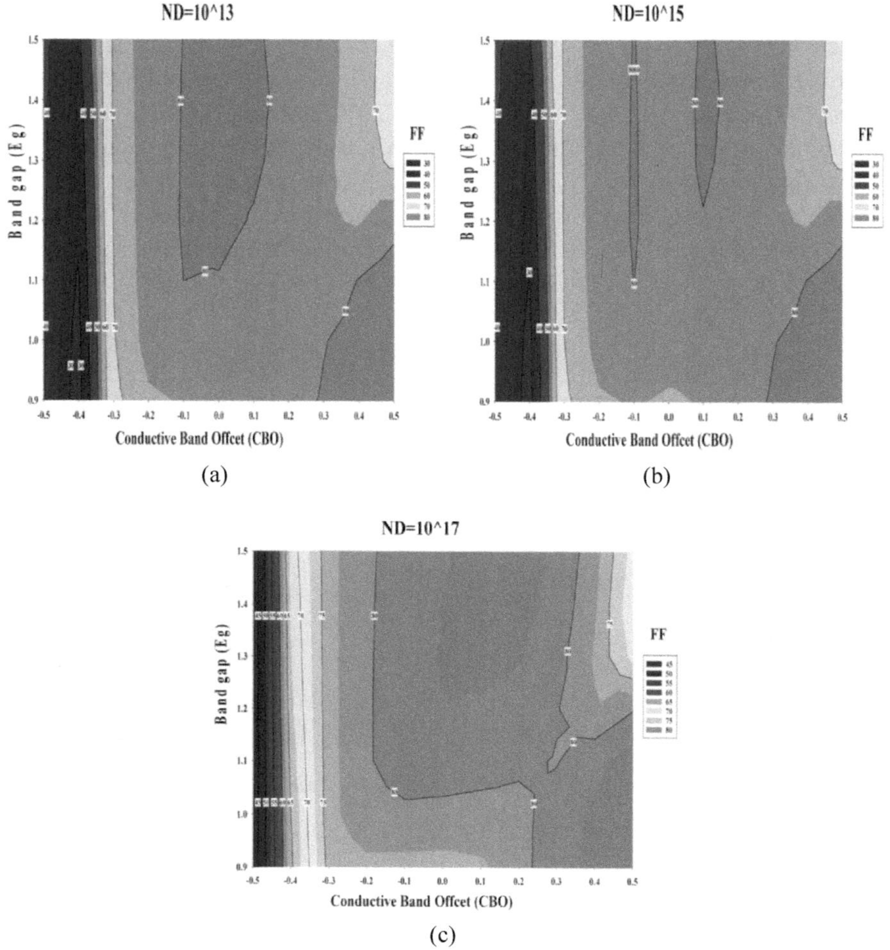

FIGURE 8.4 Photovoltaic performance parameters for FF at; (a) ND: 10^{13} cm^{-3} (b) ND: 10^{15} cm^{-3} (c) and ND: 10^{17} cm^{-3}

8.4 CONCLUSION

In conclusion, the performance of solar cells is numerically simulated via the SCAPS-1D software. The numerical parameters have been adjusted by changing the carrier density, bandgap, and conductive band offset of the X absorber layer to analyze their effects on the cell performance. It has also been proven in this research that the conversion efficiency is affected by the difference in the bandgap of the absorber layer and the carrier density in the p-X layer. The reduction in the bandgap decreases the device's performance, improving V_{OC} from 0.48 V to 0.90 V when the bandgap increases from 0.9 eV to 1.5 eV. The opposite trend has been noticeable in both J_{SC} and conversion efficiency as it is reported dropped from 47.18 (mA/cm^2) to 26.10 (mA/cm^2) and 30.94% to 17.33% for J_{SC} and conversion efficiency, respectively. This can be explained by the increased energy-level division between Fermi and the CDs buffer layer.

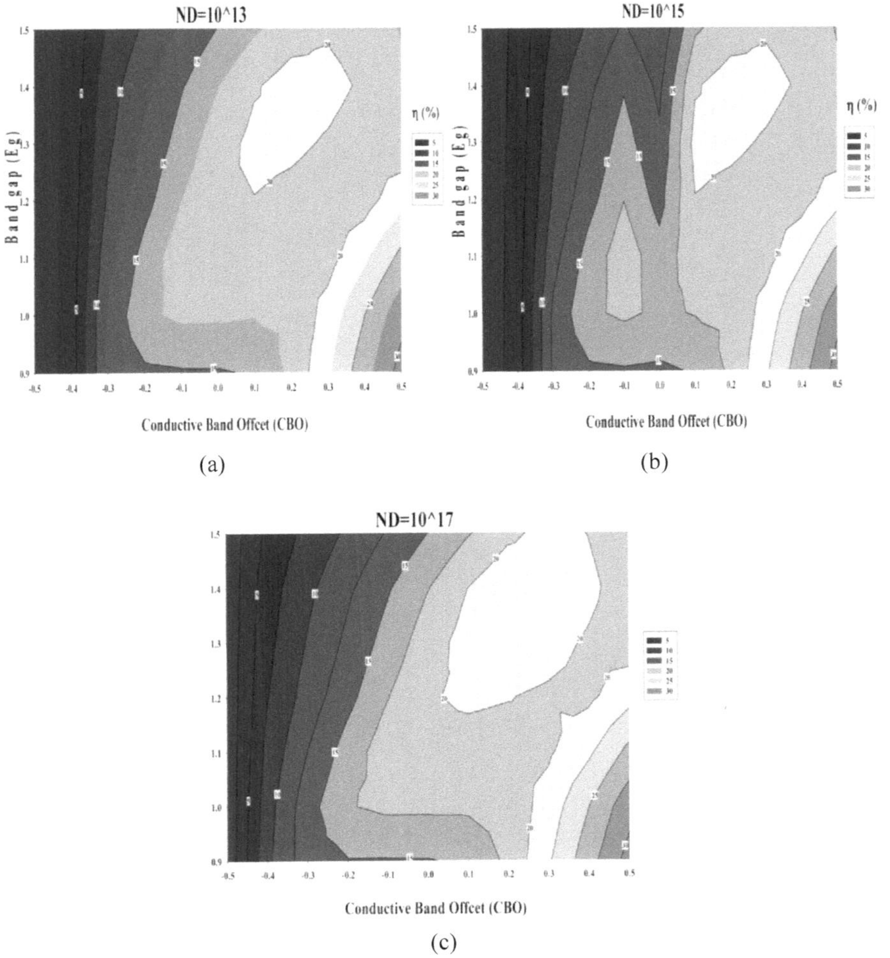

FIGURE 8.5 Photovoltaic performance parameters for PEC at; (a) N_D: 10^{13} cm^{-3} (b) N_D: 10^{15} cm^{-3} (c) and N_D: 10^{17} cm^{-3}.

TABLE 8.2 Examples of Varying of Solar Cells

Solar Cell Type	Bandgap (eV)	Electron Affinity, (eV)	Conductive Band Offset CBO (eV)
Si	1.12	4.05	–0.15
CIGS	1.15	4.5	0.3
CZTS	1.5	4.5	0.3
CdTe	1.5	3.9	–0.3
FeS_2	0.95	4.71	0.51
Perovskite	1.5	3.93	–0.27

TABLE 8.3 Solar Cells Parameters with Different Donor Concentration

Solar Cell	VOC (V)	JSC (mA/cm^2)	FF	η (%)
a) ND=10^{13}				
Si	0.65	45.17	89.60	21.28
CIGS	0.65	41.78	80.01	19.89
CZTS	0.98	26.17	76.60	19.71
CdTe	0.41	26.20	71.67	7.70
FeS_2	0.73	44.41	80.15	29.31
Perovskite	0.36	23.58	64.50	6.93
b) ND=10^{15}				
Si	0.65	45.17	89.64	21.29
CIGS	0.65	41.78	80.05	19.90
CZTS	0.98	26.17	76.67	19.72
CdTe	0.41	26.20	71.90	7.73
FeS_2	0.73	44.41	80.15	29.31
Perovskite	0.36	23.58	64.71	6.95
c) ND=10^{17}				
Si	0.65	44.59	90.94	21.30
CIGS	0.65	41.26	83.48	20.49
CZTS	0.98	25.62	82.17	20.70
CdTe	0.40	25.85	76.76	8.11
FeS_2	0.73	44.30	80.34	29.31
Perovskite	0.36	23.27	69.08	7.30

KEYWORDS

- **bandgap**
- **buffer layer**
- **cadmium sulfide**
- **carrier concentration**
- **CdS thin film**
- **conductive band offset**
- **copper indium gallium disulfide**

REFERENCES

Abdulghani, Z. R., Najm, A. S., Holi, A. M., Al-Zahrani, A. A., Al-Zahrani, K. S., & Moria, H. (2022). Numerical simulation of quantum dots as a buffer layer in CIGS solar cells: A comparative study. *Scientific Reports, 12*, 8099. https://doi.org/10.1038/s41598-022-12234-0.

Adewoyin, A. D., Olopade, M. A., & Chendo, M. (2017). Enhancement of the conversion efficiency of Cu_2ZnSnS_4 thin film solar cell through the optimization of some device parameters. *Optik, 133*, 122–131. https://doi.org/10.1016/j.ijleo.2017.01.008.

Ananthakumar, S., Ram Kumar, J., & Moorthy Babu, S. (2016). Synthesis of $Cu_2ZnSnSe_4$ hierarchical nanostructures by colloidal method. *Optik, 127*(22), 10360–10365. https://doi.org/10.1016/j.ijleo.2016.08.058.

Basak, A., & Singh, U. P. (2021). Numerical modeling and analysis of earth-abundant Sb_2S_3 and Sb_2Se_3-based solar cells using SCAPS-1D. *Solar Energy Materials and Solar Cells, 230*, 111184. https://doi.org/10.1016/j.solmat.2021.111184.

Belghachi, A., & Limam, N. (2017). Effect of the absorber layer band-gap on CIGS solar cell. *Chinese Journal of Physics, 55*(4), 1127–1134. https://doi.org/10.1016/j.cjph.2017.01.011.

Boutebakh, F. Z., Zeggar, M. L., Attaf, N., & Aida, M. S. (2017). Electrical properties and back contact study of CZTS/ZnS heterojunction. *Optik, 144*, 180–190. https://doi.org/10.1016/j.ijleo.2017.06.080.

Feurer, T., Reinhard, P., Avancini, E., Bissig, B., Löckinger, J., Fuchs, P., Carron, R., Weiss, T. P., Perrenoud, J., Stutterheim, S., Buecheler, S., & Tiwari, A. N. (2016). Progress in thin-film CIGS photovoltaics – Research and development, manufacturing, and applications. *Progress in Photovoltaics: Research and Applications, 25*(7), 645–667. https://doi.org/10.1002/pip.2811.

Green, M. A., Hishikawa, Y., Dunlop, E. D., Levi, D. H., Hohl-Ebinger, J., Yoshita, M., & Ho-Baillie, A. W. Y. (2018). Solar cell efficiency tables (Version 53). *Progress in Photovoltaics: Research and Applications, 27*(1), 3–12. https://doi.org/10.1002/pip.3102.

Green, M. A., Hishikawa, Y., Dunlop, E. D., Levi, D. H., Hohl-Ebinger, J., & Ho-Baillie, A. W. Y. (2018). Solar cell efficiency tables (Version 52). *Progress in Photovoltaics: Research and Applications, 26*(7), 427–436. https://doi.org/10.1002/pip.3040.

Gunawan, O., Pae, S. R., Bishop, D. M., Virgus, Y., Noh, J. H., Jeon, N. J., Lee, Y. S., Shao, X., Todorov, T., Mitzi, D. B., & Shin, B. (2019). Carrier-resolved photo-Hall effect. *Nature, 575*, 151–155. https://doi.org/10.1038/s41586-019-1632-2.

He, X., Wu, L., Hao, X., Zhang, J., Li, C., Wang, W., Feng, L., & Du, Z. (2019). The band structures of Zn_{1-x}MgxO (In) and the simulation of CdTe solar cells with a Zn_{1-x}MgxO(In) window layer by SCAPS. *Energies, 12*(2), 291. https://doi.org/10.3390/en12020291.

Holi, A. M., Al-Zahrani, A. A., Najm, A. S., Chelvanathan, P., & Amin, N. (2020). PbS/CdS/ZnO nanowire arrays: Synthesis, structural, optical, electrical, and photoelectrochemical properties. *Chemical Physics Letters, 750*, 137486. https://doi.org/10.1016/j.cplett.2020.137486.

Hossain, E. S., Chelvanathan, P., Shahahmadi, S. A., Bais, B., Akhtaruzzaman, Md., Tiong, S. K., Sopian, K., & Amin, N. (2019). Fabrication of Cu_2SnS_3 thin film solar cells by sulphurization of sequentially sputtered Sn/CuSn metallic stacked precursors. *Solar Energy, 177*, 262–273. https://doi.org/10.1016/j.solener.2018.10.081.

Islam, M. T., Kumar, A., & Thakur, A. K. (2021). Defect density control using an intrinsic layer to enhance conversion efficiency in an optimized SnS solar cell. *Journal of Electronic Materials, 50*, 3603–3613. https://doi.org/10.1007/s11664-021-08881-0.

Lindahl, J., Keller, J., Donzel-Gargand, O., Szaniawski, P., Edoff, M., & Törndahl, T. (2016). Deposition temperature-induced conduction band changes in zinc tin oxide buffer layers for $Cu(In,Ga)Se_2$ solar cells. *Solar Energy Materials and Solar Cells, 144*, 684–690. https://doi.org/10.1016/j.solmat.2015.09.048.

Najm, A. S., Chelvanathan, P., Tiong, S. K., Ferdaous, M. T., Shahahmadi, S. A., Yusoff, Y., Sopian, K., & Amin, N. (2021). Numerical insights into the influence of electrical properties of n-CdS buffer layer on the performance of SLG/Mo/p-Absorber/n-CdS/n-ZnO/Ag configured thin film photovoltaic devices. *Coatings, 11*(1), 52. https://doi.org/10.3390/coatings11010052.

Najm, A. S., Chowdhury, M. S., Munna, F. T., Chelvanathan, P., Selvanathan, V., Aminuzzaman, M., Techato, K., Amin, N., & Akhtaruzzaman, M. (2020). Impact of cadmium salt concentration on CdS nanoparticles synthesized by chemical precipitation method. *Chalcogenide Letters, 17*(11), 537–547.

Najm, A. S., Naeem, H. S., Alabboodi, K. O., Hasbullah, S. A., Hasan, H. A., Holi, A. M., Al-Zahrani, A. A., Sopian, K., B., Sh. Majdi, H., & Sultan, A. J. (2022). New systematic study approach of green synthesis CdS thin film via Salvia dye. *Scientific Reports, 12*, 12521. https://doi.org/10.1038/s41598-022-16733-y.

Najm, A. S., Naeem, H. S., Alwarid, D. A. R. M., Aljuhani, A., Hasbullah, S. A., Hasan, H. A., Sopian, K., Bais, B., Al-Iessa, H. J., Sh. Majdi, H., Sultan, A. J., & Moria, H. (2022). Mechanism of chemical bath deposition of CdS thin films: Influence of Sulphur precursor concentration on microstructural and optoelectronic characterizations. *Coatings, 12*(10), 1400. https://doi.org/10.3390/coatings12101400.

Ostadebrahim, M., & Dehghani, H. (2020). Improving the photovoltaic performance of $CdSe_{0.2}S_{0.8}$ alloyed quantum dot sensitized solar cells using CdMnSe outer quantum dot. *Solar Energy, 199*, 901–910. https://doi.org/10.1016/j.solener.2019.10.036.

Priya, A., & Singh, S. N. (2021). Enhancement of efficiency and external quantum efficiency of CIGSSe solar cell by replacement and inserting buffer and Cu_2O ER-HTL layer. *Superlattices and Microstructures, 152*, 106840. https://doi.org/10.1016/j.spmi.2021.106840.

Slonopas, A., Ryan, H., Foley, B., Sun, Z., Sun, K., Globus, T., & Norris, P. (2016). Growth mechanisms and their effects on the opto-electrical properties of CdS thin films prepared by chemical bath deposition. *Materials Science in Semiconductor Processing, 52*, 24–31. https://doi.org/10.1016/j.mssp.2016.05.011.

Voznyi, A., Kosyak, V., Opanasyuk, A., Tirkusova, N., Grase, L., Medvids, A., & Mezinskis, G. (2016). Structural and electrical properties of SnS_2 thin films. *Materials Chemistry and Physics, 173*, 52–61. https://doi.org/10.1016/j.matchemphys.2016.01.036.

Weiss, T. P., Bissig, B., Feurer, T., Carron, R., Buecheler, S., & Tiwari, A. N. (2019). Bulk and surface recombination properties in thin film semiconductors with different surface treatments from time-resolved photoluminescence measurements. *Scientific Reports, 9*, 5385. https://doi.org/10.1038/s41598-019-41716-x.

Zhou, H., Hsu, W. C., Duan, H. S., Bob, B., Yang, W., Song, T. B., et al. (2013). CZTS nanocrystals: a promising approach for next generation thin film photovoltaics. *Energy & Environmental Science, 6*(10), 2822. https://doi.org/10.1039/c3ee41627e.

Zhuk, S., Kushwaha, A., Wong, T. K. S., & Masudy-panah, S. (2016). Critical review on sputter-deposited Cu_2ZnSnS_4 (CZTS) based thin film photovoltaic technology focusing on device architecture and absorber quality on the solar cells performance. *Solar Energy Materials and Solar Cells, 171*, 239–252. https://doi.org/10.1016/j.solmat.2017.05.064.

PART II
Emerging Materials for Sustainable Energy

CHAPTER 9

Optimization of Spectral and Energy Efficiency for MIMO with Grey Wolf Optimizer and BSNOMA Using IoT

R. SURESH,[1] S. DOMI EVANGELINE,[2] S. DOMI CAROLINE,[1] THERESA MICHEAL,[3] and V. MAHESH KUMAR REDDY[4]

[1]*Department of Electronics and Communication Engineering, Kings Engineering College, Irungattukottai, Chennai, Tamil Nadu, India*

[2]*Department of Computer Science and Engineering, Kings Engineering College, Irungattukottai, Chennai, Tamil Nadu, India*

[3]*Department of Computer Science and Engineering, Panimalar Engineering College, Varadharajapuram, Poonamallee, Chennai, Tamil Nadu, India*

[4]*Chaitanya Bharathi Institute of Technology, Proddatur, Andhra Pradesh, India*

ABSTRACT

Huge MIMO is a central application in 5G networks. The massive Multiple Input and Multiple Output (MIMO) scheme inspired the MIMO design. It comprises facilities with multiple antennas which form an adaptive beam that enhances the system performance while serving many customers simultaneously on the very same frequency blocks. Compared to traditional MIMO-based mobile networks, the use of firm MIMO technologies would increase both spectral efficiency and energy efficiencies considerably. Capital must be spread to take benefit of the stated Huge MIMO benefits. It refers to the proper allocation of resources for cellular communications in terms of time, frequency, space, and power. One method for allocating capital for

Engineering Design and Technical Applications of Physical Science: An Integrated Approach for Sustainable Development. Chin Hua Chia, Sarani Zakaria, Ruey Shan Chen, & Md. Shahariar Chowdhury (Eds.)

providing large MIMO networks with great spectral efficiency and energy efficiency is power management. A feature that gives data transmitters the power to optimize the device's output signal measures is called power control. The optimal approach is implemented such that the resource utilization metric formula is used for maximal spectral efficiency and energy efficiency. Two Grey Wolf Optimizers and (Base Station Non-Orthogonal Multiple Access) BSNOMA set of rules are created to frame useful research for the beam shaping vectors and power allocation.

9.1 INTRODUCTION

A short-lived dot line Enormous MIMO is perceived as an exciting prototype for 5G communications because it offers higher spectral efficiency (Asadi, Wang, & Mancuso, 2014) and energy efficiency (Tan, Xie, Xia, Tan, Fan, & Jin, 2018; Boyd & Vandenberghe, 2004; Rusek, Persson, Lau, Larsson, Marzetta, & Tufvesson, 2013). In the massive MIMO model, the BS is equipped with a large number of antennas. A dedicated radio frequency (RF) is distributed to each activated antenna, which includes a digital-to-analog converter (DAC), low-noise amplification, high-resolution analog-to-digital converter (ADC), and other components. When a large number of antennas are enabled in the BS, this standard leads to an increase in energy consumption while still maximizing the hardware cost (Fan & Zhang, 2019; Fan, Zhao, Gong, Yang, & Karagiannidis, 2017; Fan, Li, & Zhang, 2018). While the spectral efficiency of MIMO systems is significantly enhanced by allowing a large number of antennas and connecting more users to the BS within the same radio station, energy consumption rises, reducing the energy efficiency of MIMO systems (Gandomi, Yang, Talatahari, & Alavi, 2013). Most recently, existing systems have examined whether the introduction of low-resolution/mixed ADC was useful in reducing energy consumption (Zhang, Dai, Sun, & Wang, 2016a; Zhang, Chen, Shen, & Xia, 2016b). Due to D2D's proximity, it facilitates direct communications without the use of BS or core networks. Spectral efficiency and energy efficiency can be increased as a result, while D2D reduces the delay (He, Wang, Chen, Wong, & Elkashlan, 2017). Distance is the most critical factor in D2D transmission. Interference occurs between a MIMO cellular network and a D2D network using the same frequency bands, causing the D2D network's efficiency to be lower compared to MIMO cellular networks. According to theory (Hu, Liu, Xie, Gao, & Yang, 2014), the value of energy efficiency becomes subjectively high when the number of antennas in MIMO systems is small, and the spectral efficiency value is irrelevant to MIMO. The energy efficiency worth is unwaveringly related to consumer transmits power, according to an investigation published by Huang, He, Wang, & Zhu (2018). In functional MIMO devices, however, the energy usage must include both the power expended in the radio frequency circuit and the power consumed in the antenna. When a hypothetical examination is conducted between spectral efficiency and energy efficiency in the context of massive MIMO, a profound insight into massive MIMO is discovered (Jin, Wen, & Jiang, 2017). Based on the findings of Lee & Lee, (2017), it was suggested that time and power resources be allocated to maximize the energy efficiency benefit, (Lu, Li, Swindlehurst, Ashikhmin, & Zhang, 2014; Marzetta, 2010) calculated the energy efficiency and spectral efficiency trade-off. Furthermore, it is understood from all systems that proper beam-forming architecture aids in improving energy efficiency value but decreases spectral efficiency value (Mirjalili, Mirjalili, & Lewis, 2014). Overall, it has been concluded that, even in large antenna systems, an increase in energy efficiency value can result in a reduction in spectral efficiency. As a result, in large MIMO systems, a novel revolutionary technique is needed to achieve a fair trade-off between spectral efficiency and energy efficiency (Ngo, Larsson, & Marzetta, 2013; Pan, Lei, Deng, Fan et al., 2016; Patcharamaneepakorn, Wu, Wang, Aggoune, Alwakeel, Ge, & Renzo, 2016).

The main attributes of the paper are described floatingly below.

1. Build an enriched, huge MIMO exemplary with a poised trade-off between spectral efficiency and energy efficiency by means of a smart set of rules, including massive GWO, BS NOMA, BS OMA, and FA.
2. With the Huge GWO and BS NOMA strategies, optimize power allotments and vectors shaping beams to generate trade-offs between spectral efficiency and energy efficiency in a massive MIMO scheme.
3. Boost parameters of the latest Massive GWO, BS NOMA variant so that the projected set of rules can be beneficially impacted.

The following describes the layout of the paper: The review of the literature of Sect 2 includes the Master Model Sector 3 and the Analytical Model for massive MIMO Technology Sect 4, Efficacy and Comparative Exchange Section 5, the effect on the trade-off balance between spectral efficiency and energy efficiency of optimization set of rules, and Sect 6, the final comment.

9.2 RELATED WORKS

Fan & Zhang (2019) surveyed and optimized the energy efficiency of multi-user MIMO in their downlink multi-cell networks. Initially, the Downlink Cell Network (DRN) tested the MIMO Multi-User spectral efficiency with a wide range of BS antennas when the number of antennas (ZF) precoders was used. During the evaluation process, a larger amount of data in the BS was considered, differing from the current evaluation. Subsequent iterative optimization algorithms were developed to achieve the best energy efficiency, taking into account details on the conduit state of the converter. The results from the CPU phase indicated that the evaluated outputs were accurate and that more than 2 active BS antennas were required to achieve optimal energy efficiency for each served customer. Huang et al. (2018) also demonstrated that emphasis is added to enhance the energy efficiency of Beam directly by using ample derivations, creating optimization risks. An optimization algorithm for creating spectral efficiency and energy efficiency in a multi-user beam was implemented using abrupt or geometric conduit information (CSI). In conclusion, the results from the experiments show that the evolved CSI-based algorithm can achieve the same effectiveness as the immediate CSI with substantially less work required. Zhang et al. (2016) reported a full-document broadcast conducted by a large antenna to numerous pairs of fully documented consumers to convey data over the multi-pair two-way communication channel. To recover Eng., four different power-scaling replicas based on the maximal ratio combining (MRC)/maximal ratio transmission (MRT) and zero-forcing receiver (ZFR)/zero-forcing transmitter (ZFT) by trace were introduced. The asymptotic spectral efficiency and energy efficiency of the proposed power scaling method are defined as the number of antennas approaches infinity. The loop's distortion has been shown, indicating that big relay antennas are used to reduce transmission power. It also contrasted the utility of MRC/MRT, and ZFR/ZFT, and described the influence on customer couples' spectral efficiency. The impact of cellphone antennas on energy efficiency. The power consumption of the system service was also used to calculate performance. The established fields where full-duplex systems outperformed half-duplex systems became later available. The proposed solution is the strongest between spectral efficiency and energy efficiency, according to the experimental findings. He et al. (2017) developed a power management strategy, D2D, which could be easily adapted to circumstances of high transmission and no D2D connections. Under the power regulation in these particular networks, a hierarchical approach was accessible to legalize the spectral efficiency and the energy efficiency consequently; confident spectral efficiency statements were received for a cellular network customer or D2D transmitter in which the results of the primary computer parameters such as large MIMO antennas and D2D were collected. D2D scale characteristics have also been achieved, and include the requirements necessary for achieving the planned spectral efficiency. The statistical effects disclosed that the proposal proposed was effective. Zhang et al. (2016a) propose a set of exposure criteria for wide MIMO systems based on linear dispensation actions. Analyzed and exposed as a quality indicator the overachiever performance of economic efficiency (EE), which refers to economic feasibility, has enabled a hidden equilibrium between the spectral efficiency and the energy efficiency for the spectral efficiency energy efficiency, Yang et al. (2017) have looked at a massive MIMO two-way communication amplifier, multi-pair customer, in which data is transmitted using large antennas as contact sites. The conduit State information was supposed to be insufficient, and hence, the maximum ratio of combining/maximum transmission ratio was chosen for the touch point. Under the power-scaling systems that diffuse influence at each customer and communicate station, the sum of spectral efficiency and energy efficiency asymptotic was reduced to a number of relay antennas, as well as the transmission power of the pilot sequence, which is compatible or scaled. Furthermore, a spectral efficiency declaration was approximated in the closed form. The findings revealed that a non-disappearing signal was used to increase the power transmission of large antennas, as well as a ratio between any consumer and the transmitting station (SINR). Xin et al. also used viral contagion to establish massive MIMO cellular lines (Xin, Wang, Li, Zhu, Wang, & You, 2016). By using a permanently dispensed form of market location; a functional power utilization skill has been cast off to achieve the field of energy efficiency in a cell. The spectral efficiency and AE.EFF tests were used. A few constraints have been taken into consideration for the final analysis, including the number of symbols, antenna numbers in the BS, and the number of users. The optimal data-pilot power ratio (PDPR) was investigated to improve spectral efficiency and energy efficiency. Tan et al. examined the power spectral efficiency and energy efficiency of the massive Hybrid MIMO in 2018 (Tan, Xie, Xia, Tan, Fan, & Jin, 2018). The phase shifters were simulated using ZF precoding to create this model. The hybrid form of ideal phase shifters has a rough upper bound on the spectral efficiency that can be achieved. To obtain the essential minimal response, an approach was proposed for constructing the analog quantified matrix and for the investigation of hybrids with quantified phase shifters. Hybrid structural design has the highest performance compared to full digital structural design.

The output of any network is greatly influenced by security. The latest wireless networks have been found to have a significant effect on a variety of applications. A network with no infrastructure through which each node can send and receive data is more vulnerable to multiple security attacks in different layers due to its high mobility, infrastructure-less functionality, and lack of centralized structure (Tan, Xie, Xia, Tan, Fan, & Jin, 2018).

9.3 PROPOSED SYSTEM

9.3.1 GWO-BASED CONDUIT ESTIMATION

MIMO devices have attracted a lot of attention over the past two decades. They are used in a variety of wireless protocols because they increase the ability and reliability of contact networks. Early MIMO research focused on MIMO end-to-end networks, in which data is transmitted between two systems with multiple antennas. For the stint and incidence domains of MIMO-OFDM, the conduit transmission function seems inferior as opposed to other subcontractors. As a result, estimating the dynamic conduit is important. GWO determines the right conduit in this case. This set of rules simulates the hunting and chasing abilities of a GWO. In total, a population of wolves has five to twelve members. Pack is the name of the band. The faction is in control of the hierarchy. Men and women will be the pack's leaders. Alpha (α), beta (β), delta (Δ), and omega (Ω) are the four classifications for the packs. The numerical articulation of the GWO set of rules is given in the proposed component. Fitness solutions should be equally, the second-best solution, and the third-best solution as delta (based on the architecture of the GWO set of rules). Ω also has other up-and-coming arrangements. The intention of this set of rules is to choose the best conduit based on mean squared error (MSE). To quantify conduits based on GWO, follow the steps in subsections (Figure 9.1).

9.3.1.1 INITIALIZATION

Solutions initialization is one of the key techniques of optimization. Initialization parameters are available for the Bit Error Rate, Minimum time, PAPR, Specific Reliability, Uplink Downlink, and download downlinks. To maximize these parameters, GWO develops a random solution population. The candidate solution is the canal matrix here. Null equation (1).

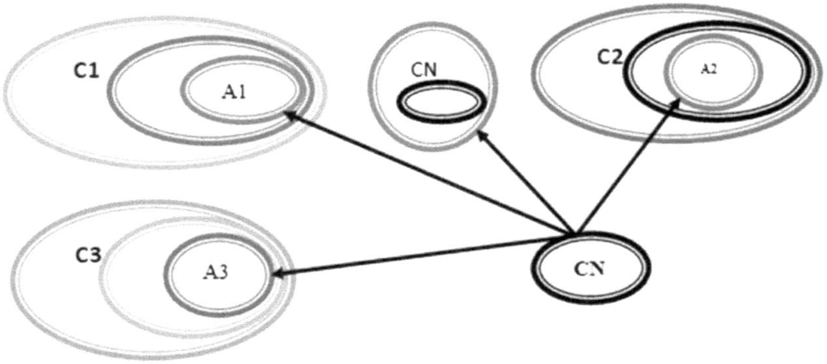

FIGURE 9.1 Gray wolf optimizer network.

$$Z = Z_1 + Z_2 + Z_3 \ldots\ldots + Z_N Z = Z_1 + Z_2 + Z_3 \ldots\ldots + Z_N \quad (1)$$

Here, Z represents the candidate solution.

9.3.1.2 *FITNESS EVALUATIONS*

The fitness is assessed after the solution is initialized. The specification of the necessary fitness parameter is one of the key aspects of any optimization set of rules. A vigor exam is used to assess the candidate's strategic abilities. For all the solutions, fitness is measured. The Mean Fitness Assessment Criterion is known as the key criterion here. MSE is determined based on equation (2) and exercise equation (3).

$$MSE = \frac{1}{m}\sum_{N=1}^{M}\left(T_N - \hat{T}_N\right)^2 MSE = \frac{1}{m}\sum_{N=1}^{M}\left(T_N - \hat{T}_N\right)^2 \quad (2)$$

Where represents the motion sent from the correspondent and represents the signal at the system's receiver end.

$$FIT_N = Min\ \{MSE\} FIT_N = Min\{MSE\} \quad (3)$$

a. **Updating:** Solution focused on fitness. After the fitness estimation, the solutions are isolated. Let J& α be the first best exercise options, J β the second best, and J μ the third best solution for fitness.
b. **Encircling Prey:** The aspirants are pursued by, and the chasing mechanism is pushed by, and. The prey is fat at first, causing the pack to become uneasy. The solution is chosen using an equation in this phase. (4)

$$\mathrm{U(s + 1) = U(S) - \bar{Z}\bar{G}U(s + 1) = U(S) - \bar{Z}\bar{G}} \tag{4}$$

$$\mathrm{\bar{G} = \bar{G} = |\bar{B}\bar{B}\bar{U}(s + 1) - U(s)|} \tag{5}$$

$$\mathrm{\bar{Z} = 2\bar{z}q_1 - \bar{z}\,\bar{Z} = 2\bar{z}q_1 - \bar{z}} \tag{6}$$

$$\mathrm{And\ \bar{Z} = 2q_2\,\bar{Z} = 2q_2} \tag{7}$$

At this point, the repetition number is s, the position of the prey is U(s), the coefficient vectors are Z, B, and the random vector [0, 1] is exemplified by S1, S2.

c. **Hunting:** The alpha, beta, and delta provide effective data on the position of the potential proxy to replicate biologically the gray wolf's hunting. In order to be able to find the first 3 best options, we need key search agents to examine their locations to find the best search officer. The new solution is evaluated for recurrence using the equation (7).

$$\bar{K}^{\propto} = |\bar{M}_1.S_{\propto} - S|, \quad \bar{K}^{\beta} = |\bar{M}_1.S_{\beta} - S|, \quad \bar{K}^{\gamma} = |\bar{M}_1.S_{\gamma} - S|$$
$$\bar{K}^{\propto} = |\bar{M}_1.S_{\propto} - S|, \quad \bar{K}^{\beta} = |\bar{M}_1.S_{\beta} - S|, \quad \bar{K}^{\gamma} = |\bar{M}_1.S_{\gamma} - S| \tag{8}$$

$$M_1 = M_{\propto} - \bar{Z}_1(\bar{K}^{\propto}), \quad M_2 = M_{\beta} - \bar{Z}_2(\bar{K}^{\beta}), \quad M_3 = M_{\gamma} - \bar{Z}_3(\bar{K}^{\gamma})$$
$$M_1 = M_{\propto} - \bar{Z}_1(\bar{K}^{\propto}), \quad M_2 = M_{\beta} - \bar{Z}_2(\bar{K}^{\beta}), \quad M_3 = M_{\gamma} - \bar{Z}_3(\bar{K}^{\gamma}) \tag{9}$$

$$M(S+1) = \frac{M_1 + M_2 + M_3}{3}\, M(S+1) = \frac{M_1 + M_2 + M_3}{3} \tag{10}$$

The result would be that the final position in the quest space will be random, like a single circle, using the alpha, beta, and delta spots. In addition, alpha, beta, and delta predict the position of the prey and the remaining wolves randomly report their positions near the prey.

d. **Attacking Prey (Exploitation) and Examine for Prey (Exploration):** Exploration and exploitation are ensured with the aid of bounds a and A. The adaptive parameters a and A enable GWO to travel in a random way between exploitation and exploration. With Z decrease, one-half of the replicates will be explored ($|A| \geq 1$), and the remainder of them will be used ($|Z| < 1$). The other half will be explored. The GWO only has 2 key parameters to be used (A and B). There is evidence that the GWO set of rules is smaller than the humblest operators. The development will stay pending until the highest amount of iteration is completed. Lastly, the optimum solutions are selected according to the fitness value.

9.3.1.4 TERMINATION CRITERIA

The output of the set of rules is stopped when you reach the maximum amount of iteration, and the best fitness solution is chosen as the optimal conduit.

- **Set of Rules 1:**
  Estimation of conduits in MIMO using GWO
- **Input:** Random conduit states (Ei)
- **Output:** Selected optimal conduit in MIMO
  1. Initialization of the first conduit population at random 1.
  2. Equal fitness assessment in equation. (3)
  3. Parameters z, Z, and B are initialized.
  4. Although not met by the end criteria
  5. Scan agent for the foresight
  6. Position updating of new search agent counterparts to (4) (9)
  7. Varnish with 7.
  8. Z, Z and B update
  9. Compute the health of the conduits
  10. Modernize the right resolution for the first, second, and third.
  11. Exit at 12. Optimum conduit return
- **Efficiency Metrics:** The effects of the following output metrics are calculated and analyzed in comparison to the current process.
- **BER:** The extent of bit errors is set to the bit error rate per unit (BER). The percentage of BER can be determined by dividing the total number of bits diffused over the measurable time period. BER is measured in proportions has no unit, and is stated as a measurement. Noise from the propagation of the conduit, vibrating, interruption, wireless multipath, damping, bit synchronization problems, etc., will affect the BER.
- **Spectral Efficiency:** The rate of data that can be conveyed in the bandwidth provided in the structure of demanding communication determines spectral adeptness. As indicated in the equation, it is evaluated in bits per symbol (11).

$$SE = \frac{CE}{BW * CV} SE = \frac{CE}{BW * CV} \tag{11}$$

Where;

SE= spectral efficiency
CE= the communication system's overall data channels
BW= bandwidth
CV= coverage

The spectral efficiency depends on the data rate transmitted by the bandwidth given by the exact transmission structure. It is assessed as seen in the equation "s" in bits per symbol. (11).

9.4 METHOD

This paper explores the energy-efficient allocation of vast MIMO power mixing relay systems during conduit aging. In a number of simulation tests involving various device parameters, the stability of the proposed systems was tested extensively, demonstrating the supremacy of the proposed method specifically in contrast with other literature studies. Significantly higher energy efficiency was found in all different device settings with numerous implementations of the suggested set of rules, allowing for high levels of QoS to be maintained (Figure 9.2).

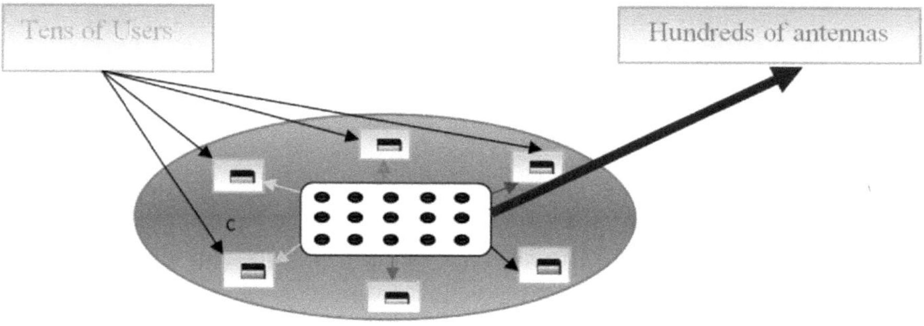

FIGURE 9.2 Massive MIMO.

Next, to illustrate the basics, there are more specifics of solitary user and multiple user MIMO cases. Both examples are shown in Figure 9.3. The number of transmitted antennas is Nt, and the number of established antennas is NR, whereas the number of spatial data streams transmitted to all users is M.

9.4.1 EXPERIMENTAL ANALYSIS

The graphs show and explain both of these methods' comparative outcomes, as well as the performance of all their performance measures in this field. Figure 9.4 depicts the suggested and current energy efficiency vs. user methods.

The delay is initially the same as it is in the new processes. However, in terms of increasing consumption prices, the suggested solution is more reliable than the existing alternatives. As shown in Figure 9.5, the energy efficiency analysis of big MIMO is carried out using several sets of rules, as well as the suggested set of rules, in relation to various enhancements considered in the BS. Figure 9.6 shows the energy efficiency approximation using several sets of criteria, with the number of antennas in the BS adjusted to examine how the output changes. The expected massive GWO exceeds BS NOMA, Existing, BS OMA, and FA as the quantity of antennas in the BS grows to 20. However, the current system performed admirably since 125 antennas were used in BS. BS OMA and FA, followed by Massive GWO and BS NOMA, were in second and third, respectively. Moreover, when the number of antennas is set to 4, Major GWO is 87%, 70%, 25%, and 20% higher for BS NOMA, BS OMA, and FA. Figure 9.7 displays the variations in the per-BS performance of the various sets of rules for finding energy efficiency (Table 9.1).

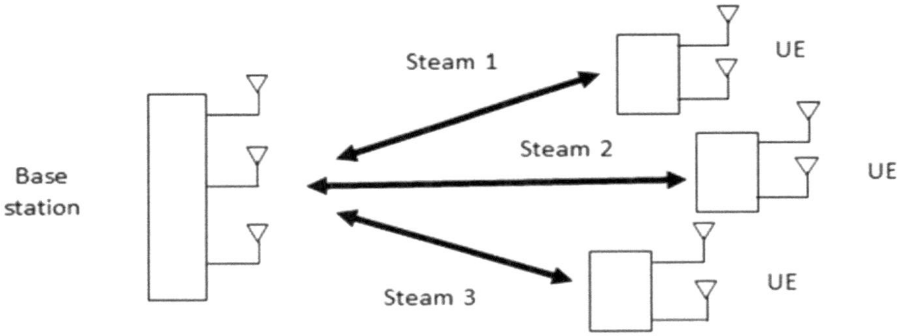

FIGURE 9.3 Generic single and multiple-user MIMO system.

TABLE 9.1 Comparison of Performance Parameter Strategies Based on SNR Value

Performance Parameters	Techniques			
	FD	**BS-OMA**	**MIMO-GWO**	**BS-NOMA Energy Efficiency**
Efficiency	20%	25%	87%	70%
Capacity	12%	60%	90%	60%

9.5 CONCLUSION

Using smart methodologies and a set of rules, this work utilizes a large MIMO model to manage the trade-off between spectral efficiency and

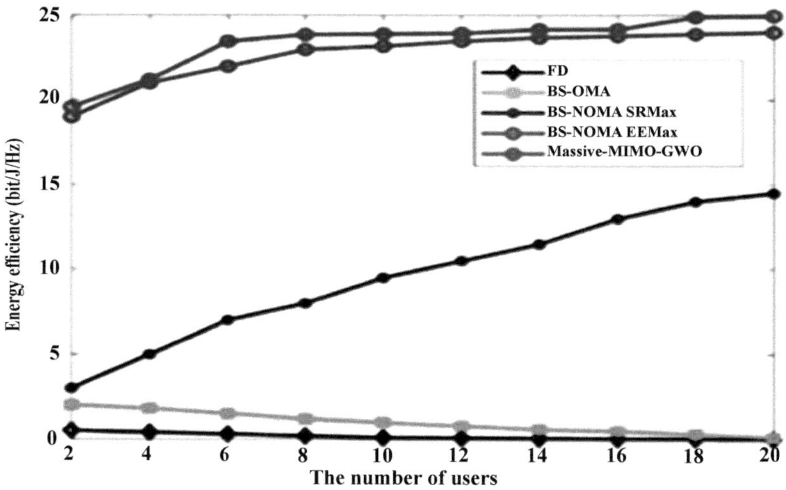

It indicates the energy usage of number of users
Existing Method = 2.5-2/2.5 * 100 = 20%
Proposed Method = 19.5-2.5/19.5 * 100 = 87.17%

FIGURE 9.4 Energy efficiency vs No of the user.

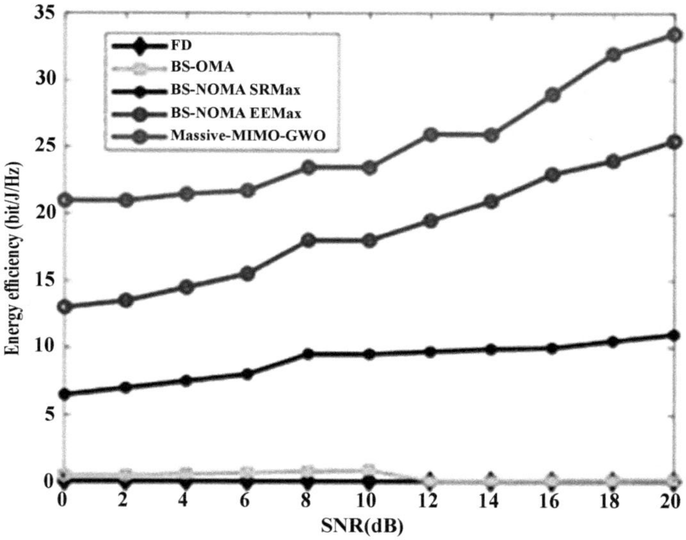

Existing Method = 34-21/34 * 100 = 38.23%
Proposed Method = 13.5-6.5/13.5 * 100 = 51.87%

FIGURE 9.5 Energy efficiency Vs SNR.

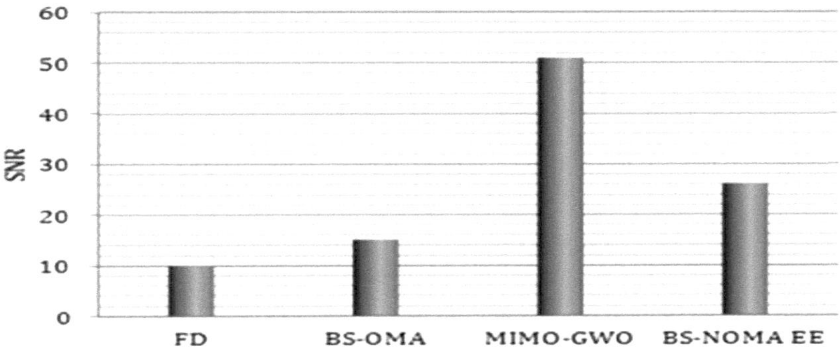

(a) Number of BS antennas

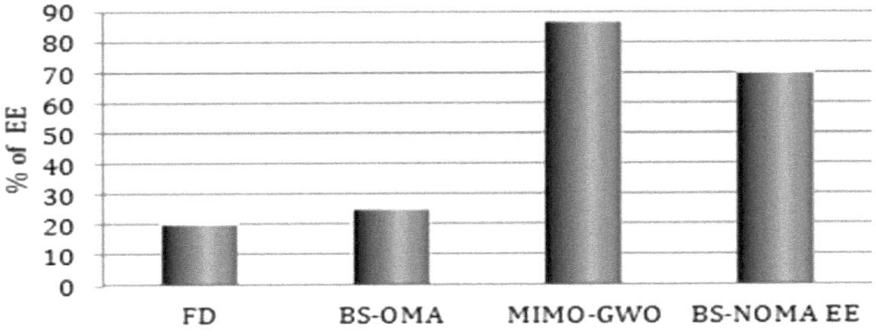

(b) Number of consumer devices per BS

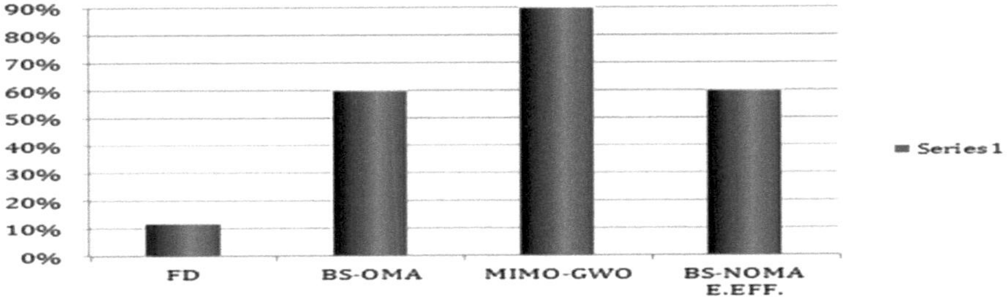

(c) % value of Capacity vs SNR

FIGURE 9.6 Huge MIMO systems energy efficiency review in relation to (a) number of BS antennas, (b) number of consumer devices per BS (c) percent value of capacity vs. SNR.

energy efficiency. The key objective of the proposed model was to find the best methods for the formation of vectors and the transfer of force. The optimum solution was found through the RE metric model to ensure that both optimum spectral efficiency and energy efficiency were achieved. In the

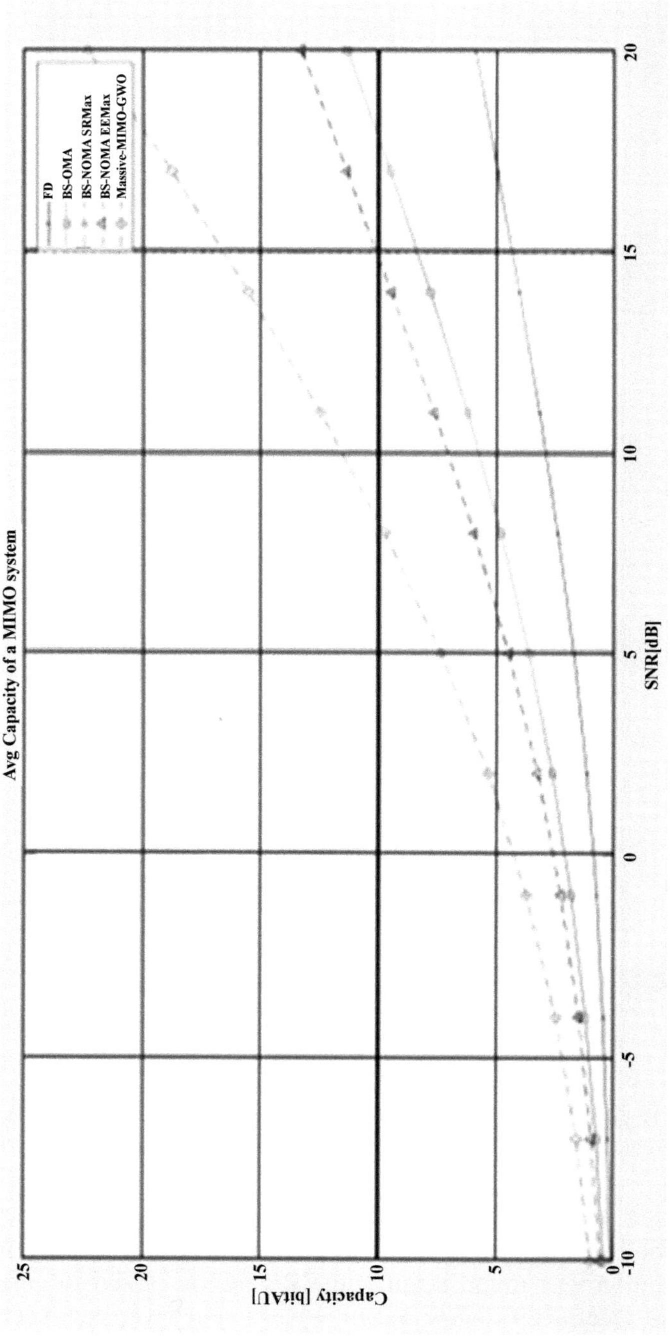

FIGURE 9.7 Capacity VS SNR.

current paper, the Gray Wolf Optimization set of rules offers a new conduit estimating strategy for large MIMOs. Relevant conduits are analyzed, and MIMO systems estimate their capacity. The proposed solution was implemented using the MATLAB method. The proposed MIMO-GWO massive approach is compared to BS NOMA, BS OMA, and FA in terms of their specified performance metrics. The performance of the proposed approach's conduit figures being higher than those of current approaches suggests that the stability and data capacity of massive 16QAM network MIMO systems have been improving. Both sets of rules have been effective at resolving complex optimization problems in various applications, improving the convergence rate. In this scenario, the overall performance of Massive GWO is 87%, 70%, 25%, and 20% higher than that of BS NOMA, BS OMA, and FA. The proposed sets of rules have therefore been found to provide a better balance in broad MIMO technology between spectral efficiency and energy efficiency.

KEYWORDS

- **downlink cell network**
- **IoT**
- **maximal ratio combining**
- **maximal ratio transmission**
- **Mimo Energy Efficiency**
- **Spectral Efficiency**
- **zero-forcing receiver**
- **zero-forcing transmitter**

REFERENCES

Asadi, A., Wang, Q., & Mancuso, V. (2014). A survey on device-to-device communication in cellular networks. *IEEE Communications Surveys & Tutorials, 16*(4), 1801–1819.

Boyd, S., & Vandenberghe, L. (2004). *Convex Optimization.* Cambridge University Press.

Fan, J., & Zhang, Y. (2019). Energy efficiency of massive MU-MIMO with limited antennas in downlink cellular networks. *Digital Signal Processing, 86*, 1–10.

Fan, J., Li, W., & Zhang, Y. (2018). Pilot contamination mitigation by fractional pilot reuse with threshold optimization in massive MIMO systems. *Digital Signal Processing, 78*, 197–204.

Fan, L., Zhao, R., Gong, F., Yang, N., & Karagiannidis, G. K. (2017). Secure multiple amplify-and-forward relaying over correlated fading conduits. *IEEE Transactions on Communications, 65*(7), 2811–2820.

Gandomi, H., Yang, X.-S., Talatahari, S., & Alavi, A. H. (2013). Firefly algorithm with chaos. *Communications in Nonlinear Science and Numerical Simulation, 18*, 89–98.

He, A., Wang, L., Chen, Y., Wong, K., & Elkashlan, M. (2017). Spectral and energy efficiency of uplink D2D underlaid massive MIMO cellular networks. *IEEE Transactions on Communications, 65*(9), 3780–3793.

Hu, B., Liu, Y., Xie, G., Gao, J. C., & Yang, Y. L. (2014). Energy efficiency of massive MIMO wireless communication systems with antenna selection. *Journal of China Universities of Posts and Telecommunications, 21*(6), 1–8.

Huang, Y., He, S., Wang, J., & Zhu, J. (2018). Spectral and energy efficiency tradeoff for massive MIMO. *IEEE Transactions on Vehicular Technology, 67*(8), 6991–7002.

Jin, W. T. S., Wen, C. K., & Jiang, T. (2017). Spectral efficiency of multi-user millimeter wave systems under single path with uniform rectangular arrays. *EURASIP Journal on Wireless Communications and Networking, 181*, 1–13.

Lee, J. H., & Lee, J. Y. (2017). Optimal beam forming-selection spatial preceding using population-based stochastic optimization for massive wireless MIMO communication systems. *Journal of the Franklin Institute, 354*(10), 4247–4272.

Lu, L., Li, G. Y., Swindlehurst, A. L., Ashikhmin, A., & Zhang, R. (2014). An overview of massive MIMO: Benefits and challenges. *IEEE Journal of Selected Topics in Signal Processing, 8*(5), 742–758.

Marzetta, T. L. (2010). Non-cooperative cellular wireless with unlimited numbers of base station antennas. *IEEE Transactions on Wireless Communications, 9*(11), 3590–3600.

Mirjalili, S., Mirjalili, S. M., & Lewis, A. (2014). Grey wolf optimizer. *Advances in Engineering Software, 69*, 46–61.

Ngo, H. Q., Larsson, E., & Marzetta, T. (2013). Energy and spectral efficiency of very large multiuser MIMO systems. *IEEE Transactions on Communications, 61*(4), 1436–1449.

Pan, G., Lei, H., Deng, Y., Fan, L., Yang, J., Chen, Y., & Ding, Z. (2016). On secrecy performance of MISO SWIPT systems with TAS and imperfect CSI. *IEEE Transactions on Communications, 64*(9), 3831–3843. https://doi.org/10.1109/TCOMM.2016.2573822.

Patcharamaneepakorn, P., Wu, S., Wang, C.-X., Aggoune, M. E.-H., Alwakeel, M. M., Ge, X., & Di Renzo, M. (2016). Spectral, energy, and economic efficiency of 5G multicell massive MIMO systems with generalized spatial modulation. *IEEE Transactions on Vehicular Technology, 65*(12), 9715–9731. https://doi.org/10.1109/TVT.2016.2526628.

Pedersen, M. E. H., & Chipperfield, A. J. (2010). Simplifying particle swarm optimization. *Applied Soft Computing, 10*(2), 618–628. https://doi.org/10.1016/j.asoc.2009.08.029.

Rusek, F., Persson, D., Lau, B. K., Larsson, E. G., Marzetta, T. L., & Tufvesson, F. (2013). Scaling up MIMO: Opportunities and challenges with very large arrays. *IEEE Signal Processing Magazine, 30*(1), 40–60. https://doi.org/10.1109/MSP.2011.2178495.

Sanguinetti, L., Zappone, A., & Debbah, M. (2019). Deep learning power allocation in massive MIMO. *Electrical Engineering and Systems Science, 1257–1261,* 1–5.

Tan, W., Xie, D., Xia, J., Tan, W., Fan, L., & Jin, S. (2018). Spectral and energy efficiency of massive MIMO for hybrid architectures based on phase shifters. *IEEE Access, 6*, 11751–11759.

Tang, J., So, D., Alsusa, E., & Hamdi, K. (2014). Resource efficiency: A new paradigm on energy efficiency and spectral efficiency tradeoff. *IEEE Transactions on Wireless Communications, 13*(8), 4656–4669.

Xin, Y., Wang, D., Li, J., Zhu, H., Wang, J., & You, X. (2016). Area spectral efficiency and area energy efficiency of massive MIMO cellular systems. *IEEE Transactions on Vehicular Technology, 65*(5), 3243–3254.

Yang, J., Wang, H., Ding, J., Gao, X., & Ding, Z. (2017). Spectral and energy efficiency analysis for massive MIMO multi-pair two-way relaying networks under generalized power scaling. *Science China Information Sciences, 60*(10), 102303. https://doi.org/10.1007/s11432-016-9007-2.

Zhang, J., Dai, L., Sun, S., & Wang, Z. (2016a). On the spectral efficiency of massive MIMO systems with low-resolution ADCs. *IEEE Communications Letters, 20*(5), 842–845.

Zhang, Z., Chen, Z., Shen, M., & Xia, B. (2016b). Spectral and energy efficiency of multi-pair two-way full-duplex relay systems with massive MIMO. *IEEE Journal on Selected Areas in Communications, 34*(4), 848–863.

CHAPTER 10

A Review of Generating Biofuels Through Syngas Fermentation Toward Sustainable Energy Production, Its Challenges, and Opportunities

MINHAJ UDDIN MONIR,[1,2] AZRINA ABD AZIZ,[3] SHAHARIAR CHOWDHURY,[4] KHAMPHE PHOUNGTHONG,[4] and KUAANAN TECHATO[4]

[1]*Department of Petroleum and Mining Engineering, Jashore University of Science and Technology, Jashore, Bangladesh*

[2]*Energy Conversion Laboratory, Department of Petroleum and Mining Engineering, Jashore University of Science and Technology, Jashore, Bangladesh*

[3]*Faculty of Civil Engineering Technology, Universiti Malaysia Pahang, Gambang, Malaysia*

[4]*Faculty of Environmental Management, Prince of Songkhla University, Songkhla, Thailand*

ABSTRACT

The purpose of this review is to provide an outline to produce clean and sustainable biofuels from lignocellulosic biomass through syngas fermentation. A hybrid thermochemical and biochemical process is discussed systematically, whereby syngas is produced from lignocellulosic biomass, and is then subjected to fermentation using different biocatalysts of yeast and bacteria, to produce biofuels. The most feasible experimental setup for

Engineering Design and Technical Applications of Physical Science: An Integrated Approach for Sustainable Development. Chin Hua Chia, Sarani Zakaria, Ruey Shan Chen, & Md. Shahariar Chowdhury (Eds.)

both processes is critically reviewed. This review also reflects on existing scientific advances in industrial techniques in areas such as thermochemical and biochemical pathways. Finally, this review provides a conventional method to produce biofuels and explores the problems and challenges that need to be addressed in biofuel processing.

10.1 INTRODUCTION

Recently, the ever-diminishing reserves of fossil fuels globally and the growing need for energy security have become matters of public concern (Monir, Azrina, Kristanti, & Yousuf, 2018). The two main sources of energy are renewable and non-renewable energy, and much of the energy that is being used by humans is produced from fossil fuels such as coal, natural gas, and petroleum (Güney, 2019). Over a period of many decades, fossil fuels like coal and hydrocarbons have emerged as the leading form of energy resources. Consequently, there is a great need for alternative energy sources due to uncertainties regarding fossil fuels and energy shortages. These worldwide issues may be effectively addressed using low-carbon biofuels, which are appealing fuels for transportation and industrial activities. The global usage of biofuels is expected to expand as much as four times in the next 45 years (Newell, Raimi, & Aldana, 2019).

Biofuels are fuels that have been extracted from organic materials (biomass) and then liquefied or gasified to be used as fuels (bioethanol, biobutanol, bio-methane, and biodiesel) for transportation as well as for heating and power production. They are often mixed with fossil fuels in proportions ranging from 5% to 85% by volume, with the normal range being somewhere in the middle (Newell, Raimi, & Aldana, 2019; Alves, Laviola, Formighieri, & Carels, 2015). In addition, biofuels are generated by biological processes or derived from the biomass of living creatures, including bacteria, plants, and microalgae. It is possible for biofuels to be created by several kinds of algae, and they have the greatest potential to produce other forms of green energy. Gasoline, alcohols, biodiesel, triglycerides, fatty acids, lipids, carbohydrates, cellulose, and biological biomass are the primary components that may be converted into biofuels (Rodionova, Poudyal, Tiwari, Voloshin et al., 2017).

According to Sikarwar, Zhao, Fennell, Shah, & Anthony (2017), biofuels are classified into first-generation, second-generation, third-generation, and fourth-generation biofuels, depending on the kinds of biomass and the procedures used to produce them. They can be converted into biomethanol, bioethanol, biopropanol, biobutanol, and fatty acid esters using simple sugars, starch, lipids, and vegetable oils as the starting materials (Naik, Goud, Rout, & Dalai, 2010). Inderwildi & King (2009) reported that second-generation biofuels such as ethanol are produced from lignocellulosic biomass, while algae can be converted to produce third- and fourth-generation biofuels. Third-generation biofuels are obtained from the processing and refining of algae biomass, while algae microbial engineering is employed to generate fourth-generation biofuels from photosynthetic microbes with oxygen and to create artificial carbon sinks (Lü, Sheahan, & Fu, 2011). Bioethanol is cost-effective compared to fossil fuels (Kennes, Abubackar, Diaz, Veiga, & Kennes, 2016). Rastogi & Shrivastava (2017) recorded an increase in the daily usage of lignocellulosic biomass to generate different sources of energy (second-generation biofuels) such as biogas, biodiesel, bioethanol, etc. Multiple forms of reactors are used to produce bioethanol. The bioreactors that are often used include the continuous stirred tank reactor (CSTR), bubble column reactor (BCR), monolithic biofilm reactor (MBR), trickle bed reactor (TBR), microbubble dispersion stirred-tank reactor (MDSR), and so on (Acharya, Roy, & Dutta, 2014). The yield of ethanol can be increased by having a well-designed reactor, an adequate transfer time, choosing a biocatalyst that can optimize yield, and having an efficient recovery method (Shen, Brown, & Wen, 2014). Shen, Brown, & Wen (2017) stated that a horizontal rotary packed bed (h-RPB) reactor is more efficient than a CSTR for the generation of bioethanol. Recently, Monir, Abd. Aziz, Yousuf, & Alam (2020) and Monir, Aziz, Khatun, & Yousuf (2020) stated that yeast and bacteria-based syngas fermentation using a tar-free bioreactor is more effective for the production of clean and sustainable bioethanol.

10.2 ENERGY RESOURCES FOR BIOFUEL PRODUCTION

In general, sugar (such as that found in sugar cane, sugar beet, and sweet sorghum), starch (such as that found in maize and cassava), and oil (such as soybean, rapeseed, coconut, sunflower, and palm oil) are the primary components of biofuels (Singh, Pandey, & Gnansounou, 2016; Konur, 2018). Biofuels (ethanol, methanol, hydrogen, dimethyl ether, synthetic natural gas) are generated through the fermentation of syngas, which is produced by biomass gasification (Molino, Larocca, Chianese, & Musmarra, 2018; Monir, Yousuf, & Aziz, 2020). The commonly used biofuels are bioethanol and biodiesel. Biofuels are categorized as primary and secondary biofuels, depending on their type, origin, and processing technique. The major sources of biofuels, including wood chips, pellets, and landfill gas, are organic and unprocessed waste. They are mostly used for heating by direct combustion and to produce electricity. Secondary biofuels such as bioethanol, biodiesel, methanol, biogas, etc., are generated from biomass (Nigam & Singh, 2011). Usually, biofuels are categorized based on their generation.

10.2.1 FIRST-GENERATION BIOFUELS

First-generation biofuels (conventional biofuels) are derived from the fermentation of simple sugars or starch into biological fuels, as well as the hydrogenation or transesterification of raw vegetable oil (Callegari, Bolognesi, Cecconet, & Capodaglio, 2020). Advanced technologies have resulted in the production of second-generation biofuels (crop residues, milling and pulping residues, seasonal plants, forestry, drainage, farm, and organic wastes) and increased biofuel output that will minimize the rivalry between land and food. Edible biomass is used to refine ethanol from first-generation ethanol as a substrate. Sugarcane, maize, whey, sugar beet, wheat, potato, and other wastes are biomass that are used to produce alcohol at first generation. Their use in the generation of electricity has numerous drawbacks in that they directly affect food prices, quality production, and the use of soil for crops, and result in greenhouse gas emissions.

10.2.2 SECOND-GENERATION BIOFUELS

Second-generation biofuels contribute primarily to the production of ethanol through the utilization of lignocellulosic biomass, and farm and municipal solid waste as its energy sources. Lignocellulosic feedstock, which is the most abundant and energy-intensive biomass on earth, enables modern technology to manage all biofuel power (cellulose, hemicellulose, and lignin). However, it is difficult to successfully transform this feedstock into biofuel by means of pre-treatment and hydrolysis to extract the fermentable sugars (Alves, Laviola, Formighieri, & Carels, 2015). The use of feedstocks reduces greenhouse gas emissions and, at the same time, does not conflict with the cultivation of agricultural food crops. The selected plant types are cultivated on marginal land that is not appropriate for use in food production but is suitable for bioenergy production (bioenergy cultures). The parts of plants and forest trees that cannot be used for food are then used to produce ethanol. The biggest challenge to the use of feedstocks for second-generation biofuels is the high cost of therapy recipes, the collection of biomass, and the inefficient and difficult multi-stage conversion technology.

Second-generation biofuels can be generated through biochemical or thermochemical processes. Biochemical processes enable biofuels to be developed by microbial fermentation, where these products may also serve as a catalyst for the processing of biofuels, such as biodiesel (non-fermenting biofuels) (Elshahed, 2010). Pyrolysis or gasification techniques are used to produce syngas (mostly a mixture of carbon monoxide and hydrogen) (Monir, Abd Aziz, Kristanti, & Yousuf, 2020), which is then transformed into biofuels, e.g., by Fisher-Tropsch synthesis (Luque, & Clark, 2010), or fermented into biofuels utilizing biochemical processes (Monir, Abd Aziz, Yousuf, & Alam, 2020). The biochemical conversion process yields less expensive biofuels than the thermochemical process, as it is possible to reduce costs by means of the convergence process, successful pre-treatment techniques, improved fermentation enzymes, the use of lignin, enzyme recycling, and the construction of cost-effective bioreactors (Sims, Mabee, Saddler, & Taylor, 2010).

The priority of second-generation biofuels, which have the advantage of lower GHG emissions, is to use sustainable biomass that does not compete with grains and land for cultivation. Bioethanol, depending on the feedstock used, will minimize GHG emissions by around 85% compared to gasoline (Saini, Saini, & Tewari, 2015). Due to supply, environmental issues, and cost-effectiveness, the future lies in second-generation feedstocks for biofuel production (bioethanol, biomethanol, biobutanol, and biohydrogen) (Monir, Yousuf, & Aziz, 2020). However, advanced technology is required to ferment these feedstocks into biofuels to sustain the conversion process (Monir, Aziz, Khatun, & Yousuf, 2020). These feedstocks can be categorized as residues of industrial crops, seasonal crops, municipal solid waste (MSW), and forest residues (Monir, Abd Aziz, Kristanti, & Yousuf, 2020). This segment addresses the global supply of these feedstocks, their potential, and related challenges.

10.2.3 THIRD-GENERATION BIOFUELS

Third-generation biofuels are generated from the production of algae oil, also known as "oilgae." It is supposed to be generated at a low cost and give a high performance of up to around 30 times the energy per unit area extracted compared to conventional first-generation feedstocks. It is obtained mainly from agricultural feedstocks owing to the inadequate photosynthetic effectiveness of terrestrial plants since they consume less than 0.5% of solar energy at middle latitudes. Moreover, microalgae can be converted by more than solar energy, suggesting that they can be converted into biofuels through commercial processes (Su, Song, Zhang, Su, Cheng, & Chen, 2017). Because of its high lipid content, ease of culture, and rapid growth, microalgae is emerging as a potential contender for use in the manufacture of biofuels (Alam, Mobin, & Chowdhury, 2015). In most cases, microorganisms should have a substantial amount of lipids for effective algae production. Species like chlorella are often targeted because they have a high lipid content (between 60 and 70%) and high productivity (7.4 g/L/d for chlorella protothecoides) (Lee & Lavoie, 2013). Tsuchida, Tsukagoshi, & Ueyama (2017) documented microbial methods for the processing of bioethanol. Moreover, different types of yeast strains are responsible for better bioethanol output by fermentation processes that take into account factors that influence bioethanol production and yeast immobilization (Mohd Azhar, Abdulla, Jambo, Marbawi et al., 2017).

10.3 HYBRID PROCESS FOR BIOFUEL PRODUCTION

A hybrid process to produce clean biofuels from lignocellulosic biomass includes both thermochemical and biochemical processes. Firstly, syngas are produced through the gasification (thermochemical process) of different tyspes of biomasses using various gasifiers (Monir, Abd Aziz, Kristanti, & Yousuf, 2020). Secondly, clean biofuels (bioethanol) are produced through syngas fermentation using a tar-free bioreactor, where tar and small particles are removed from the syngas (Monir, Abd Aziz, Yousuf, & Alam, 2020). A flowchart for the generation of clean biofuels is presented in Figure 10.1.

10.3.1 THERMOCHEMICAL PROCESS

A thermochemical pathway is used for complex processes involving limited biomass heating. Some combustion processes yield one or more solid, liquid, or gaseous end products, known as biochar, pyrolytic oil or bio-oil, and syngas, which is composed of carbon monoxide (CO), hydrogen (H_2), short-chain alkanes, and carbon dioxide (CO_2). Pyrolysis, combustion, and gasification are used in the thermochemical process.

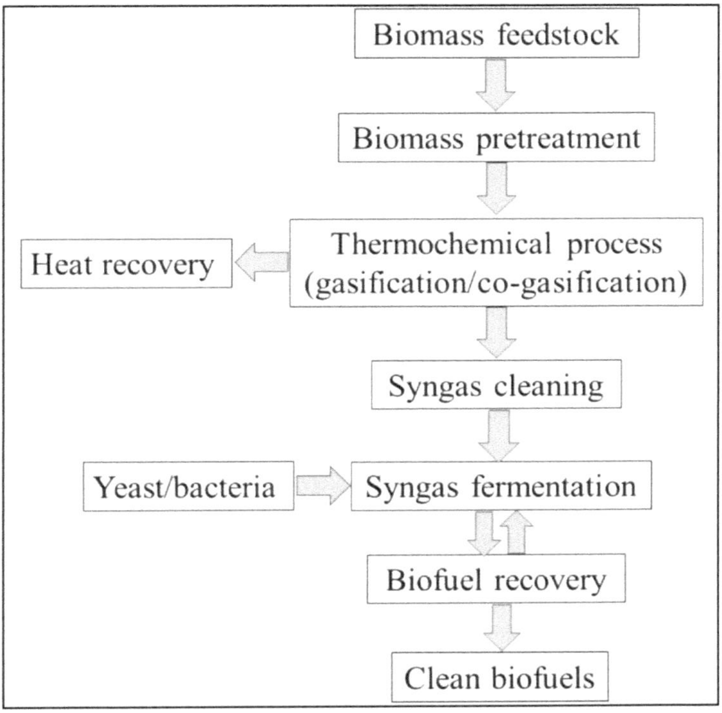

FIGURE 10.1 Process flow for the generation of biofuels utilizing sustainable biomass.

10.3.1.1 PYROLYSIS

Pyrolysis results in the production of bio-oil, liquid products, solid char, and gaseous compounds. Bio-oil is another type of oil that is used for various products. Pyrolysis is a process where biomass is decomposed at high temperatures (375°C–525°C) and a pressure of 5–10 bar, in the absence of air or oxygen. A major product of pyrolysis is bio-oil, which is produced at high temperatures with limited oxygen input. Consequently, syngas and biochar are mainly produced as by-products. The thermal process is somewhat self-sufficient in terms of energy as the biomass can be burned to provide heat. During this process, biomass is converted into solid fuel, liquid oils, and gaseous products through thermal drying (100°C), the initial phase (100–300°C), the intermediate phase (>200°C), and the final phase (300–900°C), as described by Basu (2010). Carbon dioxide is the primary gas produced during pyrolysis, and hydrogen generation is not considered. The gaseous co-products of pyrolysis, based on the organic composition of the biomass, include H_2, CO, CH_4, CO_2, and other gases. Wang, Dai, Yang, and Luo (2017) discussed their approaches to pyrolyzing lignocellulosic biomass for biofuel and bio-based chemical processing. They highlighted the properties of cellulose, hemicellulose, and lignin that influence their behavior during pyrolysis, as well as the thermochemical characteristics of the building blocks and their functional groups. They also emphasized biofuel production through catalytic pyrolysis, aromatic or aliphatic carbons, and chemical additives such as anhydro sugars and furans.

10.3.1.2 COMBUSTION

The most important conversion chain for biomass is to convert it into CO_2 and steam through temperatures ranging from 800 to 1000°C (Basu, 2018), where the biomass becomes CO, H_2, CO, CH_4, and CO_2. It also decomposes chemical energy from wood for heating and electricity (Kuo, Wu, & Chen, 2014). It has also been reported that certain co-fired boilers consumed the biomass and charcoal separately when the biomass ratio reached 10%. This process marked the first use of biomass, and mankind has been exploring fire since. Wood is burnt for cooking and heating (Basu, 2018). The combustion of biomass involves a chemical reaction between oxygen and hydrocarbons. The biomass is converted into two stable H_2O and CO molecules. In this phase, some biomass reactions release around 90% of heat energy. It is used for heat output in rural cooking. Steam also produces industrial ventilation in biomass boilers. Both these modern economic activities are focused on biomass-based energy (Ruiz, Juárez, Morales, Muñoz, & Mendívil, 2013). The most common method is by burning biomass in a furnace and producing electricity from a steam turbine. In certain countries, the combustion of biomass generates electricity (Sikarwar, Zhao, Clough, Yao et al., 2016).

10.3.1.3 GASIFICATION

Gasification involves the conversion of biomass into oxygen-deficient gas or synthetic gas (Aydin, Yucel, & Sadikoglu, 2018). Syngas are mostly carbon monoxide and hydrogen. Various types of gasifiers can be used for the gasification of biomass, the most popular being fixed bed, fluidized bed, and entrained flow gasifiers (Basu, 2018b). Baruah & Baruah (2014) also reported on the structures for bio-gasification, namely, fixed bed gasification (down-drafting and up-drafting), fluidized bed gasification (circulating and bubbling), and entrained flow gasification (Oakey, 2015). The choice of a gasifier usually depends on the design, height, humidity, and end product of the steam/oxygen/air interaction device (Sansaniwal, Pal, Rosen, & Tyagi, 2017). Different gasifiers are used commercially for the generation of electricity. Downdraft gasifiers produce hazardous tar by-products, impurities, and ash (Samiran, Jaafar, Ng, Lam, & Chong, 2016). Patra & Sheth (2015) stated that downdraft gasifiers are able to achieve a maximum capacity of 5 MW. Gao, Zhang, Li, and Yu (2016) developed a biomass gasification model with an intrinsic gasifier reaction rate, whereby the relative volumetric error levels were 1–18%, and the heating, power storage, cold gas, and carbon conversion efficiency were 1–13%, 1–8%, 1–12%, and 1–11%, respectively. Figure 10.2 shows a schematic diagram of a downdraft gasifier used for biomass gasification. The body of the downdraft reactor (c) is usually made of a sheet of mild steel (MS). The height and diameter of the reactor are around 1000 mm and 400 mm, respectively (Monir, Aziz, Kristanti, & Yousuf, 2018). A neck or throat, with an angle of slope of almost ~70°, is created inside the reactor near the grid to preserve the natural flow of the feedstock due to the attraction of gravity (Atnaw, Sulaiman, & Yusup, 2013).

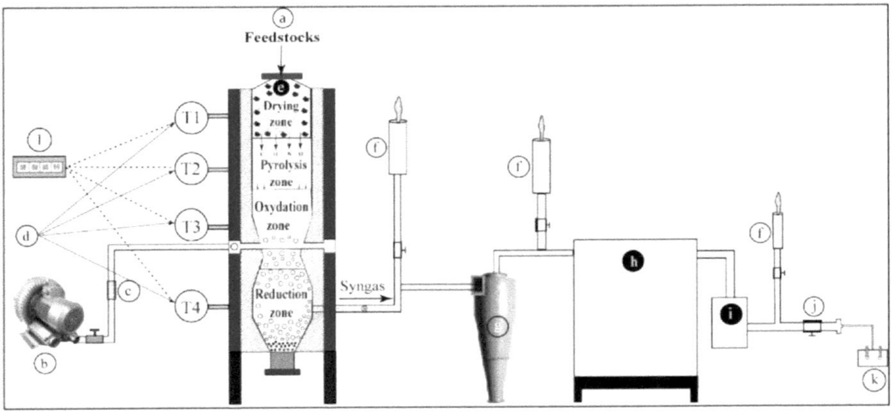

FIGURE 10.2 Schematic diagram of a downdraft gasifier: (a) feedstock, (b) air blower, (c) rotameter, (d) K-type thermocouples, (e) reactor, (f) gas flare points, (g) cyclone separator, (h) heat exchanger, (i) small particle filter, (j) clean gas sampling point, (k) syngas bag (Monir, Abd Aziz, Kristanti, & Yousuf, 2020; Monir, Abd Aziz, & Yousuf, 2022).

The reactor consists of four zones: (a) drying zone; (b) pyrolysis zone; (c) oxidation zone; and (d) reduction zone. These zones are linked to the data acquisition data logger in real-time by means of K-type thermocouples, and the temperature profile is recorded distinctly. The K-type thermocouples are mounted at intervals of 200 mm along the gasifier sheet in the middle section. The reactor is also attached to an air blower with an ambient air inlet device, a rotameter to regulate the quantity of airflow, gas flare points to examine the syngas composition, a cyclone separator to remove smaller particles from the gas produced, and a heat exchanger to cool the hot syngas. An inlet air pipe is attached to the reactor to enable air to flow through the air blower into the oxidation region. The cyclone separator, which is a device that is located between the reactor and the heat exchanger, is used to purify the system by eliminating fine carbon particles left by the manufactured syngas (Monir, Azrina, Kristanti, & Yousuf, 2018). Finally, the heated syngas is cooled to room temperature. The syngas heating value is the main optimized output parameter for the key operating parameters (i.e., temperature, strain, feedstock ratio, and moisture content).

During the gasification phase, the feedstock is moved from the drying zone across four zones within the downstream reactor. Evaporation takes place during the drying phase. When the temperature is above 100°C during gasification, part of the vapor is transformed into hydrogen and ends up as moisture in the syngas that is formed. Subsequently, the feedstock is pyrolyzed at ~200°C (Basu, 2018). The biomass begins to undergo thermal disintegration at this stage to become reactive gases and carbon, where the chemical composition of the material is influenced by the proportion of these materials. The working conditions in the reactor also change. Subsequently, an oxidation zone is created to induce the pyrolyzed products into the hotter zone. Atmospheric air is pumped into the oxygen-starved zone. Oxidation occurs between 700 and 1000°C (Basu, 2018), and possible reactions take place during gasification. The material is then moved to the reduction zone, where there is a deficit of oxygen, resulting in decreased reactions between the hot gases and the biomass. The heated responsive gases and carbon that have been transformed into syngas accumulate chemical energy in this zone, and the gas temperature is reduced.

10.3.2 BIOCHEMICAL PROCESSES

Biochemical processes are chemical processes that occur in the biomolecules of living organisms. Alcoholic fermentation is one such biochemical process, where carbohydrates are converted into cellular energy, with ethanol (alcohol) and carbon dioxide as the by-products. Various forms of biocatalysts are used in the biochemical processing of liquid biofuels.

10.3.2.1 BIOCATALYSTS FOR SYNGAS FERMENTATION

Various forms of biocatalysts are used in the fermentation of syngas (Sun, Atiyeh, Kumar, Zhang, & Tanner, 2018). Yeast and bacteria are the common biocatalysts used to ferment chemical fuels or biofuels (Sayed, & Abdelkareem, 2017; Foo, Susanto, Keasling, Leong, & Chang, 2017; Davis, Molloy, Quigley, Nikodinovic-Runic et al., 2018). Microorganisms cannot fully degrade lignocellulosic biomass by the direct fermentation of the lignin-containing biomass. However, microorganisms can produce biofuels from biomass (Asimakopoulos, Gavala, & Skiadas, 2018; Molitor, Marcellin, & Angenent, 2017). Since syngas also includes impurities such as tar (higher hydrocarbons) and fine particulate matter (Kuba, & Hofbauer, 2018; Moon, Jo, Jeong, Bang, Choi, Hwang, & Lee, 2017; Al-Rahbi, & Williams, 2017), it cannot be used as a transport fuel without first being purified, and hence, current engines must be modified or improved (Mansfield, & Wooldridge, 2015; Xu, Tree, & Lewis, 2011). The conversion of syngas to fuels and other chemicals by such particular types of microorganisms is thus an effective option for the energy market of the future (Henstra, Sipma, Rinzema, & Stams, 2007).

Mesophilic and thermophilic microorganisms are used in the syngas fermentation process. The popular microorganisms that are widely used for the fermentation of syngas are *Clostridium autoethanogenum, Clostridium carboxidivorans, Clostridium ragsdalei, Clostridium ljungdahlii*, and others. The use of biological catalysts (microorganisms) (*Acetobacterium Woodii* and *Clostridium Carboxidivorans*) for the fermentation of syngas has been documented to be more efficient than chemical catalysts (copper, cobalt, or iron). Microorganisms (yeasts, bacteria, and fungi) have been reported to be used for the conversion of lignocellulosic biomass into biofuels. *Escherichia coli, Bacillus sp., Clostridium* sp., etc., are the common biocatalysts used for biofuel production. *S. cerevisiae, Fusarium oxysporum*, and *Trichoderma reesei* have been recorded as being the most effective microorganisms. However, the yield of ethanol from the fermentation of xylose is less than that obtained from the fermentation of glucose using *S. cerevisiae*, which is suitable for producing clean bioethanol. The production of the next generation of biofuels directly related to microalgae is based on the ability of lignocellulose bacteria and fungi to transform CO_2. The fermentation of yeast is the simplest and most conventional method (Monir, Abd Aziz, Yousuf, & Alam, 2020). It is economically competitive with other fermentation agents, has a recognized capacity for fermentation, and produces zero chemical waste. Another study by Akhtar (2016) reported that the final product (ethanol) in biofuel processing has been found to be environmentally friendly (Shen, Brown, & Wen, 2017). *C. Carboxidvorans* and *C. Butyricum* play a vital role in the production of bioethanol and biohydrogen, respectively, through the fermentation process (Kanchanasuta, Prommeenate, Boonapatcharone, & Pisutpaisal, 2017). Figure 10.3 shows a bioreactor where the biomass-based syngas is converted into liquid biofuels.

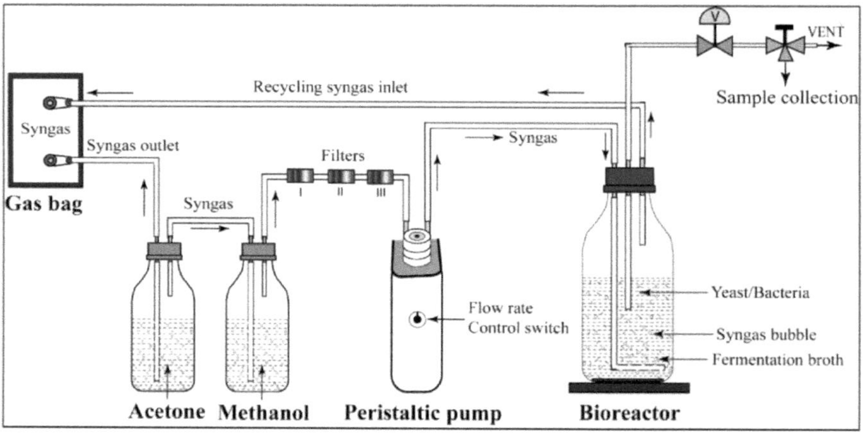

FIGURE 10.3 Schematic diagram of bioreactor for the conversion of syngas into biofuels using a biocatalyst (yeast/bacteria) (Monir, Abd Aziz, Yousuf, & Alam, 2020; Monir, Abd Aziz, & Yousuf, 2022).

In this bioreactor, the syngas derived from the gasification of lignocellulosic biomass are usually fermented in batches or a fed-batch mode. A fermentation broth needs to be prepared separately for the fermentation of the yeast and bacteria-based syngas. During the syngas fermentation process, a peristaltic pump is also required to facilitate the flow of the syngas through the reactor. The medium is autoclaved for 20 minutes after the combination. The syngas are also passed through acetone and methanol to filter out the tar compounds. A cotton filter is often used to remove fine particles from the syngas. The temperature, pH, colony-forming unit (CFU), total organic carbon (TOC), and syngas impurity are needed to investigate the growth of microbial cells during syngas fermentation.

10.3.2.2 YEAST-BASED SYNGAS FERMENTATION

A bioreactor is required to carry out yeast-based syngas fermentation with different varieties of yeast. The yeast must be freshly cultured to ensure that the biofuel is produced at maximum efficiency. Figure 10.4 shows yeast under a field emission scanning electron microscope. A fermentation medium (broth) must also be prepared for the fermentation of syngas. The fermentation broth is loaded to fill 80% of a 500/1000-mL impinger bottle, with the remaining 20% being the working volume. According to a study by Monir, Abd Aziz, Yousuf, and Alam (2020), the prepared medium includes (1) yeast extract (0.2 gm), (2) peptone (0.8 gm), (3) KH_2PO_4 (0.4 gm), (4) $MnSO_4.7H_2O$ (0.2 gm), (5) $(NH_4)2SO_4$ (0.8 gm), (6) carbon dioxide by-product (0.7 gm), and (7) DI water (400 mL). Yeast is inoculated into the biomass-based syngas. The pH is controlled within the range of 4.0–6.0. The experiment is carried out within a shaking incubator. The temperature is controlled at 37°C with a shaking rate of 200 rpm. It is necessary to run the experiment for 16 days in order to produce clean bioethanol Monir, Abd Aziz, Yousuf, & Alam, (2020).

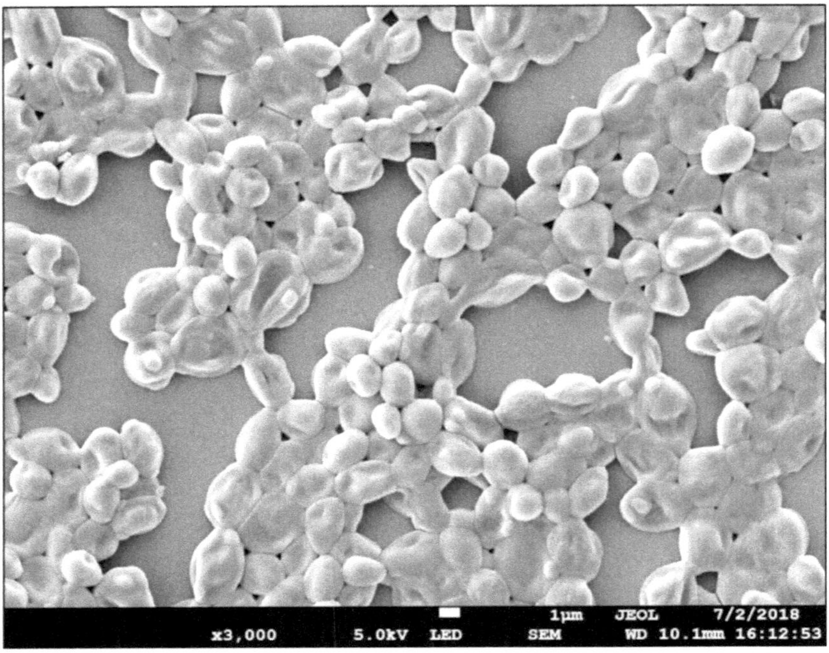

FIGURE 10.4 FESEM image of yeast used for the conversion of bioethanol through syngas fermentation.

10.3.2.3 BACTERIA-BASED SYNGAS FERMENTATION

Bacteria can also be used for the conversion of biomass-based syngas through fermentation. In such tests, a 500/1000-mL impinger bottle is used, where 80% of it is loaded with the fermentation broth, and the remaining 20% is regarded as the working volume. The fermentation medium is combined with the nutrient broth and 10 g of it is mixed in 400 mL of DI water. Bacteria are inoculated into the fermentation broth. The pH is controlled between 6.5 and 7.5. The experiment is performed inside a shaking incubator. The temperature and shaking rate are maintained at 37°C (for mesophilic bacteria) and 200 rpm, respectively. The fermentation process is carried out for about 16 days to obtain clean bioethanol (Monir, Aziz, Khatun, & Yousuf, 2020). Figure 10.5 shows the bacteria under a field emission scanning electron microscope.

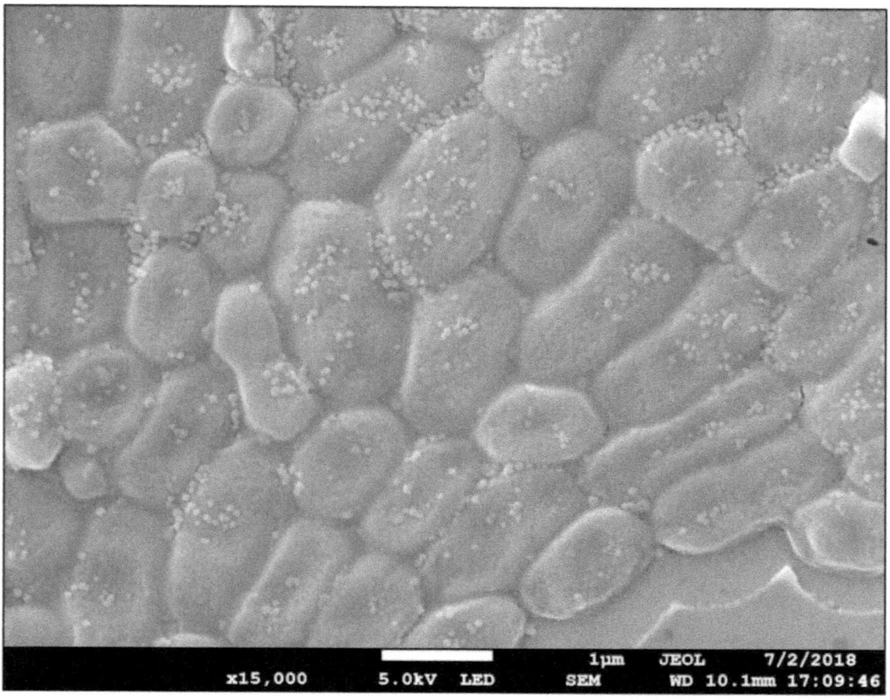

FIGURE 10.5 FESEM image of bacteria used for the conversion of bioethanol from syngas fermentation.

10.3.2.4 PROCESS OPTIMIZATION

Chemical processes are needed to optimize the operating parameters (temperature and pressure) to minimize the experimental costs and time. Various types of simulators can be used to optimize the parameters, among them, the Aspen Plus®, which is a process modeling tool for chemical and biological calculations. The Aspen Plus algorithm is more complicated than the Aspen Hysys. The Aspen Plus has simplified the simulations and updates over the years, and small parts of complicated and embroidered frameworks can be created and tested separately, as some researchers have argued (Ismail, Abbas, Azizi, & Zeaiter, 2017). This simulator can be used to model a number of methods, e.g., methanol synthesis, indirect coal liquefaction processes, advanced combined-cycle coal gasification (IGCC) power plants, environmentally friendly combustion processes, coal-gasification processes and coal-gasification simulations (Kaushal, & Tyagi, 2017). According to the existing literature, the Aspen Plus® can be used to optimize gasification (temperature and pressure). The gasification simulations are based on the required assumptions. It minimizes experimental time and expenses. Simulation tends to be an effective solution for forecasting the valuation and gasification of commercial syngas. The advantage of this simulation is that it uses the same combination of temperature, pressure, and mole fraction. The central simulation theory of Aspen Plus is to use the basic property relationships of the structural product (weight and energy balance, velocity coefficient, reaction, mass and heat transfer, and all sorts of balanced relationships) to model and simulate the flow rate, structure, and properties of the fuel, maximizing the operating conditions and design of the reactor. Various operating system templates can be used in the gasifier, depending on the conversion process, to reflect different configurations. For example, the RYield block simulates the biomass pyrolysis. The Gibbs block is used to simulate the Gibbs theory-based oxidation-reduction minimization mechanism. Then, both blocks are applied to the material transfer, thermal movement, or the origin of work to create a flow layer.

10.4 CHALLENGES IN BIOFUEL PRODUCTION

Overall, third-generation biofuels derived from microalgae have a better chance of becoming viable sources of energy. It is very necessary to enhance the species of microalgae and gain a better knowledge of the processes involved in the generation of biofuels. Full-scale industrial facilities for the production of lignocellulosic biofuels are being developed, with the majority of them in Europe, North America, South America, and Asia already functioning in some capacity (Raud, Kikas, Sippula, & Shurpali, 2019). One of the most significant difficulties is the relatively expensive capital and processing expenses, which are caused by the use of specialized technology and apparatus to facilitate the conversion phase (Jambo, Abdulla, Azhar, Marbawi, Gansau, & Ravindra, 2016). The price is expected to decrease as manufacturing moves to larger production facilities and more experience is gained. Now, research and development in this field are being financed by a range of channels, both public and private, such as funding from venture capitalists and cash from the internal coffers of big firms. Although this money has helped move things ahead, there is still a need for further investments to get this technology to a state of maturity.

10.5 CONCLUSIONS AND FUTURE TRENDS

Biofuels are extracted by the processing of syngas (methanol, gasoline, dimethyl ether, synthetic natural gas, hydrogen), which is obtained from biomass gasification. Biofuels are a renewable source of green energy relative to restricted life, geopolitical volatility, and the negative global effects of fossil fuel energy. Collectively, biofuels contain all the energy-enriched chemicals generated directly by biological processes or derived from prior living organisms via biomass conversion. Biofuels are mainly extracted from photosynthetic plants, including photosynthetic bacteria, microalgae, macroalgae, and vascular soil. The key ingredients of a biofuel may be gas, liquid, or solid, but these things can be converted by biochemical, physical, and thermochemical methods.

ACKNOWLEDGMENTS

The authors would like to thank the Energy Conversion Laboratory at the Department of Petroleum and Mining Engineering at Jashore University of Science and Technology, Jashore, Bangladesh. This research was supported by Prince of Songkhla University and the Ministry of Higher Education, Science, Research and Innovation, Thailand, under the Reinventing University Project (Grant Number 2564–2565) and funded by the Universiti Malaysia Pahang

International Publication Grant (Grant No. RDU 213305) for completing this review work.

KEYWORDS

- **biocatalyst**
- **biofuels**
- **bubble column reactor**
- **continuous stirred tank reactor**
- **lignocellulosic biomass**
- **microbubble dispersion stirred-tank reactor**
- **monolithic biofilm reactor**
- **syngas fermentation**
- **thermochemical process**

REFERENCES

Acharya, B., Roy, P., & Dutta, A. (2014). Review of syngas fermentation processes for bioethanol. *Biofuels, 5*(5), 551–564.

Akhtar, M. N. (2016). Biofuel: A stepping stone towards sustainable development. *International Journal of Research in BioSciences, 5*(3), 1–12.

Alam, F., Mobin, S., & Chowdhury, H. (2015). Third-generation biofuel from algae. *Procedia Engineering, 105*, 763–768.

Al-Rahbi, A. S., & Williams, P. T. (2017). Hydrogen-rich syngas production and tar removal from biomass gasification using sacrificial tire pyrolysis char. *Applied Energy, 190*, 501–509.

Alves, A. A., Laviola, B. G., Formighieri, E. F., & Carels, N. (2015). Perennial plants for biofuel production: Bridging genomics and field research. *Biotechnology Journal, 10*(4), 505–507.

Asimakopoulos, K., Gavala, H. N., & Skiadas, I. V. (2018). Reactor systems for syngas fermentation processes: A review. *Chem. Eng. J., 348*, 732–744.

Atnaw, S. M., Sulaiman, S. A., & Yusup, S. (2013). Syngas production from downdraft gasification of oil palm fronds. *Energy, 61*, 491–501.

Aydin, E. S., Yucel, O., & Sadikoglu, H. (2018). *Numerical Investigation of Fixed-Bed Downdraft Woody Biomass Gasification.* In *Exergetic, Energetic and Environmental Dimensions*; Dincer, I., Colpan, C.O., Kizilkan, O., Ed.; Academic Press, 323–339.

Baruah, D., & Baruah, D. C. (2014). Modeling of biomass gasification: A review. *Renewable and Sustainable Energy Reviews, 39*, 806–815.

Basu, P. (2010). *Biomass Gasification and Pyrolysis: Practical Design and Theory*. Academic Press.

Basu, P. (2018). *Biomass Gasification, Pyrolysis, and Torrefaction: Practical Design and Theory* (3rd edn.). Academic Press.

Callegari, A., Bolognesi, S., Cecconet, D., & Capodaglio, A. G. (2020). Production technologies, current role, and prospects of biofuels feedstocks: A state-of-the-art review. *Critical Reviews in Environmental Science and Technology, 50*(4), 384–436.

Davis, R., Molloy, S., Quigley, B., Nikodinovic-Runic, J., Solano, F., & O'Connor, K. E. J. A. M. (2018). Biocatalytic versatility of engineered and wild-type tyrosinase from *R. solanacearum* for the synthesis of 4-halocatechols. *Journal Name: Applied Microbiology and Biotechnology, 102, 1–11.*

Elshahed, M. S. (2010). Microbiological aspects of biofuel production: Current status and future directions. *Journal of Advanced Research, 1*(2), 103–111.

Foo, J. L., Susanto, A. V., Keasling, J. D., Leong, S. S. J., & Chang, M. W. J. B. (2017). Whole-cell biocatalytic and de novo production of alkanes from free fatty acids in Saccharomyces cerevisiae. *114*(1), 232–237.

Gao, X., Zhang, Y., Li, B., & Yu, X. (2016). Model development for biomass gasification in an entrained flow gasifier using intrinsic reaction rate sub-model. *Energy Conversion and Management, 108*, 120–131.

Güney, T. (2019). Renewable energy, non-renewable energy, and sustainable development. *International Journal of Sustainable Development & World Ecology, 26*(5), 389–397.

Henstra, A. M., Sipma, J., Rinzema, A., & Stams, A. J. M. (2007). Microbiology of synthesis gas fermentation for biofuel production. *Current Opinion in Biotechnology, 18*(3), 200–206.

Inderwildi, O. R., & King, D. A. (2009). Quo vadis biofuels? *Energy and Environmental Science, 2*(4), 343–346.

Ismail, H. Y., Abbas, A., Azizi, F., & Zeaiter, J. (2017). Pyrolysis of waste tires: A modeling and parameter estimation study using Aspen Plus®. *Waste Manage., 60,* 482–493.

Jambo, S. A., Abdulla, R., Azhar, S. H. M., Marbawi, H., Gansau, J. A., & Ravindra, P. (2016). A review on third generation bioethanol feedstock. *Renewable Sustainable Energy Rev., 65,* 756–769.

Kanchanasuta, S., Prommeenate, P., Boonapatcharone, N., & Pisutpaisal, N. (2017). Stability of Clostridium butyricum in biohydrogen production from non-sterile food waste. *Int. J. Hydrogen Energy, 42*(5), 3454–3465.

Kaushal, P., & Tyagi, R. (2017). Advanced simulation of biomass gasification in a fluidized bed reactor using ASPEN PLUS. *Renewable Energy, 101*(Supplement C), 629–636.

Kennes, D., Abubackar, H. N., Diaz, M., Veiga, M. C., & Kennes, C. (2016). Bioethanol production from biomass: Carbohydrate vs syngas fermentation. *Journal of Chemical Technology and Biotechnology, 91*(2), 304–317.

Konur, O. (2018). *Bioenergy and Biofuels*. CRC Press.

Kuba, M., & Hofbauer, H. (2018). Experimental parametric study on product gas and tar composition in dual fluid bed gasification of woody biomass. *Biomass Bioenergy, 115,* 35–44.

Lee, R. A., & Lavoie, J.-M. (2013). From first- to third-generation biofuels: Challenges of producing a commodity from a biomass of increasing complexity. *Animal Frontiers, 3*(2), 6–11.

Lü, J., Sheahan, C., & Fu, P. (2011). Metabolic engineering of algae for fourth generation biofuels production. *Energy & Environmental Science, 4*(7), 2451–2466.

Luque, R., & Clark, J. (2010). *Handbook of Biofuels Production: Processes and Technologies*. Elsevier.

Mansfield, A. B., & Wooldridge, M. S. (2015). The effect of impurities on syngas combustion. *Combustion and Flame, 162*(5), 2286–2295.
Mohd Azhar, S. H., Abdulla, R., Jambo, S. A., Marbawi, H., Gansau, J. A., Mohd Faik, A. A., & Rodrigues, K. F. (2017). Yeasts in sustainable bioethanol production: A review. *Biochemistry and Biophysics Reports, 10*, 52–61.
Molino, A., Larocca, V., Chianese, S., & Musmarra, D. (2018). Biofuels production by biomass gasification: A review. *Energies, 11*(4), 811.
Molitor, B., Marcellin, E., & Angenent, L. T. (2017). Overcoming the energetic limitations of syngas fermentation. *Current Opinion in Chemical Biology, 41*, 84–92.
Monir, M. U., Abd Aziz, A., & Yousuf, A. (2020). Hydrogen-rich syngas fermentation for bioethanol production using *Saccharomyces cerevisiae*. *International Journal of Hydrogen Energy, 45*(36), 1–9.
Monir, M. U., Abd Aziz, A., Kristanti, R. A., & Yousuf, A. (2020). Syngas production from co-gasification of forest residue and charcoal in a pilot scale downdraft reactor. *Waste and Biomass Valorization, 11*(2), 635–651.
Monir, M. U., Aziz, A. A., & Yousuf, A. (2022). Integrated technique to produce sustainable bioethanol from lignocellulosic biomass. *Materials Letters: X, 13*, 100127.
Monir, M. U., Aziz, A. A., Khatun, F., & Yousuf, A. (2020). Bioethanol production through syngas fermentation in a tar-free bioreactor using *Clostridium butyricum*. *Renewable Energy, 157*, 1116–1123.
Monir, M. U., Aziz, A. A., Kristanti, R. A., & Yousuf, A. (2018). Co-gasification of empty fruit bunch in a downdraft reactor: A pilot scale approach. *Bioresource Technology Reports, 1*, 39–49.
Monir, M. U., Azrina, A. A., Kristanti, R. A., & Yousuf, A. (2018). Gasification of lignocellulosic biomass to produce syngas in a 50 kW downdraft reactor. *Biomass and Bioenergy, 119*, 335–345.
Monir, M. U., Yousuf, A., & Aziz, A. A. (2020). Syngas fermentation to bioethanol. In A. Yousuf, D. Pirozzi, & F. Sannino (Eds.), *Lignocellulosic Biomass to Liquid Biofuels* (pp. 195–216). Academic Press.
Moon, J., Jo, W., Jeong, S., Bang, B., Choi, Y., Hwang, J., & Lee, U. (2017). Gas cleaning with molten tin for hydrogen sulfide and tar in producer gas generated from biomass gasification. *Energy, 130*, 318–326.
Naik, S. N., Goud, V. V., Rout, P. K., & Dalai, A. K. (2010). Production of first- and second-generation biofuels: A comprehensive review. *Renewable and Sustainable Energy Reviews, 14*(2), 578–597.
Newell, R., Raimi, D., & Aldana, G. (2019). Global energy outlook 2019: The next generation of energy. *Resources for the Future, 1*(8), 8–19.
Nigam, P. S., & Singh, A. (2011). Production of liquid biofuels from renewable resources. *Progress in Energy and Combustion Science, 37*(1), 52–68.
Oakey, J. (2015). *Fuel Flexible Energy Generation: Solid, Liquid and Gaseous Fuels*. Woodhead Publishing.
Patra, T. K., & Sheth, P. N. (2015). Biomass gasification models for downdraft gasifier: A state-of-the-art review. *Renewable and Sustainable Energy Reviews, 50*, 583–593.
Patra, T. K., & Sheth, P. N. (2015). Biomass gasification models for downdraft gasifier: A state-of-the-art review. *Renewable and Sustainable Energy Reviews, 50*, 583–593.
Rastogi, M., & Shrivastava, S. (2017). Recent advances in second generation bioethanol production: An insight to pretreatment, saccharification, and fermentation processes. *Renewable and Sustainable Energy Reviews, 80*, 330–340.

Raud, M., Kikas, T., Sippula, O., & Shurpali, N. (2019). Potentials and challenges in lignocellulosic biofuel production technology. *Renewable Sustainable Energy Rev., 111,* 44–56.
Rodionova, M. V., Poudyal, R. S., Tiwari, I., Voloshin, R. A., Zharmukhamedov, S. K., Nam, H. G., & Allakhverdiev, S. I. (2017). Biofuel production: Challenges and opportunities. *International Journal of Hydrogen Energy, 42*(12), 8450–8461.
Saini, J. K., Saini, R., & Tewari, L. (2015). Lignocellulosic agriculture wastes as biomass feedstocks for second-generation bioethanol production: Concepts and recent developments. *3 Biotech, 5*(4), 337–353.
Samiran, N. A., Jaafar, M. N. M., Ng, J. H., Lam, S. S., & Chong, C. T. (2016). Progress in biomass gasification technique–With a focus on Malaysian palm biomass for syngas production. *Renewable and Sustainable Energy Reviews, 62,* 1047–1062.
Samiran, N. A., Jaafar, M. N. M., Ng, J.-H., Lam, S. S., & Chong, C. T. (2016). Progress in biomass gasification technique–With a focus on Malaysian palm biomass for syngas production. *Renewable and Sustainable Energy Reviews, 62,* 1047–1062.
Sansaniwal, S. K., Pal, K., Rosen, M. A., & Tyagi, S. K. (2017). Recent advances in the development of biomass gasification technology: A comprehensive review. *Renewable and Sustainable Energy Reviews, 72,* 363–384.
Sansaniwal, S. K., Pal, K., Rosen, M. A., & Tyagi, S. K. (2017). Recent advances in the development of biomass gasification technology: A comprehensive review. *Renewable and Sustainable Energy Reviews, 72,* 363–384.
Sayed, E. T., & Abdelkareem, M. A. (2017). Yeast as a Biocatalyst in Microbial Fuel Cell. *In Old Yeasts-New Questions*. InTech. Sayed, E. T., & Abdelkareem, M. A. (2017). Yeast as a biocatalyst in microbial fuel cells. In *Old Yeasts-New Questions,* InTech.
Shen, Y., Brown, R. C., & Wen, Z. (2017). Syngas fermentation by *Clostridium carboxidivorans* P7 in a horizontal rotating packed bed biofilm reactor with enhanced ethanol production. *Applied Energy, 187*, 585–594.
Shen, Y., Brown, R., & Wen, Z. (2014). Syngas fermentation of *Clostridium carboxidivorans* P7 in a hollow fiber membrane biofilm reactor: Evaluating the mass transfer coefficient and ethanol production performance. *Biochemical Engineering Journal, 85*, 21–29.
Sikarwar, V. S., Zhao, M., Fennell, P. S., Shah, N., & Anthony, E. J. (2017). Progress in biofuel production from gasification. *Progress in Energy and Combustion Science, 61*, 189–248.
Sims, R. E., Mabee, W., Saddler, J. N., & Taylor, M. (2010). An overview of second-generation biofuel technologies. *Bioresource Technology, 101*(6), 1570–1580.
Singh, R. S., Pandey, A., & Gnansounou, E. (2016). *Biofuels: Production and Future Perspectives.* CRC Press.
Su, Y., Song, K., Zhang, P., Su, Y., Cheng, J., & Chen, X. (2017). Progress of microalgae biofuel's commercialization. *Renewable and Sustainable Energy Reviews, 74*, 402–411.
Sun, X., Atiyeh, H. K., Kumar, A., Zhang, H., & Tanner, R. S. J. B. (2018). Biochar enhanced ethanol and butanol production by Clostridium carboxidivorans from syngas. *Bioresource Technology, 265*, 128–138.
Tsuchida, Y., Tsukagoshi, N., & Ueyama, M. (2017). Growth method for microbe and bioethanol production method. *Google Patents*, U.S. Patent 9,732,363.
Wang, S. R., Dai, G. X., Yang, H. P., & Luo, Z. Y. (2017). Lignocellulosic biomass pyrolysis mechanism: A state-of-the-art review. *Progress in Energy and Combustion Science, 62*, 33–86.
Xu, D., Tree, D. R., & Lewis, R. S. (2011). The effects of syngas impurities on syngas fermentation to liquid fuels. *Biomass Bioenergy, 35*(7), 2690–2696.

PART III
Case Studies

CHAPTER 11

Geostatistical Analysis to Determine Underground Coal Gasification (UCG) Site for Exploiting Jamalganj Coalfield in NW Bangladesh

ARUP KUMAR BISWAS,[1,2] MD. RAFIQUL ISLAM,[3] MD. FUAD HASAN,[4] JAFAR SHARAFELDEEN ALBAGIR MOHAMMAD,[1,5] KUAANAN TECHATO,[1] KHAMPHE PHOUNGTHONG,[1] and MAKATAR WAE-HAYEE[6]

[1]*Faculty of Environmental Management, Prince of Songkla University, Hat Yai, Songkhla, Thailand*

[2]*Hydrocarbon Unit, Energy and Mineral Resources Division, Government of the People's Republic of Bangladesh, Dhaka, Bangladesh*

[3]*Energy and Mineral Resources Division, Ministry of Power, Energy, and Mineral Resources, Government of the People's Republic of Bangladesh, Dhaka, Bangladesh*

[4]*Mineral Processing Center (MPC), Institute of Mining, Mineralogy, and Metallurgy (IMMM), Bangladesh Council of Scientific and Industrial Research (BCSIR), Joypurhat, Bangladesh*

[5]*Oil Exploration and Production Authority, Ministry of Energy and Petroleum, Republic of the Sudan*

[6]*Department of Mechanical and Mechatronics Engineering, Faculty of Engineering, Prince of Songkhla University, Hat Yai, Songkhla, Thailand*

Engineering Design and Technical Applications of Physical Science: An Integrated Approach for Sustainable Development. Chin Hua Chia, Sarani Zakaria, Ruey Shan Chen, & Md. Shahariar Chowdhury (Eds.)

ABSTRACT

Geostatistical analysis can depict the subsurface conditions and geological features in such a way that an analogous decision can be made regarding mining feasibility. Jamalganj Coalfield in North-western (NW) Bangladesh was discovered in 1962 but remains unexploited to date. Adoption of any conventional mining techniques (i.e., underground or opencast) has been difficult due to greater depth (~1100 m). However, in recent years, advances in unconventional coal mining methods such as Underground Coal Gasification (UCG) have demonstrated greater prospects to extract the untapped resources in a sustainable manner. In this study, we employed the Inverse Distance Weighted (IDW) method of interpolation for interpreting ten (10) available borehole logging data. Subsurface geological modeling and isopach mapping were conducted in the study for locating an appropriate site in this coalfield for UCG exploitation by interpolating depth and thickness data of the seven (07) coal seams. This reveals that the eastern part of the Jamalganj coal basin can be initially targeted for UCG piloting, where the thickest Seam III merges with Seam IV. This work can assist further initiatives by providing a reasoned approach for siting the first UCG trial in Bangladesh.

11.1 INTRODUCTION

Bangladesh has been placed among the few fastest-growing GDPs in the world. To support economic growth, the rapidly growing energy demand is frequently surpassing the supply capacity. Consequently, energy importation is growing inexplicably, leaving the country in a vulnerable state regarding energy security. Yet, the country does not have a comprehensive plan to exploit the vast coal resources (first discovered during the 1960s) due to various techno-environmental issues (Ball et al., 1983; Ghose & UNDP, 2009). All exploitable coal reserves of Bangladesh are in the Permian-aged Gondwana sedimentary basins located in NW Bangladesh. Permian coals are intersected at 118 m to 1150 m in small graben structures found in Phulbari, Barapukuria, Dighipara, Khalashpir, Jamalganj, and other places in Dinajpur and Rangpur districts as High Volatile C to A type bituminous coal (Islam, Hayashi, & Kamruzzaman, 2009; Imam, 2013; Farhaduzzaman et al., 2015). The total coal reserves are about 8 billion tons in five coalfields so far discovered (BCMCL, 2019).

Among the five discovered fields, coal from four deposits (depths ranging from 118 m to 509 m) excluding the Jamalganj deposit, may be extractable at present by conventional methods (i.e., underground or opencast mining). Only the Barapukuria coalfield has been operated through underground mining methods since 2005, while the others are awaiting any kind of suitable mining operation to be undertaken (Sajjad & Rasul, 2015). It is noteworthy that Jamalganj Coalfield alone holds two-thirds of the total still-exploitable coal resources in Bangladesh, being the largest (5.45 billion tons) and deepest (over 1000 m) coal deposit in the country. The resource is roughly equivalent to more than 109 Tcf gas (U.S. EIA, 2021). In the wake of global warming anxiety, UCG can be a viable clean coal technology as this can effectively reduce many obstacles to conventional mining straightaway. By doing so, many of the prevailing hindrances can be overcome, when the energy markets call on these resources. UCG does not bring coal to the surface, thus there is no coal handling/washing/transportation. Also, there is no surface burning, so no ash problem happens, and there is no need for any surface or subsurface excavation (Beath et al., 2004; Burton, Friedmann, & Upadhye, 2006). The feasibility of UCG deployment highly depends on some critical factors which are mainly associated with the site-specific coalfield characteristics, e.g., burial thickness of coal seam, depth, hydraulic head, seam dipping, scale of deposit, etc. (Biswas et al., 2023; Perkins, 2018b). Geostatistical analysis can depict the subsurface by modeling those critical factors and delineating the pre-feasibility for UCG application in a coalfield. Geostatistics offers a wide range of analysis and stochastic modeling tools that have been developed over the years to solve practical problems in the mining industry (Journel & Huijbregts, 1978), and it is now commonly used in combination with deterministic methods; also widely used for mineral resource evaluation and production. A non-exhaustive list of common applications of geostatistics in the mining industry is presented in (Rambert, 2005).

11.1.1 GEOLOGIC AND STRATIGRAPHIC OVERVIEW

The Gondwana basin of the Jamalganj coal area lies on the Bogra slope of the stable platform zone. This intra-cratonic basin is an E-W elongated asymmetric half-graben with an E-W trending northern boundary fault (Buzrak-Durgadah Fault). The southern part of the basin is dipping at a shallow angle (3° to 4°) towards the Continental slope (Imam, 2013). Previously, the contacts of Gondwana (Barakar) with the basement granite were found to be intersected in the deeper exploration wells, and the total thickness of Gondwana (Damuda) was found to be 494 m to ~1200 m. Noticeably, the thickness of Gondwana intersected in the recent drilling program of 3 core holes in Jamalganj also has a similar thickness (over 450 m). The Gondwana rocks are overlain by the Jaintia Group of rock (Paleocene-Eocene) with the Tura Formation at the base, followed by the Sylhet Limestone and Kopili Formations. Surma Group (Miocene), Dupi Tila (Pliocene-Pleistocene), and Alluvium (Holocene) (at the top 20–30 m) are the upward-trending younger groups and formations of the rock sequence of the study area. A significant gap in sedimentation between Lower Gondwana (Early to Middle Permian) and Lowermost Tertiary (Early Paleocene) is found. The occurrence of volcanic rocks below the Tertiaries is common and correlated with the Rajmahal Trap of Eastern India (Imam, 2013). The contour on top of the Gondwana sequence indicates a synclinal fold of the sequence (Islam, 1998). The stratigraphic succession of the study area is given in Table 11.1.

11.1.2 INITIATIVES FOR DEVELOPING JAMALGANJ COALFIELD

Several significant attempts have been undertaken to develop coal resources in the Jamalganj area since its discovery in 1962 (Table 11.2). The main proposals have focused on conventional mining techniques, such as underground mining, but the greater depth has always been a major concern. Therefore, unconventional options, such as CBM/UCG, have gained interest as feasible alternatives.

11.1.3 OPTIMIZATION OF COAL EXTRACTION METHOD FOR STUDIED COALFIELD

Between the two conventional mining techniques (i.e., underground or opencast), the only possible underground mining method adoption in Jamalganj coalfield may be difficult due to greater depth. This is also the major reason for not taking any initiative to develop this coalfield within the last 60 years. While coal cannot be physically extracted, unconventional coal mining methods such as CBM extraction or UCG deserve to be considered as a potential energy exploitation option. Noteworthy, the latest advances in these techniques have demonstrated greater opportunities to extract coal resources sustainably. However, in 2016, Petrobangla conducted a study of this deposit assessing CBM prospects which has been deemed non-potential. Therefore,

TABLE 11.1 Generalized Stratigraphy of Jamalganj Coalfield

Age		Group	Formation	Thickness(m)	Lithology
Holocene			Alluvium/ Barind clay	20-30	Clay, silt, coarse, medium and fine sand and gravels
			~UNCONFORMITY~		
Pleistocene			Dupi Tila	12-70	Coarse to fine gravel and coarse sand
Pliocene	Late				
	Early				
~UNCONFORMITY~					
Miocene	Late	SURMA	Jamalganj	250-380	Silt, fine sand, clay, medium gravel, shaly clay, sandstone, shaly coal
	Middle				
	Early				
			~UNCONFORMITY~		
Oligocene	Late	BARAIL			Undifferentiated
	Early				
Eocene	Late	JAINTIA	Kopili	25-30	Shaly clay, sandstone
	Middle		Sylhet	15-45	Limestone, shaly clay
	Early		Tura	50-110	Sandstone, shaly clay, silt, shaly coal
Paleocene	Late				
	Early				
Cretaceous	Late				
	Early		Rajmahal		Basalt trap and associated sill/dyke
Jurassic	Late				
	Middle				Undifferentiated
	Early				
Triassic	Late				
	Middle				
	Early				

TABLE 11.1 *(Continued)*

Age		Group	Formation	Thickness(m)	Lithology
			~UNCONFORMITY~		
Permian	Middle	LOWER GONDWANA	Baren Measure	115-280	Medium to very coarse grained light greenish grey to greyish white chloritic sandstone; often interbedded with pebbles and conglomerate; matrix feldspathic; large cross bedded with minor siltstone/ chloritic shale often ferruginous
	Early		Barakar	300-400	Coarse to medium, white feldspathic sandstone, very hard and compact with seven major coal seam zone: with grey shale and carb shale. Garnet bearing cross bedded at the lower part and devoid of coal.
Precambrian				-	Granite and gneisses

**Data sources: Borehole logs acquired from the Petrobangla study in 2016 (Petrobangla Mines Division, 2016) and (Rahman & Zaher, 1980)* the UCG potential of Jamalganj Coalfield needs a detailed study program for further appraisal. The prospect of UCG in this coalfield has been studied by a few researchers. Sajjad et al. (2015) presented a review work considering the hydrology and geology of the coal-bearing region of Bangladesh (Sajjad & Rasul, 2015). Thomas Kempka et al. (2011) suggested the theoretical feasibility based on the economics of the coupled UCG-Urea process, along with the good potential of UCG covering CO_2 storage options in the already gasified zone (Kempka et al., 2011). Nakaten et al. (2014) projected a techno-economic assessment for Bangladesh to support the scarcity of electricity and fertilizer from syngas products; also, additional CO_2 utilization through the method of Enhanced Gas Recovery at one of the gas fields of Bangladesh (Nakaten, Islam, & Kempka, 2014). Constructing a subsurface model through geostatistical analysis to delineate the UCG feasibility might be a novel approach. Several studies have developed lists of the key factors for UCG suitability based on previous experiences from trials and research (Creedy et al., 2001; Beath et al., 2004; Burton, Friedmann, & Upadhye, 2006; Perkins, 2018b). These can be assessed in favor of the studied coalfield to explore the potentiality of UCG application (Table 11.3).

TABLE 11.2 Chronological Events for the Development of Jamalganj Coalfield

Year	Initiator	Details of Events
1962	Geological Survey of Pakistan (now, Geological Survey of Bangladesh)	Discovered first coalfield of Bangladesh by the UN-sponsored coal exploration program; in total, 10 wells (both open hole and core drill wells) were drilled, while 9 well-intersected coal seams; 11.7 sq. km area encountered-bounded by 5 wells (EDH 5, 6, 10, 11, and 12) was declared feasible for underground coal mining, as shown in Figure 11.1.
1966	Fried Krupp Rohstoffe	Recommended that mining is technically feasible considering the previously proposed area in the eastern part of the designated coal basin.
1969	Powell Duffryn Technical Services	
1976	Robertson Research International Ltd.	
2016	Petrobangla (Bangladesh Oil, Gas & Mineral Corporation)	A feasibility study focusing on Coal Bed Methane (CBM) prospects in 2016 found it non-profitable but recommended further feasibility studies for deep underground mining in the NW corner; about 15 sq. km is bounded by 3 wells (EDH 7, 8, and 9), as shown in Figure 11.1. Three (03) core holes were drilled in the coalfield where all of them encountered significant coal occurrences.

Data sources:(Ball et al., 1983; Imam, 2013; Petrobangla Mines Division, 2016).

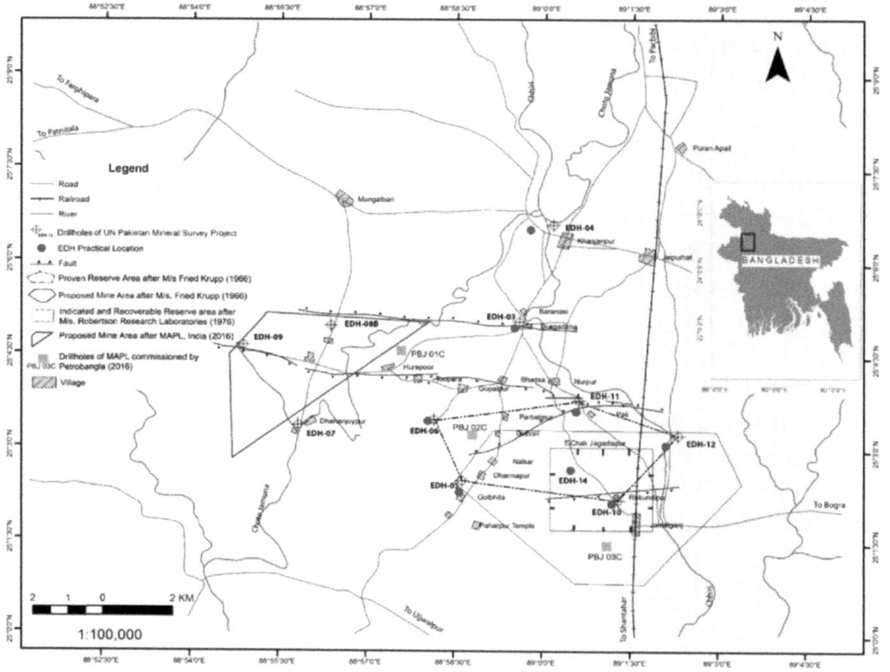

FIGURE 11.1 Jamalganj Coalfield–a general map with previous development proposals and drilling points. Also, the E-W trending northern boundary fault (Buzrak-Durgadah fault) is a major geological feature.

This study primarily assumes that the UCG potential of Jamalganj coalfield is positive through the integration of multi-factors from previous experiences of worldwide trials or demonstration projects. Apparently, the favorable geochemical properties of high-volatile bituminous (HV-B type) coal in the Jamalganj deposit are suitable for UCG development in terms of depth of occurrence, seam thickness, reserve of coal, areal coverage, and a workable overburden. The favorable factors are very high thickness (over 46 m) as well as depth (within 1000 m) of target seams, the chemical properties of high-volatile bituminous (HV-B type) coal with low ash content (7%–16% in the top 4 seams), very low sulfur content (~0.3%), a positive gross calorific value (GCV) of about 28 MJ/kg with a massive deposit of 5.45 billion tons and a structurally robust overburden of over 200 m to guard the regional aquifer. One of the biggest concerns is that extraction of this solid fuel resource through conventional mining methods can be catastrophic, particularly for the north-western region,

TABLE 11.3 Assessment of Physical Factors for Considering UCG Potential of Jamalganj Coalfield

Factors	Attributes of Jamalganj Coalfield+	Recommended Limit	Reference
Burial depth	>1000 m (Thickest Seam III = 660 m to 977 m)	> 200 m preferred	(Beath, 2006; Perkins, 2018a)
The thickness of the coal seam	Up to 46 m (Thickest Seam III = 8 m to 47 m)	> 6 m preferred	(Creedy et al., 2001; Beath et al., 2004; Perkins, 2018b)
Ash content (air-dried basis)	7–16% (seam I-IV);23–28% (seam V-VII)	<40%	(Beath et al., 2004; Beath, 2006; Perkins, 2018b)
Coal rank	Seams I-V: HV-B; seams VI-VII: HV-A bit.	Lignite or Bituminous with high volatiles	(Beath et al., 2004; Perkins, 2018b)
Discontinuities	E-W trending boundary fault in the north and another 2 minor faults in the south	should be away from any major discontinuities (faults/fractures)	(Beath et al., 2004; Perkins, 2018b)
Closeness to aquifers	No aquifer in the vicinity; safer distance (~600 m) from the overlying Dupi Tila aquifer	no aquifer within a distance of 25 times the seam height	(Vyas & Singh, 2015; Perkins, 2018b)
Overburden/Roof materials	150–200 m cover of hard, compacted, impervious Gondwana sandstone	Compacted, highly thick (>40 m preferred), essentially impervious	(Perkins, 2018b)
Seam dipping	5°–15°	not steeply, <30°	(Creedy et al., 2001; Beath et al., 2004; Perkins, 2018b)
Hydraulic head	108 bar	High (preferred >20 bar)	(Beath et al., 2004; Perkins, 2018b)
Scale of deposit	Not estimated in terms of UCG exploitation; but this field has 5.45 billion tons of total deposit	>100 MT preferred	(Perkins, 2018b)

+*Imam, Rahman, & Akhter (2002); Imam (2013); and Petrobangla Mines Division (2016).*

and generally for the whole country, due to the dewatering of arsenic-free sources of drinking and irrigation groundwater from the overlying Dupi Tila formation. In this context, the Jamalganj coal deposit offers great advantages, as there is considerable overburden (about 200 m) of sufficiently impermeable strata. However, the concerned authority may undertake a detailed hydrogeological study-investigation on the major potential aquifers in the NW region of the country, to support decisions for selecting a suitable mining method.

11.1.4 UCG – AT A GLANCE

The UCG is an advanced industrial and globally employed technique of harnessing energy from coal and provides an alternative to conventional underground mining for seemingly un-mineable coal (from the point of being uneconomical, especially when the deposit is too deep, of low grade, or very thin) with minimal surface disturbance (Khadse et al., 2007). The thermochemical process converts coal into a synthesis gas/fuel through partial oxidation. It involves the drilling of injection and production wells from the surface to the target coal seam followed by creating a highly permeable channel between those two wells to develop a flow path within the coal seam (Garkusha et al., 1984) (Figure 11.2). The process requires injecting steam and an oxidant (either oxygen or air and steam) which are the only feed materials to sustain the gasification process, and their effective proportion maneuvers the gasification performance (Ranade, Mahajani, & Samdani, 2019). As coal is not excavated, the associated coal wastes and combustion residuals, e.g., the ashes remain underground. The UCG objective is to produce syngas with a high percentage of CO and H_2, while the coal characters-quality-geology and the gasification parameters strongly directly impact the composition of produced substances. The produced syngas (H_2, CO, and CO_2 with minor and trace quantities of CH_4, hydrocarbons, and tars and traces of pollutants) can be utilized-recycled as a feedstock for chemical products like hydrogen, methanol, ammonia, fertilizers, and synthetic natural gas, or for power generation.

In the rising concern of global warming, UCG can be a viable clean coal technology as this can effectively reduce many of the impediments to conventional mining straightaway. It does not bring the coal to the surface, thus there is no coal handling/washing/transportation, no ash problem, and no mine working underground.

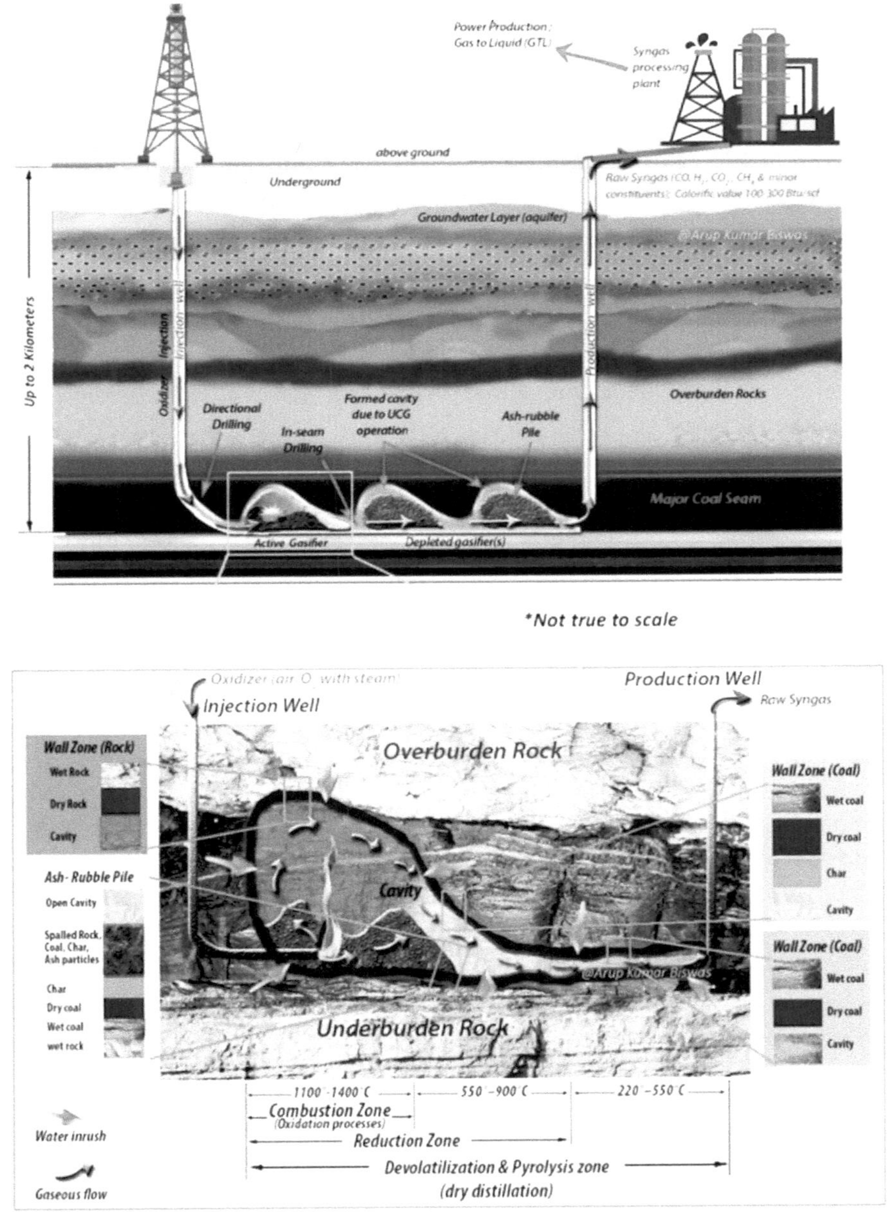

FIGURE 11.2 Schematic of UCG, aboveground and underground setup with a formed cavity inside the coal seam showing distinct physical/chemical-process domain.

Source: Adapted from Biswas et al., 2021

11.2 METHOD

Determination of the appropriate site for UCG in the targeted coalfield requires information on subsurface strata and lateral continuity of the coal beds. It is also significant to delineate areas of the thickest intersection of target seam(s). Moreover, the local geological setting needs to be considered (e.g., the presence/connection of the hydrogeological layer, any fault-induced strata disruption, etc.).

11.2.1 BOREHOLE LOG DATA COMPILATION AND CORRELATION

Borehole logging data can depict the subsurface. Lithological logs and geophysical logs of core holes PBJ01C, PBJ02C, and PBJ03C have facilitated coal seam correlation (Figures 11.3 & 11.4). The other available borehole logs, retrieved in 1962, were appended with the latest drilled data. Moreover, all available papers/notes and a few confidential reports relevant to this study have assisted in integrating the necessary information for building the concept.

11.2.2 METHOD OF GRIDDING BOREHOLE LOG

Geostatistical analysis might be an effective tool for resolving pre-mining considerations (Balić, Velić, & Malvić, 2008; Husanovi´c & Malvi´c, 2014). Geostatistical simulation explicitly portrays spatial variation, which is considered superior to other methods for building earth models. The main focus of geostatistical data analysis is understanding and describing the spatial patterns in variables like coal thickness (Srivastava, 2013). Hence, an analysis of available borehole logs data has been done geostatistically with the aid of commercial software, Rockworks 2020. For assessing the thickness of each coal seam alongside its lateral continuity, a special geostatistical analysis method, Inverse Distance Weighting (IDW), has been employed in this study. IDW is one of the most commonly used deterministic models, and its calculation accuracy is affected by two parameters: search radius and inverse-distance weight power value (Liu et al., 2020).

The assumption is based on a weighting of each data point value inversely according to its distance (d) from the grid node, in order to assign the node values, taken to the nth power

$$Z-node = \frac{\sum\left(\frac{\text{Z}-\text{point}}{d^n}\right)}{\sum\left(\frac{1}{d^n}\right)} \quad (1)$$

Values of distant data points will exert less influence than those from near the grid node by increasing the distance factor in the denominator to a power >1. The greater the exponent value, the lesser influence is exhibited on the distant points. The weighting exponent (which determines the extent of gridding process execution – local/global) of 3.8 was adopted in this study for confining the IDW results within an assumed buffer zone, which appears to have a realistic coal seam extent. This is again examined upon careful inspection of the available borehole log data.

11.2.3 GEOSTATISTICAL ANALYSIS FOR DOWNHOLE DATA INTERPRETATION

Spatial data along with the borehole logging data has been plotted, mainly the coal-bearing (hosted) successions. Using the Inverse Distance gridding method, several visual representations of the coal seams, such as 2D isopach maps, and coalbed successions (Figures 11.5–11.7) have been derived. The developed 3D interpretative geological models of the considered coalfield improved the understanding of the subsurface geological conditions – structure and the extent of the deposit – spatial distribution pattern of the coal deposits. This, in turn, rationalizes the selection of the most favorable site for initiating a pilot UCG project in the coal basin, based on required depth, thickness, and lateral continuity.

11.3 RESULTS AND DISCUSSION

11.3.1 BOREHOLE CORRELATION – 2D AND 3D BOREHOLE LOGS

A correlation among the neighboring boreholes is drawn emphasizing the coal intersections, and a simple two-dimensional (2D) view is presented in Figure 11.3. A topographic elevation on average 25 m above mean sea level (MSL) was adopted to plot, discuss, and interpret the borehole logs information. The available borehole logging data have been compiled

(Appendix A). The depths of the coal seams range from 641 to 1126 m below the surface. The thickest coal seam III lies at depths from 659 m (EDH-9) to 1032 m (EDH-14). Coal seam IV gradually merges with seam III in the eastern part of the basin (Imam, Rahman, and Akhter, 2002; Imam, 2013).

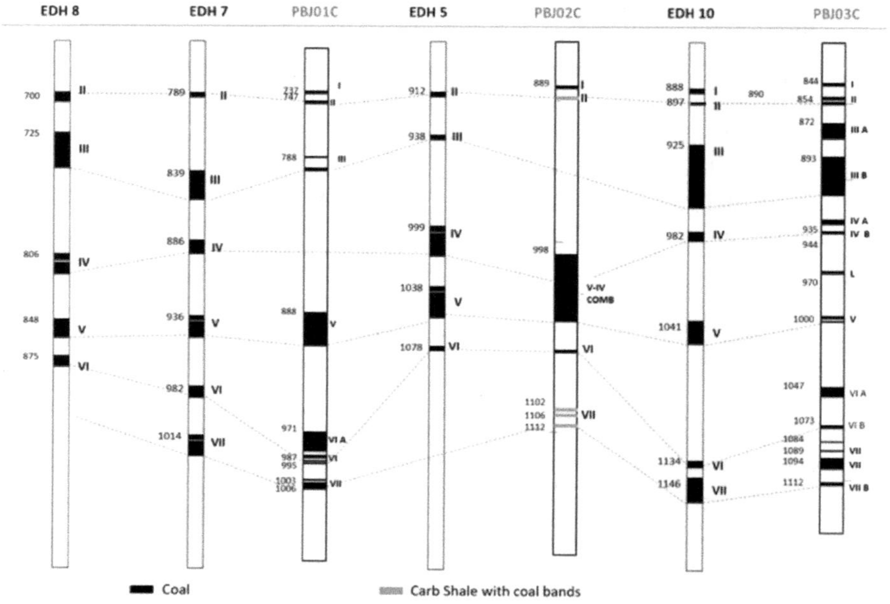

FIGURE 11.3 Borehole correlation depicting the coalbed intersections.

For a better understanding, a three-dimensional (3D) downhole log plot has been constructed (Figure 11.4). This fundamentally builds the first step to simulate the lateral extent of coal seams over the region. Thus, an isopach mapping of the coal seams has been constructed. Furthermore, these coal intersections have been interconnected to represent the stratigraphic correlation of the coal seams.

11.3.2 *ISOPACH MAPPING*

Figure 11.5 presents a 2-D isopach mapping of the basin exhibiting the thickness of the coal seams in place. It is worth noting that the overall isopach mapping of the studied coalfield does not depict any subsurface occurrence of coal seam extension in the SW area of the basin, as no borehole logs were recorded in that part of the basin. From Figure 11.5(a), it can be assumed that Seam I is not widely extended throughout the basin. It is almost absent in the western half of the coalfield. The thickest section (~13 m) is found around the middle of the eastern part (near borehole EDH-10). Figure 11.5(b) presents seam II, which is somewhat more extended than the topmost seam. The center of the basin has the thickest intersection of Seam II, about 12 m (around borehole EDH-6). To the Northwest, it thins out gradually, while towards the NE it diminishes very sharply.

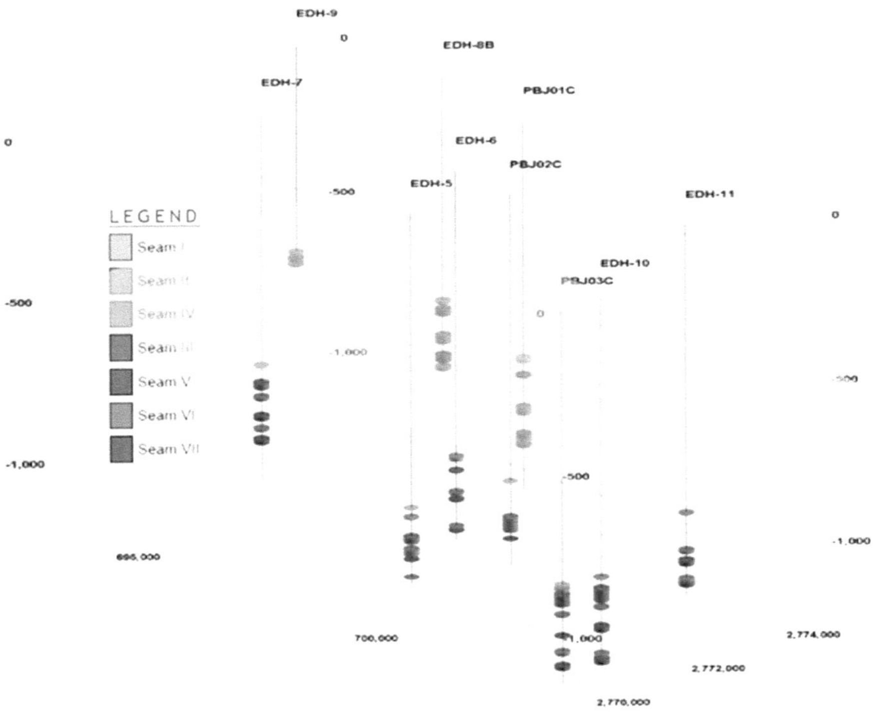

FIGURE 11.4 A 3-dimensional representation of borehole logs featuring coal intersections at depth in Jamalganj coalfield (isometric view).

Figure 11.5(c) represents Seam III, which is widely developed throughout the whole extent of Jamalganj coal basin and very thick by worldwide standards, and also it has appreciable lateral continuity. Seam III occurs at variable thicknesses, being around 5 m thick in the western part and around 45 m thick in the middle of the Eastern to Northeastern part. This seam, being the thickest, has been the main zone of interest to initiate mining. Seam IV is presented in Figure 11.5(d), which also shows another widely developed seam; thickest in the middle of the basin, gradually thinning out towards both east and west, yet maintaining a good average of over 5 m. The coal seam IV gradually merges with seam III in the Eastern part of the basin forming the thickest coal intersection within the basin (Imam, Rahman, & Akhter, 2002;

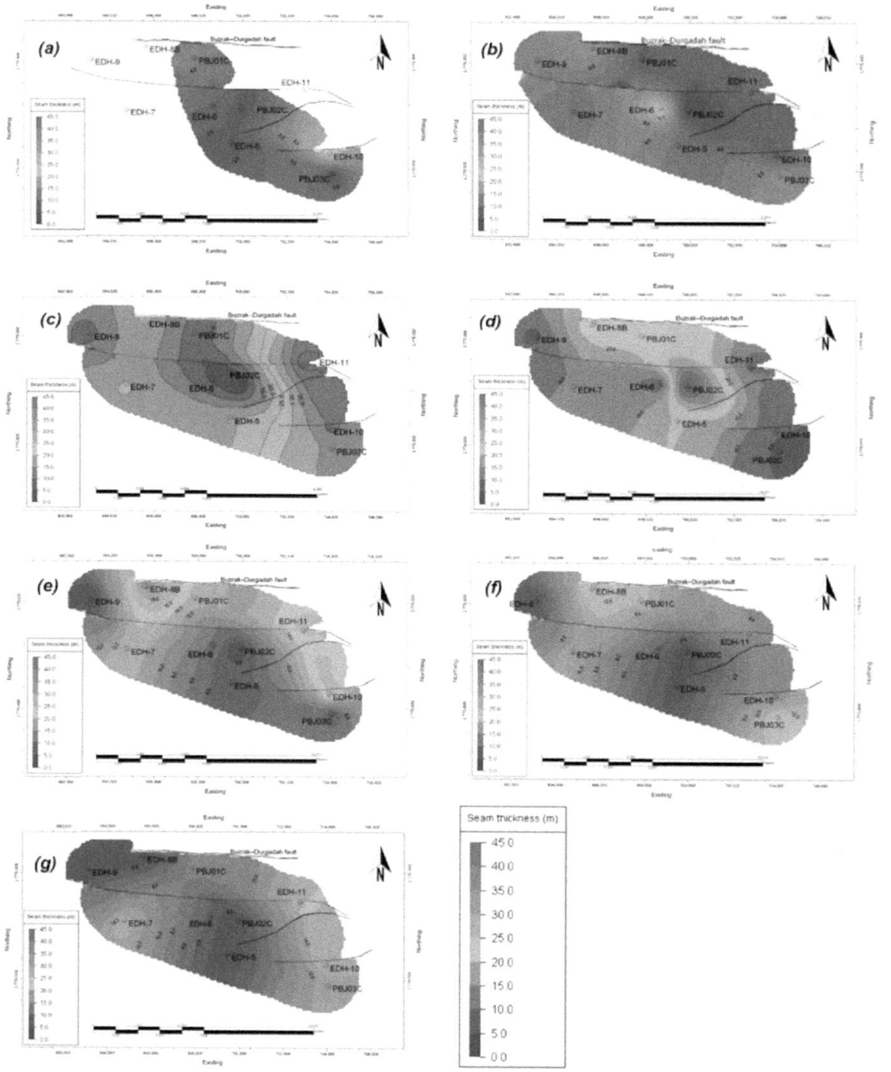

FIGURE 11.5 Isopach maps of Jamalganj coal seams; (a) Seam I (b) Seam II (c) Seam III (d) Seam IV (e) Seam V (f) Seam VI and (g) Seam VII.

Imam, 2013). The Northern and northeastern corners of the basin show considerable thickness, about 15 m, in Seam V shown in Figure 11.5(e), but this may not be widely distributed. From Figure 11.5(f), Seam VI might be absent from the central part. It only sporadically appears in a few areas of the Northern and South-eastern corners. Figure 11.5(g) presents Seam VII, which might have notable lateral continuity and considerable thickness (around 15 m) as encountered in the borehole logs drilled in the Eastern part of the basin. Imam et al. found that seam VII contains 35% of the total reserve in the Jamalganj coalfield, while the thickest seam III occupies 50% (Imam, Rahman, & Akhter, 2002).

11.3.3 UNDERGROUND COALBED SUCCESSION

A 3-D stratigraphic correlation of underground coalbed occurrences marked from the borehole logging data has been drawn (Figure 11.5). This has been achieved by implementing the geostatistical IDW extrapolation method as mentioned earlier. The method is carefully deployed to represent a nearly realistic coal seam extent throughout the basin. This apparently replicates the subsurface layout of the coal deposit. Thereby, seven (07) different coal layers (seam I to seam VII) have been stacked to represent the underground coalbed succession in the Jamalganj basin. In the illustrations of Figure 11.6(a), the isometric view presents the stratification pattern of the coal seams. This, along with the opposite North-western corner view in Figure 11.6(b), exhibits that the Jamalganj coal basin is sunken in the central part, as also proven by the deepest coal intersections in some drilled boreholes (EDH-5, PBJ02C, etc.). Figure 11.6(c) represents an oblique/tilted view from the North-eastern corner providing a better understanding of the coal seam occurrences throughout the extent. It can be seen in this view that the thickest Seam III gradually merged with Seam IV towards the East. Regarding the study goal to find an appropriate site for the UCG pilot test, these illustrations contain important information. Though the coal seams appear nearest to the surface (as shallow as 641 m in borehole EDH 9) in the northwestern part of the basin, the lateral continuity of the coal seams is likely to be disrupted by the faults. From available borehole logs, it is evident that some of the coal seams are non-existent or have thinned out. In contrast, the middle part seems to be sunken, and prospective coal seams appear at greater depths (mostly from 900 to 1100 m). Moreover, towards the east, this bends upward again and the most promising Seam

III gradually merges with Seam IV to make it further productive for UCG exploitation: the thicker the coal seam, the better; and no discontinuity is preferable (Beath et al., 2004). In this case, fewer wells are required, thus drilling costs are reduced.

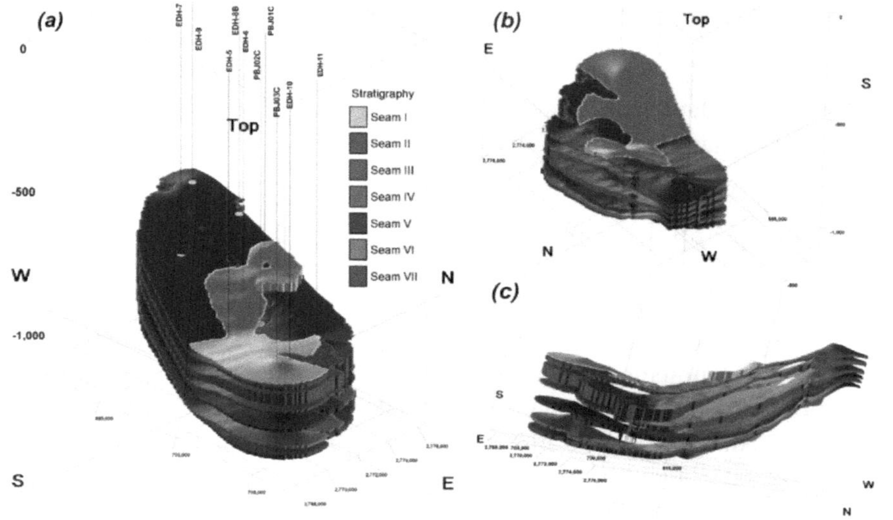

FIGURE 11.6 Coalbed succession – layout of the coal seam occurrences. (a) Isometric view with the boreholes, (b) N-W view, and (c) N-E tilted view.

11.3.4 FAULTING IN THE ZONE

From old data sources, only the locations of the existing faults have been retrieved. The dipping, fault throw, or vertical displacement data were not available during this study. In Figure 11.7, the projected displacement (as marked in a rectangular block) has been extrapolated methodically by using IDW based on the available borehole logs. Therefore, fault-induced strata disorientation cannot be measured accurately from the data. However, faulting seems to have less influence in the eastern part than in the northwestern part, as perceived from the borehole logs. This is certainly a positive decisive factor in the selection process of a pilot site for UCG operation. Supposedly, it will not affect the UCG operation if there are only minor faulting-induced displacements. Studies have documented that minimal faulting or fractures in both coal seam and surrounding formations are acceptable for UCG siting (Beath et al., 2004; Perkins, 2018b).

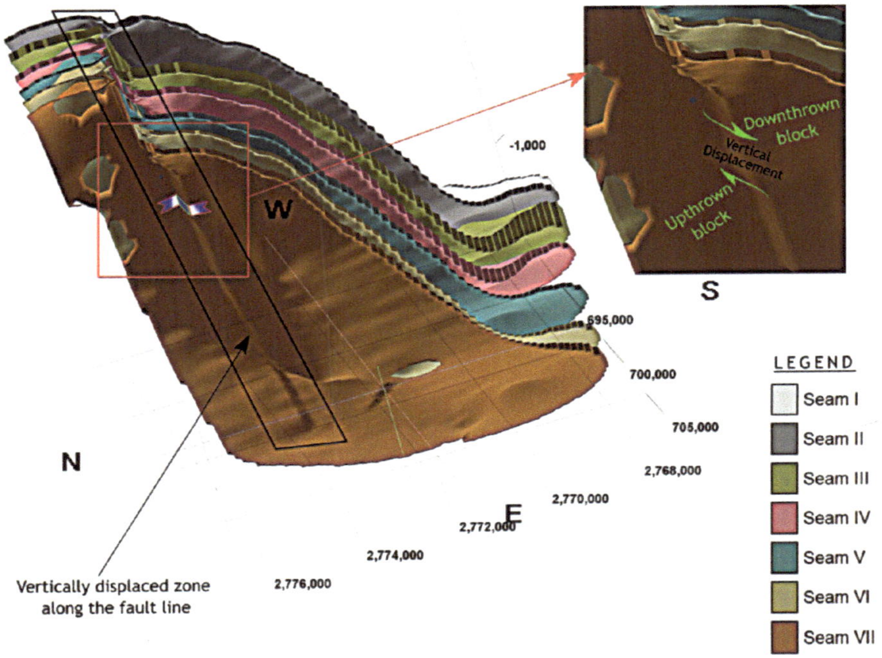

FIGURE 11.7 Coalbed succession showing displacements along the fault planes (bottom view).

11.3.5 ESTABLISHING A PROSPECT AREA FOR UCG SITING

The isopach mapping indicated that the occurrences of coal seams in the eastern part are much thicker than those in the western part of the basin. This is also backed up by the borehole log data, corroborating the reliability of this deduction. It appears that the coal seams in the western zone of interest are highly faulted, as the major Buzrak–Durgadah Boundary Fault in the Jamalganj coal basin runs alongside the boundary. In contrast, the zone of interest in the middle of the eastern part has been affected only by local faulting, and the coal seams have better lateral continuity throughout this part. Thus, the available data on Jamalganj coal deposits points towards a prospect for this unconventional mining approach to be pilot-tested in the middle of the eastern part of the basin (Figure 11.8). Hence, an area of 11.7 sq. Km bounded by 5 wells (EDH 5,6, 10, 11, & 12) among the total of 14 UN-PAK drill holes, appears the best feasible spot for pilot testing UCG (Figure 11.8). It is worth noting that this area was previously recommended for underground coal mining by Fried Krupps Rohstoffe (1966).

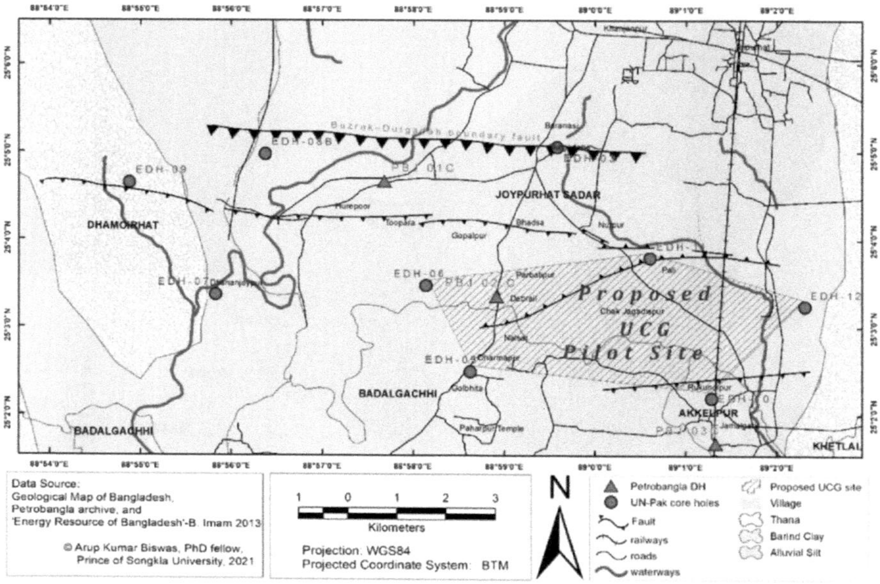

FIGURE 11.8 Proposed UCG site in the Jamalganj coalfield.

11.4 CONCLUSION

In Bangladesh, coal resources can be the key to future energy security and a significant fuel source alternative to natural gas, the reserves of which are depleting very fast. The country has started relying heavily on imported fuel. Nevertheless, most of the known coal deposits in the country remain inaccessible due to disputes in adopting a proper conventional mining method (open-pit or underground). Especially, the Jamalganj coalfield containing the biggest coal deposit (5.45 billion tons) among the five known coalfields in the country has been left unattended for over half a century due to its greater depth of occurrence (around 1100 m). While the depth of Jamalganj coal deposits appears not to be within any commercially viable conventional mining methods compatible with the techno-socio-economic-environmental aspects of Bangladesh, this still bears the promise to carry out a pilot test through underground coal gasification (UCG). UCG can be a potential clean coal technology to harness energy from coal resources in an environmentally sustainable manner. In terms of the UCG prospects, Jamalganj Coalfield has the advantage of a very high net coal seam thickness. Other favorable factors for UCG development include the basic (geo)chemical properties of high-volatile bituminous (HV-B type) coal with low ash (<20%) and sulfur content (<1%), in conjunction with appreciable depth of occurrence (zone of interest lies within 1000 m), a positive GCV of ~28 MJ/kg, and a concrete overburden keeping a safe distance from the regional major aquifer. In this study, a geostatistical analysis using the Inverse Distance Weighting (IDW) method has modeled the subsurface and created isopach maps of the coal intersections in the Jamalganj basin. These findings reveal that the very thick Seam III can presumably be the prime target for a UCG pilot test, especially where it merges with Seam IV in the eastern part of the Jamalganj coal basin. This study could facilitate the development of a viable operational strategy for venturing UCG piloting in the Jamalganj coalfield, the first of its kind in Bangladesh.

ACKNOWLEDGMENT

The authors would like to extend their gratitude to Dr. Mohammad Moshiur Rahman (Associate Professor, Department of Environmental Science and Management, North South University, Bangladesh), Dr. Dhiman R. Mondal (Research Scientist, MIT Haystack Observatory, USA) for their meaningful criticism in formulating this document. The authors highly appreciate Md. Shahariar Chowdhury (Lecturer, Faculty of Environmental Management, Prince of Songkla University, Hat Yai, Songkhla 90110, Thailand), Dr. Md. Ahosan Habib (Deputy Director, Geological Survey of Bangladesh), and Dr. Md. Hafiz Al Amin (Professor, Dept. of Agroforestry and Environment, Faculty of Agriculture, Hajee Mohammad Danesh Science and Technology University, Bangladesh) for helping the thought process while constructing the manuscript.

KEYWORDS

- **Bangladesh coal**
- **Geostatistical analysis**
- **Jamalganj coalfield**
- **UCG**
- **Unconventional coal mining**
- **Underground coal gasification**

REFERENCES

Balić, D., Velić, J., & Malvić, T. (2008). Selection of the most appropriate interpolation method for sandstone reservoirs in the Kloštar oil and gas field. *Geologia Croatica, 61*(1), 27–35. https://doi.org/10.4154/gc.2008.05.

Ball, M. M., Landis, E. R., & Woodside, P. R. (1983). Geologic assessment of the fossil energy potential of Bangladesh. USGS Publications Warehouse. https://doi.org/10.3133/ofr83280.

BCMCL. (n.d.). Barapukuria Coal Mining Company, Dinajpur, Bangladesh. https://bcmcl.org.bd/coal-mining-development/ (accessed on 5th June 2024).

Beath, A. (2006). Underground Coal Gasification Resource Utilisation Efficiency. Technical Report.

Beath, A. C., Craig, S., Littleboy, A., Mark, R., & Mallett, C. W. (2004). Underground coal gasification: Evaluating environmental barriers. *Prog. Energ. Combust., 39,* 189–214. https://doi.org/10.4225/08/59a9ae521cd41.

Biswas, A. K., Suksuwan, W., Phoungthong, K., & Wae-hayee, M. (2021). Effect Of Equivalent Ratio (ER) on the flow and combustion characteristics in a typical Underground Coal Gasification (UCG) cavity. *Journal of Advanced Research in Fluid Mechanics and Thermal Sciences, 86*(2), 28–38. https://doi.org/10.37934/arfmts.86.2.2838.

Burton, E. A., Upadhye, R., & Friedmann, S. J. (2019). Best practices in underground coal gasification. No. LLNL-TR-225331. Lawrence Livermore National Lab. (LLNL), Livermore, CA (United States).

Creedy, D. P., Gamer, K., Holloway, S., Jones, N., & Ren, T. X. (2001). Review of underground coal gasification technological advancements. Technical report ETSU-COAL-R-211, Energy Technology Support Unit (ETSU), Harwell Oxfordshire, OX11 0RA, UK.

Farhaduzzaman, M., Abdullah, W. H., Islam, M. A., & Sia, S. G. (2015). Quality assessment of the Permian coals from Dighipara Basin, Bangladesh based on proximate, ultimate, and microscopic analyses. *J. Bangladesh Acad. Sci., 39*(2), 177–194. https://doi.org/10.3329/jbas.v39i2.25951.

Garkusha, I. S., Kazak, V. N., & Kapralov, V. K. (1986). Vsesojuznoe Nauchno-Proizvod-Stvennoe Obiedinenie sojuzpromgaz. Method of underground gasification of coal seam. U.S. Patent 4,573,531.

Ghose, A. K. (2009). Contemporary issues in coal and energy in SAARC – opportunities and challenges in the backdrop of globalization. UNDP.

Husanović, E., & Malviæ, T. (2014). Review of deterministic geostatistical mapping methods in Croatian hydrocarbon reservoirs and advantages of such approach. *Nafta, 65*(1), 64–68.

Imam, B. (2013). Energy Resource of Bangladesh (2nd edn.). University Grants Commission of Bangladesh.

Imam, M. B., Rahman, M., & Akhtar, S. H. (2002). Coalbed methane prospect of Jamal Ganj coalfield Bangladesh. *Arabian Journal for Science and Engineering, Section B: Engineering, 27*(1A), 17–27.

Islam, Md. R., Hayashi, D., & Kamruzzaman, A. B. M. (2009). Finite element modeling of stress distributions and problems for multi-slice longwall mining in Bangladesh, with special reference to the Barapukuria coal mine. *Int. J. Coal Geol., 78*(2), 91–109. https://doi.org/10.1016/j.coal.2008.10.006.

Islam, Md. S.-U.-. (1998). Origin of a thick coal seam in the Gondwana sequence of Jamalgonj basin, Bangladesh. *Journal of Nepal Geological Society, 18,* 289–298.

Journel, A. G., & Huijbregts, Ch. J. (1976). *Mining Geostatistics.* The Blackburn Press.
Kempka, T., Plötz, M.-L., Schlüter, R., Hamann, J., Deowan, S. A., & Azzam, R. (2011). Carbon dioxide utilization for carbamide production by application of the coupled UCG-urea process. *Energy Procedia, 4,* 2200–2205. https://doi.org/10.1016/j.egypro.2011.02.107.
Khadse, A., Qayyumi, M., Mahajani, S., & Aghalayam, P. (2020). Underground coal gasification: A new clean coal utilization technique for India. *Energy, 32*(11), 2061–2071. https://doi.org/10.1016/j.energy.2007.04.012.
Liu, H., Chen, S., Hou, M., & He, L. (2020). Improved inverse distance weighting method application considering spatial autocorrelation in 3D geological modeling. *Earth Sci. Inform., 13,* 619–632. https://doi.org/10.1007/s12145-019-00436-6.
Nakaten, N., Islam, R., & Kempka, T. (2014). Underground coal gasification with extended CO_2 utilization – An economic and carbon neutral approach to tackle energy and fertilizer supply shortages in Bangladesh. *Energy Procedia, 63,* 8036–8043. https://doi.org/10.1016/j.egypro.2014.11.840.
Perkins, G. (2018). Mathematical modeling of in situ combustion and gasification. *P. I. Mech. Eng. A-J. Pow., 232*(1), 56–73.
Perkins, G. (2018). Underground coal gasification – Part I: Field demonstrations and process performance. *Prog. Energy Combust. Sci., 67,* 158–187.
Petrobangla Mines Division. (2016). Drilling and testing of three core holes for a feasibility study for the extraction of coal bed methane at Jamal Ganj [Restricted access]. Dhaka, Bangladesh.
Rahman, R. R., & Zaher, M. A. (1980). Jamal Ganj Coal – its quality, quantity, and mine ability. In *Petroleum and Mineral Resources of Bangladesh. Seminar and Exhibition,* 41–53. Dhaka.
Rambert, F. (2005). Introduction to mining geostatistics. In *67th EAGE Conference & Exhibition,* 1–4. https://doi.org/10.3997/2214-4609.201405201. European Association of Geoscientists & Engineers.
Ranade, V., Mahajani, S., & Samdani, G. (2019). *Computational Modelling of Underground Coal Gasification.* CRC Press.
Sajjad, M., & Rasul, M. G. (2015). The prospect of underground coal gasification in Bangladesh. *Procedia Eng., 105,* 537–548. https://doi.org/10.1016/j.proeng.2015.05.087.
Srivastava, R. M. (2013). Geostatistics: A toolkit for data analysis, spatial prediction, and risk management in the coal industry. *Int. J. Coal Geol., 112,* 2–13. https://doi.org/10.1016/j.coal.2013.01.011.
U.S. Energy Information Administration (n.d.). *U.S. Energy Information Administration.* https://www.eia.gov/energyexplained/units-and-calculators/. (accessed on 5th June 2024).
Vyas, D. U., & Singh, R. P. (2015). Worldwide developments in UCG and Indian initiative. *Procedia Earth Planet. Sci., 11,* 29–37.

APPENDIX A

Depths and Thicknesses of the Coal Seams (Islam, 1998; Imam, Rahman, & Akhter, 2002; Petrobangla Mines Division, 2016).

Coal Seams in Jamalganj field	UN-sponsored core holes									
	EDH 5		EDH 6		EDH 7		EDH 8B		EDH 9	
	Depth from surface (m)	Thickness (m)	Depth from surface (m)	Thickness (m)	Depth from surface (m)	Thickness (m)	Depth from surface (m)	Thickness (m)	Depth from surface (m)	Thickness (m)
Seam I	912.9	1.52	-	-	-	-	-	-	-	-
Seam II	940	4.56	886	12.46	786.4	4.26	700	7.9	641.2	5.17
Seam III	999.7	19.41	929.8	4.26	838.4	20.37	725	20.67	659.6	8.87
Seam IV	1037.5	20.22	995.2	7.9	882.7	10.34	806	24.78	679.4	4.55
Seam V	1070.3	2.36	1018	5.17	941.8	13.68	865.9	20.98	-	-
Seam VI	1126.4	-	1102.5	2.56	981.6	7.6	902.5	10.99	-	-
Seam VII	-	-	-	3.19	1013.9	15.05	-	-	-	-

APPENDIX A *(Continued)*

Coal Seams in Jamalganj field	UN-sponsored core holes				Petrobangla core holes					
	EDH 10		EDH 11		PBJ01C		PBJ02C		PBJ03C	
	Depth from surface (m)	Thickness (m)	Depth from surface (m)	Thickness (m)	Depth from surface (m)	Thickness (m)	Depth from surface (m)	Thickness (m)	Depth from surface (m)	Thickness (m)
Seam I	867.1	12.68	-	-	737.1	5.6	888.72	1.43	844.5	0.95
Seam II	876.1	2.56	892.2	3.19	747.7	1.7	-	-	854.3	7.25
Seam III	908.7	40.76	976.9	46.82	788.8	7.2	-	-	872.77	35.44
Seam IV	966.7	5.22	1005.6	8.97	888.5	24.2	998.29	44.59	935.56	4.06
Seam V	1024.6	16.42	1036.2	16.42	971	13.4	1067.52	0.4	999.97	3.13
Seam VI	1108.7	6.04	1093.6	6.04	987.85	9.25	-	-	1047.22	10.45
Seam VII	1124.1	15.81	1101.2	15.81	1003.3	8.5	-	-	1089.59	12.69

CHAPTER 12

Preliminary Assessment of Water-Related Policies and Plans in Davao, Philippines

AUGIE FUENTES,[1] NOOR HASHIMAH HASHIM LIM,[2] ANTHONY SALES,[3] JEFF CAMKIN,[4] HEMANT OJHA,[4] BASUNDHARA BHATTARAI,[4] RODCEAL MALINAO,[1] ROEL APAS,[1] ROY MOLINA,[1] JOVANIE PUTONG,[1] SIVERLYN CAMPOSANO,[1] JARSEY KRIM TAUB,[1] ALBERTO BANDIOLA,[5] ARCHIELITO MALLARI,[5] DELLA GRACE BACALTOS,[1] and SHAHARIAR CHOWDHURY[6,7,8]

[1]*Davao del Sur State College, Matti, Digos City, Philippines*

[2]*Department of Urban and Regional Planning, Faculty of Built Environment, Universiti Malaya, Kuala Lumpur, Malaysia*

[3]*Department of Science and Technology XI, Davao City, Philippines*

[4]*Institute for Study and Development Worldwide, Sydney, Australia*

[5]*Provincial Environment and Natural Resources Office, Digos City, Philippines*

[6]*Faculty of Environmental Management, Prince of Songkhla University, Songkhla, Thailand*

[7]*Faculty of Health and Life Sciences, INTI International University, Persiaran Perdana BBN, Putra Nilai, Nilai, Negeri Sembilan, Malaysia*

[8]*Research Center on Industrial Ecology in Energy Business, Faculty of Environmental Management, Prince of Songkhla University, Songkhla, Thailand*

Engineering Design and Technical Applications of Physical Science: An Integrated Approach for Sustainable Development. Chin Hua Chia, Sarani Zakaria, Ruey Shan Chen, & Md. Shahariar Chowdhury (Eds.)

ABSTRACT

The global situation of watersheds is currently in a state of decline. The quality and condition of these water systems are on a worrisome downward trend in most countries around the world. Threats to watersheds are manifold and include problems like soil erosion, agricultural encroachment, a decrease in forest cover, and a loss of biodiversity. These issues have become increasingly noticeable in recent years, and they are contributing to a worldwide water crisis that is affecting both the quantity and quality of water available. This information sets the context for a focused study conducted in Davao, Philippines. The main goal of this study was to identify the specific problems and shortcomings in water policies and planning within the region. To gather data, questionnaires were distributed to various groups, including stakeholders, government officials, policymakers, and residents. The results of the study revealed several significant deficiencies. One of the primary findings was that policy documents related to water management were unclear. Moreover, there was an unexpected shortage of human resources, indicating a potential lack of capacity or expertise to effectively manage water resources. Additionally, there was a noticeable lack of data regarding water management or policy, suggesting a need for more robust data collection and analysis methods. These shortcomings pose significant challenges to the efficient and sustainable management of water resources in Davao.

12.1 INTRODUCTION

The global state of watersheds is deteriorating, with quality and conditions showing an alarming downward trend in most countries. Endangered watersheds consistently face issues including but not limited to soil erosion, agricultural encroachment, as well as declining forest cover, and biodiversity is now becoming evident in watersheds in most countries. The accumulation of these issues contributes to a water crisis in terms of quantity and quality, which is considered a critical challenge in the urbanizing global southern region. The crisis is further aggravated by other externalities such as the effects of climate change, mismanagement of water resources, poor water governance, and increasing water demands (Devkota et al., 2020). The issues mentioned are also observed in a watershed in Kapatagan; a small village in Davao del Sur, Philippines used as a demonstration site for the implementation of an Integrated Water Resource Management Project. Barangay Kapatagan is situated on the island of Mindanao. The elevation of its coordinates is estimated at 1,296.0 meters or 4,251.9 feet above mean sea level (Figure 12.1). Based on the population census of the Philippines, Barangay Kapatagan is estimated to have about 15,070 inhabitants in the year 2020, which represents 7% of the total population of Digos City. Issues and threats identified in the project site, hereinafter known as the study area of this paper, include: a) increasing agricultural practices; b) increasing population due to migration; c) lack of zoning policies especially in protected areas; d) conversion of forest to agriculture and industrial uses; and e) solid waste pollution.

Mt. Apo National Park (MANP) shows that areas identified as important watersheds within its vicinity are now in a critical state of denudation. This causes concerns as the same areas are important water resources to support the surrounding demand for domestic, agricultural, and industrial purposes. The incidence of soil erosion, landslides, sedimentation, and siltation is becoming more prevalent in areas of limited forest cover.

The issue of land conversion to other uses, especially agricultural activities mainly due to the expansion of banana plantations, is increasingly worrying in the areas of Barangay Kapatagan, Digos City. Moreover, the expansion of vegetable growing ventures has also caused alarming concerns in the upland slopes. The problem of unregulated land conversion in MANP stems from deeply rooted socioeconomic factors compounded by conflicting policies from time to time. The massive agricultural development in MANP and the extensive use of inorganic or chemical fertilizers and pesticides may also affect the freshwater ecosystems in the area, particularly the numerous river networks. Run-offs from agricultural areas may cause water contamination and pollution which further threatens species inhabiting in and around the freshwater ecosystems of MANP. Inappropriate agricultural practices also resulted in the depletion of soil nutrients, which further triggered low farm productivity. Pursuant to the socioeconomic issues mentioned earlier, this disrupts the activities that support the livelihood of the MANP's occupants. There are settlers who are still relying on natural resource extraction for subsistence, such as gathering timber and non-timber forest products for kaingin and charcoal production. On the other hand, agricultural practices, especially high-value cash crops, tend to use the lands intensively, especially with the application of inorganic or chemical fertilizers and pesticides. Certain agricultural developments do not even have proper provisions for soil and water conservation measures according to the requirements stated in related policies. These problems can be addressed through a comprehensive research and development project such as the establishment of an Integrated Water Resource Management (IWRM), focused on implementation at the

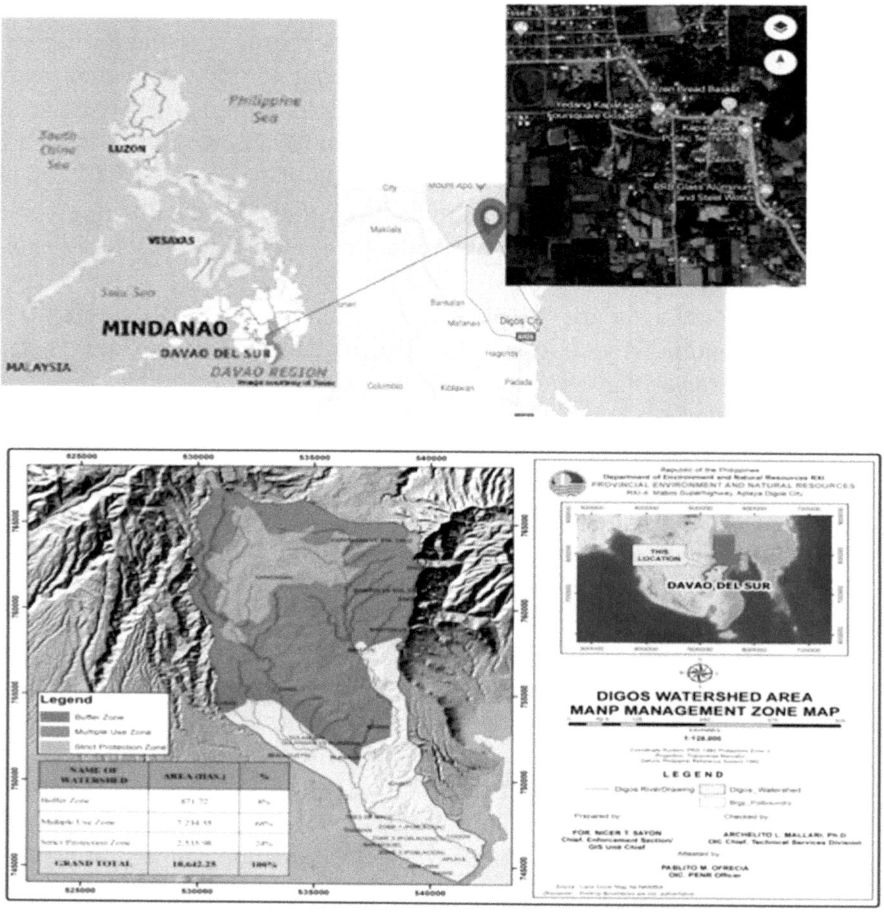

FIGURE 12.1 Maps showing the location of barangay Kapatagan, Digos City, Philippines.
Source: Courtesy of Provincial Environment and Natural Resources Office (PENRO), Department of Environment and Natural Resources (DENR)

local level with support from policymakers and water resource managers from the Davao Region, national agencies, and international partners. The concept of Integrated Water Resource Management (IWRM) emerged in the 1990s (Liu et al., 2008), where it is defined as a "process which promotes the coordinated development and management of water, land, and related resources, in order to maximize the economic benefit and social welfare in an equitable manner without compromising the sustainability of vital ecosystems" (Global Water Partnership). IWRM is now widely recognized by many countries around the world as an important component of their approach to water governance. The Asian Development Bank (ADB) recognizes that Asia's river basins also have issues related to flooding, agricultural encroachment, and declining biodiversity, among others. The role of IWRM aligns with the ideals that watersheds and water crises need to be managed in an integrated manner that promotes equitable sharing of water resources while preserving the environment. Furthermore, the UNESCO Agenda 2030 puts emphasis on a global network for sustainable development, such as the management of water resources and the conduct of the United Nations World Water Assessment Program that measures progress in achieving Sustainable Development Goal 6 (SDG 6: Clean water and sanitation) as it relates to other SDGs. Although IWRM serves as a strategy for sustainable water management, it is said that there is no blueprint or any "one-size-fits-all" strategy for IWRM; it has to be built to suit varying political, socio-economic, and cultural conditions and contexts. The main aim of the IWRM is to provide a broader systematic approach to water management which; (i) supports more coordinated management and use of land and water, surface, and groundwater, as well as up and downstream needs; and (ii) it is an ongoing process of learning. There are five key items shown in the IWRM Guidelines at River Basin Level (2009) through a spiral and continuing process of learning; (i) recognizing and identifying a need to adopt or modify IWRM; (ii) conceptualizing the changes required; (iii) coordinating and planning change strategies; (iv) implementing, monitoring and evaluating the change strategies; (v) then recognizing and identifying again the need for further adoption or modification of the IWRM approach being taken. As governments and communities cycle through this process, they are exposed to trying new solutions, respond to changing social, economic, and environmental circumstances and different needs better, and they are able to update the visions and agreements on trade-offs. All these create an ongoing process of learning and capacity building for IWRM (Figure 12.2).

The implementation of the IWRM is most needed at the local level. This collaborative project, which focuses on the assessment of water-related policies and plans, is an initial step in implementing IWRM locally. However, the study did come with a limitation where not all related policies were included in the review process due to accessibility. However, the project has addressed Phase 1 of the IWRM spiral and process, which is recognizing and identifying issues. It also serves as an effort to support the Sustainable Development Goals (SDGs), many of which rely on good water management. As an important component of a future IWRM program, the preliminary assessment of water-related policies and plans has determined specific policy provisions from each level, the extent of the stakeholders' knowledge and awareness, as well as their extent of implementation and impact on water and land resources in the Kapatagan Watershed and the whole Davao Region.

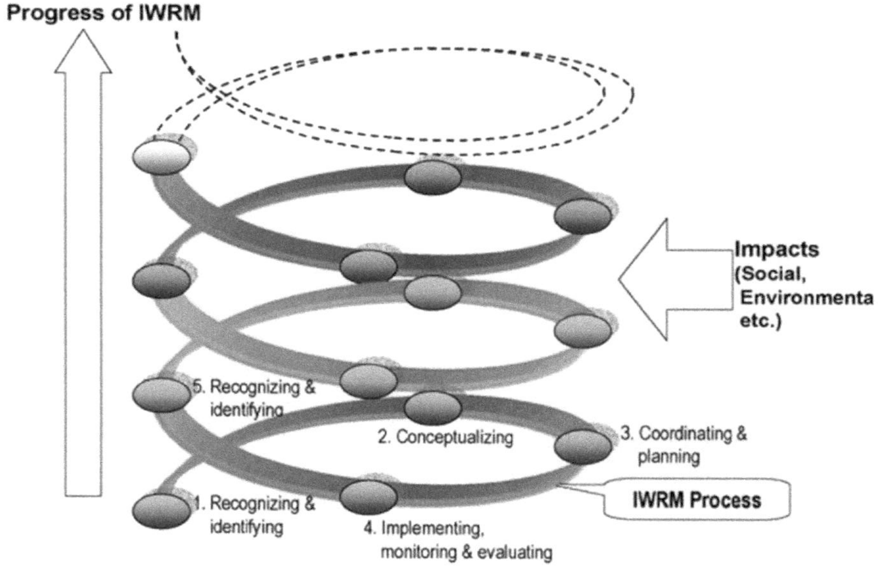

FIGURE 12.2 The IWRM spiral and process.

Source: Reprinted from Programme Hydrologique International, 2009. Courtesy of Integrated Water Resources Managment (IWRM).

12.2 METHODS

12.2.1 ENTRY PROTOCOLS

As an act of courtesy to the stakeholders, various entry protocols were done. These include the preparation and sending of Letters of Intent to the Protected Area Management Board (PAMB), the Mt. Apo National Park (MANP) Office, the Provincial Environment and Natural Resource Office, as well as the Local Government Unit at the city and barangay levels. For the stakeholders, a Free Prior and Informed Consent was accomplished. There were also preliminary meetings scheduled to present the plans and proposed activities for the project.

12.2.2 DATA COLLECTION

- **Secondary Data Gathering:** Important references to be used in the project were gathered from various official sources and were reviewed as secondary data. These comprised national policies, local ordinances, technical reports, land use plans, socio-economic profiles, and resource maps that were available for public access (Table 12.1). All available national and local policies, including ordinances related to water, were gathered and reviewed. Apart from reviewing the content of the documents, the extent to which those documents were used by stakeholders was also assessed through interviews. Some of the policies and structures being reviewed are as follows:

TABLE 12.1 List of Related Policies and Plans Reviewed in the Study

A. National-Level	
A.1	Philippine Clean Water Act of 2004
A.2	Expanded National Integrated Protected Areas System Act of 2018
A.3	Mt Apo Protected Area Act of 2003
A.4	The Indigenous Peoples Rights Acts of 1997
B. Regional, Provincial, and City-Level	
B.1	Davao Water Partnership
B.2	Water Vision for Davao Region
B.3	Regional Development Council Resolution for the Adoption of IWRM Davao Water Partnership Action Plan and Guidelines
B.4	Ancestral Domain Sustainable Development and Protection Plan (ADSDPP)
B.5	Ordinances (solid waste, greening ordinance re 1 million trees, local septage)
B.6	Comprehensive Land Use Plan
C. Local-Level (Barangay)	
C.1	Kapatagan Water Services Cooperative (KWASECO)

- **Stakeholder Interview:** A total of 28 representatives from various sectors including farmers, resort owners, and barangay officials were interviewed using a semi-structured questionnaire. Most farmers are members of the Kapatagan Water Service Cooperative, Spring Development and Water Association, and Tribal Water System (Figure 12.3). During the first part of the interview, the stakeholders were asked about the water-related problems and issues that they face within the vicinity, as well as the extent to which the requirements of the related policies were being implemented in the area. For the second part of the interview, the stakeholders were asked to report a subjective score on their level of awareness and knowledge towards water-related policies and plans, the extent of information that they know, and the impacts of their activities on the surrounding water and land resources.

FIGURE 12.3 Photo documentation during the interview.

- **Focused Group Discussion:** Officials from the Provincial Environment and Natural Resources Office (PENRO) of Davao del Sur, and the Planning Officers of the Local Government Units at the provincial and city levels were invited to participate in the Focused Group Discussion (FGD). Preliminary findings of the interviews conducted were presented to the FGD participants right after the first round of discussion sessions ended. Questions were raised to prompt the participants to comment and discuss with one another. The prompt questions include but are not limited to the following (Table 12.2):
- **Data Analysis:** The water-related policy review and analysis was done in three stages. Firstly, during the desk review, specific provisions related to the subject matter from the national, regional, provincial, city, and local levels were extracted and noted into a master copy. At this stage, filters were applied by using inclusion and exclusion criteria to the related documents. Then, content analysis was applied to the relevant information extracted from those documents.

Next, using the information gathered during the first stage, relevant questions were listed for the stakeholder interviews, and another set of questions was drafted to prompt conversations and discussions among the invited stakeholders for the FGD session. The questions were examined, reviewed, and amended by the team. Here, the content validity and reliability of the questionnaires were assessed and established.

During the second stage, related stakeholders were approached by the team to participate in the interviews, and their responses regarding the water-related policies and plans were collated. The interviews were recorded using a recorder, and the recordings were transcribed. With the written transcriptions, content analysis and thematic analysis were applied to identify groupings of problems and issues faced by the interviewees.

Lastly, during the third stage, the prompt questions established earlier were posed to the stakeholders during the FGD. This is where direct solicitation of opinions and ideas related to water was gauged. The participants were divided into groups, and each group was given the same set of questions to discuss and express their opinions from their points of view and positions. Similarly, the outcome of the discussions was analyzed using thematic analysis to identify common problems or issues faced by multiple parties.

TABLE 12.2 List of Questions to Prompt Conversations and Discussions Among the Stakeholders During the Focused Group Discussion (FGD)

"What are the problems and issues related to water, especially as a source for agriculture/fisheries and renewable energy?"
"What policies (national/local) are now being implemented and are contributing to water sustainability? How?"
"What are the barriers/hindrances to water policy implementation?"
"What is/are your schemes/mechanisms in monitoring and evaluating the extent of water policy implementation?"

12.3 RESULTS AND DISCUSSION

12.3.1 DESK REVIEW

Table 12.3 shows details of the various water-related policies and plans at the national, regional, provincial, and city as well as local levels. Among the

TABLE 12.3 Desk Review of Water-Related Policies and Plans

Policy Level	Title/Number	Provisions for Water Governance
National	Philippine Clean Water Act of 2004 RA 9275	Designation of Water Quality Management Areas in river basins (Article 1: Section 5)
	Expanded National Integrated Protected Areas System Act of 2018 RA 11038	Prohibiting littering on groundwater and other bodies of water (Section 25)
	Mount Apo Protected Area Act of 2003 RA 9237	Water services shall be implemented within areas designated and approved by the Protected Area Management Board (PAMB) and other appropriate government agencies (Section 25)
	The Indigenous Peoples (IPs) Rights Acts of 1997 RA 8371	IPs/ICCs have the right to claim ownership over bodies of water being occupied (Chapter 3: Section 7) It is the responsibility of IPs/ICCs to preserve the watershed (Chapter 3: Section 9)
Regional	Davao Water Partnership	Place IWRM at the helm of the development of regional land and water management strategies with due consideration to the social, economic, and environmental resources of the region, advance the development of human resources in IWRM, develop comparative analyses for the multisectoral management of water, develop policy, frameworks, and tools for IWRM, deliver capacity building for institutions and individuals, provide policy advice, benchmark performance and document social learning in the IWRM journey
	Regional Development Council Resolution for the Adoption of IWRM Davao Water Partnership Action Plan and Guidelines RDC XI Resolution No. 74 series of 2011	Davao Water Partnership action plans and guidelines are the strategies for water resource management in Davao region
Provincial/ City	Ancestral Domain Sustainable Development and Protection Plan (ADSDPP)	Prohibition of destructive activities in river basins and other water sources
	Comprehensive Land Use Plan (CLUP)	Watershed conservation and protection
Local (Barangay)	Kapatagan Water Services Cooperative (KWASECO)	Provides guidelines on the usage and distribution of potable water

national policies, there is a specific provision in Article 1, Section 5 of the Philippine Clean Water Act of 2004, which is now institutionalized by the Environmental Management Bureau (EMB) of the Department of Environment and Natural Resources (DENR). Data on water quality for the Davao River and Digos River are regularly monitored. For the Davao River Basin (DRB), there is now an effort to establish a DRB Health Scorecard. Part of the highlights in the Draft DRB Health Scorecard are shown in Figure 12.4 and Table 12.4, while the measurements for Total Suspended Solids (TSS), nutrients, and coliform for the Digos River (which partly connects to the Kapatagan Watershed) are presented in Table 12.5.

FIGURE 12.4 Image showing the condition and quality of the Davao River.

Source: Reprinted from Environmental Management Bureau (2023).

TABLE 12.4 Status of TSS and Fecal Coliform of Davao River

Parameters	Stations in DRB	Past Data (2020-WQMA)	Standard/ Normal	Status
Total suspended solids (mg/L)	Stn 9–11	7.2–2,705	50 (Class A)	Above normal
	Stn 1–3	24–3,164	60 (Class B)	
Fecal coliform (MPN/100 mL)	Stn 9–11	680–35,000	<1.1 (Class A)	Above normal
	Stn 1–3	1,300–1,600,000	100 (Class B)	

Source: Reprinted from Environmental Management Bureau (2023).

TABLE 12.5 Data on Total Suspended Solids (TSS), Phosphate, and Dissolved Oxygen of Digos Riveras (2022)

Parameters	Stations in DR	Data	Standard/Normal
Dissolved oxygen (mg/L)	Mouth of the river	5.0	5.0 (Class A)
			5.0 (Class B)
Phosphate (mg/L)	Mouth of the river	0.5	0.5 (Class A)
			0.5 (Class B)
Total suspended solids (mg/L)	Mouth of the river	80	50 mg/L (Class A)
			65 mg/L (Class B)

Source: Environment Management Bureau-XI.

12.3.2 STAKEHOLDER INTERVIEW

Table 12.6 shows the responses of 28 representatives interviewed from various sectors, including farmers, resort owners, and barangay officials, who are mostly members of the Kapatagan Water Service Cooperative, Spring Development and Water Association, and Tribal Water System.

TABLE 12.6 Data on Water-Related Issues, Types of Policies, and the Implementing Entities Gathered from the Interview Session

Segments	Frequency	Percentage (%)
Water Issues		
Accessibility	14	50
Scarcity	4	14
Quality	3	10
No response	7	25
Type of policy		
Administrative	18	64
Legislative	3	10
Regulatory	7	25
Implementing Entity		
Local Government Unit	7	25
Association	10	36
Cooperative	11	39

The majority of the respondents indicated that their problem is water accessibility, and this is represented by 50% out of the total interviewees. This is explained by the fact that most farmers cannot afford the cost of water installation required by water providers in the area like the Kapatagan Water Service Cooperative (KWASECO). There are also issues on the drying up of natural springs which are the sources of potable water. Moreover, there is reported competition on water use between the residents and the industry players such as the resort and agricultural landowners.

Types of policies implemented at the Barangay level are usually administrative, and this accounts for about 64% of the implementation percentage when compared to other related policies. This is a part of the management strategies of cooperatives and associations that hold control over the distribution of water in the Kapatagan area. However, there are issues that have surfaced; for example, the overlapping of powers among the local government units, the cooperatives, associations, and tribal leaders.

For the subjective scoring based on the perception of stakeholders on related policies and plans, the parameters measured include knowledge and awareness, the extent of implementation, impact on water and land resources, feasibility, and budgetary, and economic impact. Results showed that both the Comprehensive Land Use Plan (CLUP) and the Ancestral Domain Sustainable Development and Protection Plan (ADSDPP) have scores ranging from 2.2 to 2.7. This indicates that the stakeholders have an average or moderate level of knowledge and awareness, extent of implementation, and impact on water and land resources. On the contrary, knowledge and awareness of the Philippines Clean Water Act of 2004, the Coastal Resource Management Plan (CRMP), and the Barangay Kapatagan Development Plan (BDP) were reported to have lower scores of 1.0 to 1.8. The subjective scores are shown in Table 12.7.

12.3.3 FOCUSED GROUP DISCUSSION

There are four (4) questions being asked to respondents that represent various sectors including the government agency (Provincial Environment and Natural Resources) and the local government units both at the provincial and city levels. Table 12.8 shows the responses of the participants of the FGD.

12.4 CONCLUSION AND RECOMMENDATIONS

The study project has pointed out a common problem in water accessibility and, to a lesser degree, the issues of water scarcity and quality around the study area. Also, there is a paucity of data on the sources of water surrounding the Kapatagan watershed. This contributes to the concerns about the demand and supply of water for the users around the area. Considering that the population of the area is increasing, it is clear that the inhabitants will need water for multiple purposes.

TABLE 12.7 Data on the Subjective Scoring Based on the Perception of Stakeholders on Related Policies and Plans as Reported by Stakeholders during the Interview

Parameters/Indicators	BDP	CLUP	ADSDPP	PCWA 2004	CRMP	Mean Score
Awareness & Knowledge	1.8	2.3	2.7	1.8	1.0	1.9
Extent of implementation	1.8	2.2	2.6	1.7	1.0	1.8
Impact on Water & Land Resources	1.7	2.2	2.5	1.7	1.0	1.8
Feasibility	1.7	2.3	2.6	1.7	1.0	1.8
Economic Impact	1.7	2.3	2.5	1.7	1.0	1.8
Budgetary Impact	1.7	2.3	2.5	1.7	1.0	1.8

**BDP*: Barangay Kapatagan Development Plan; CLUP: comprehensive land use; ADSDPP: ancestral domain sustainable development and protection plan; PCWA 2004: Philippines Clean Water Act of 2004; and CRMP: coastal resource management plan.

***Subjective scoring in terms of ranking*:

1. *Low/less favorable;*
2. *Medium/favorable;*
3. *High/more favorable.*

There is also an issue with the unclear implementation of some policies, such as that can be detected in the Special Agreement in Protected Areas (SAPA). Additionally, the mean score of 1.8 to 1.9 reported by the participants of the interview on their knowledge and awareness, as well as their extent of implementation and impact on water and land resources, reveals that policies at various levels are yet to be visibly implemented.

The issues identified during the preliminary assessment may be attributed to factors shown as follows;

1. Multiple and fragmented water institutions;
2. Lack of harmonization between regional policies and local agreements;
3. Conflicts between customary rules and formal state laws; and
4. Absence of monitoring and evaluation schemes for better water governance.

TABLE 12.8 Responses of Respondents From PENRO, Provincial, and City Local Government Units Participated in the FGD

Questions	Transcription of Answers		
	PENRO	**Provincial LGU**	**City LGU**
At your level, what are the problems and issues related to water, especially as a source for agriculture/fisheries and renewable energy?	There is no data on the source of water in the Kapatagan watershed There is a need to determine the demand and supply of water in the area There is an increasing population that may demand water There is no clear implementation of some policies like Special Agreement in Protected Areas (SAPA)	There is competition on the use/demand for water between residents and industry players (resort and agricultural land owners) Drying up of natural springs which are sources of potable water Wastewater discharge to coastal areas	Conflict of uses in protected zones
What policies (National/Local) are now being implemented and are contributing to water sustainability? How?	National Greening Program General Management Plan for Mt Apo	Provincial Environment Code	Comprehensive Land Use Plan
What are the barriers/hindrances to water policy implementation?	Fragmented efforts of various agencies	Fragmented efforts of various agencies	Bureaucratic process and numerous requirements
What is your scheme/mechanism for monitoring and evaluating the extent of water policy implementation?	No monitoring scheme/mechanism	No monitoring scheme/mechanism	No monitoring scheme/mechanism

The many legal documents or policies, as well as the multi-level institutions for water, may also be a source of confusion. While there are a multitude of water agencies, these are not connected vertically nor horizontally integrated. Moreover, these various institutions do not have sufficient human and financial resources and presence at the local level to be effective in their mandates.

Also, the lack of water data for planning is insufficient, especially for Kapatagan. Preliminary findings in this project may reinforce previous studies that establish that water governance depends to a large extent on the underlying economic and social conditions (Rogers, 2002). The one-size-fits-all economic solutions are also inadequate (Sampford, 2009). While hierarchical governance seems to be the prevailing arrangement in the Philippines, polycentric governance mechanisms (Ostrom, 2009) may be contextually relevant. Instituting good water governance including responsive policy and institutional arrangements, appropriate planning, and effective implementation are the keys to addressing water issues in Kapatagan and the whole region.

Since the project is a preliminary assessment, the findings may be used as an initial basis for designing a more in-depth follow-through which will adopt the methodology for policy analysis in water resource management using the DPSIR (Drivers, Pressure, State, Impact, Response) Framework (Vazquez, 2003) and that from Pani Chautari (Devkota et al., 2020).

KEYWORDS

- **inhabitants**
- **integrated water resource management**
- **multi-level institutions**
- **policy analysis**
- **polycentric governance mechanisms**
- **preliminary assessment**
- **protected area management board**

REFERENCES

Alvarez, V. A., & Vazquez, A. (2004). Effect of water sorption on the flexural properties of a fully biodegradable composite. *Journal of Composites Materials, 38*(13), 1165–1182. https://doi.org/10.1177/0021998304042082.

Anukularmphai, A. (2010). Implementing Integrated Water Resources Management (IWRM): Based on Thailand's experience. International union for conservation of nature. https://portals.iucn.org/library/efiles/documents/Rep-2010-004.pdf (accessed on 5th June 2024).

Bruyninckx, H. (2017). The European Environment Agency. *Impact, 2017*(11), 28–29. https://doi.org/10.21820/23987073.2017.11.28.

Caspersen, O. (1999). The European Environment Agency. *Global Environmental Change, 9*(1), 71–75. https://www.academia.edu/download/48066643/s0959-3780_2898_2900025-920160815-32629-1d0k98l.pdf (accessed on 5th June 2024).

Devkota, K., Neupane, K. R., Khatri, D., Upadhyaya, R., Shrestha, S., & Ojha, H. P. (2020). Pani Chautari – A facilitation guidebook for evidence-informed dialogue on urban water issues. *Flood Resilience*. https://www.sias-southasia.org/wp-content/uploads/2021/03/Pani-Chautari-guidebook-book.pdf.

Environmental Management Bureau (2023); Annual Water Quality Assessment report; pp.02–44.

Kovacic, Z., & Benini, L. (2022). Striking the balance: Sustainability and institutional transitions in the European Environment Agency. *Futures, 141,* 102984. https://doi.org/10.1016/j.futures.2022.102984.

Li, S., Gu, S., Liu, W., Han, H., & Zhang, Q. (2008). Water quality in relation to land use and land cover in the upper Han River Basin, China. *Catena, 75*(2), 216–222. https://doi.org/10.1016/j.catena.2008.06.005.

Ostrom, E. (2009). A general framework for analyzing sustainability of social-ecological systems. *Science, 325*, 419–422. https://www.science.org/doi/abs/10.1126/science.1172133.

Programme Hydrologique International. (2009). IWRM guidelines at the river basin level. UNESCO. http://colinmayfield.com/public/PDF_files/IWRM%20Guidelines%20at%20River%20Basin%20Level%20and%20Global%20Changes.pdf (accessed on 5th June 2024).

Rana, S., & Kelly, L. (2004). The Global Water Partnership. Addressing challenges of globalization: An independent evaluation of the World Bank's approach to global programs. case study. *The World Bank, Operations Evaluation Department*, Washington, DC. https://ieg.worldbankgroup.org/sites/default/files/Data/reports/gppp_gwp_wp.pdf (accessed on 5th June 2024).

Republic Act No. 8371. (1997). *An Act to Recognize, Protect, and Promote the Rights of Indigenous Cultural Communities/Indigenous Peoples, Creating a National Commission on Indigenous Peoples, Establishing Implementing Mechanisms, Appropriating Funds Therefore, and for Other Purposes.* Republic of the Philippines.

Rogers, P., De Silva, R., & Bhatia, R. (2002). Water is an economic good: How to use prices to promote equity, efficiency, and sustainability. *Water Policy, 4*(1), 1–17. http://tarjomefa.com/wp-content/uploads/2017/10/7944-English-TarjomeFa.pdf (accessed on 5th June 2024).

Rola, A. C., Pulhin, J. M., Tabios, G. Q., Lizada, J. C., & Dayo, M. H. (2015). Challenges of water governance in the Philippines. *Philippine Journal of Science, 144*(2), 197–208. https://www.wherewaterflows.ph/wp-content/uploads/2020/04/challenges-of-water-governance-in-the-Phils_FinalCopy_05_April_2016.pdf (accessed on 5th June 2024).

Sampford, C. (2009). Water rights and water governance: A cautionary tale and the case for interdisciplinary governance. In *Water Ethics* (pp. 71–94), CRC Press. https://www.taylorfrancis.com/chapters/edit/10.1201/9780203875438-11/water-rights-water-governance-cautionary-tale-case-interdisciplinary-governance-sampford (accessed on 5th June 2024).

Temple University Ambler Campus. (2022). The UN World Day 2022. https://ambler.temple.edu/community/earthfest/celebrating-earth/world-water-day/why-water-so-important (accessed on 5th June 2024).

The Book of Ordinances 2000–2010. Sangguniang Panlungsod. City of Digos.
Vazquez, J. F., & Mattei, F. E. E. (2003). A methodology for policy analysis in water resources management. In *Resource and Environmental Economics: Political Economy of the Government,* 1–13, *J. Feás Vázquez* / EAERE FEEM VIU European Summer School 2003. http://www.feem-web.it/ess/ess03/students/feas.pdf (accessed on 5th June 2024).
Water Org. (2022). Where we work. Water Org. https://water.org/our-impact/where-we-work/philippines/ (accessed on 5th June 2024).

CHAPTER 13

Impact of Climate Change on Livestock Production and Its Adaptation in Nepal

DILLI RAM SEDAI,[1] SAROJ GYAWALI,[2] MEGH RAJ DANGAL,[3,4]
CHUMPOL YUANGYAI,[5] NOOR HASHIMAH HASHIM LIM,[6]
KUAANAN TECHATO,[1,6,7] and NUTTAYA YUANGYAI[1,6]

[1]*Program of Sustainable Energy Management, Faculty of Environmental Management, Prince of Songkhla University, Songkhla, Thailand*

[2]*Sustainable Study and Research Institute, Kathmandu, Nepal*

[3]*School of Arts, Kathmandu University, Kathmandu, Nepal*

[4]*School of Engineering, King Mongkut's Institute of Technology, Ladkrabang, Thailand*

[5]*Department of Urban and Regional Planning, Faculty of Built Environment, Universiti Malaya, Kuala Lumpur, Malaysia*

[6]*Research Center on Industrial Ecology in Energy Business, Faculty of Environmental Management, Prince of Songkhla University, Songkhla, Thailand*

[7]*Faculty of Health and Life Sciences, INTI International University, Persiaran Perdana BBN, Putra Nilai, Nilai, Negeri Sembilan, Malaysia*

ABSTRACT

Climate change directly affects livestock by increasing the temperature, which leads to increased infertility, loss of conception, poor expression of heat, repeat breeding, and sterility. This results in loss of weight gain, lower feed conversion ratio, increased transmission of vector-borne diseases, incidence and distribution

Engineering Design and Technical Applications of Physical Science: An Integrated Approach for Sustainable Development. Chin Hua Chia, Sarani Zakaria, Ruey Shan Chen, & Md. Shahariar Chowdhury (Eds.)

of external parasites, and diseases susceptible to livestock systems. Livestock can be distressed by altering external factors. A study was conducted using a review based on the Preferred Reporting Items for Systematic Reviews and Meta-Analyses (PRISMA). Environmental management is documented as a significant instrument for the review protocol. A systematic search process on the impact of climate change adaptation measures on livestock production includes enclosure and elimination criteria for suitability valuation. It is presented through quality assessment, data achievement, concepts, and investigations. Relevant literature was retrieved using Scopus, Google Scholar databases, and Web of Science. The effects of climate change on Nepalese livestock producers include reducing heat stress, risk management, adopting technology development by farmers, animal management, breeding management, feed and feeding management, and manure management for adaptation. These strategies should be applicable at the nationwide and local levels as development approaches. These conclusions and variations are relevant for improving farmers' economies and ensuring food security in the least developed and developing countries with livestock producers.

13.1 INTRODUCTION

Climate change in Nepal has resulted in adverse effects on various sectors of the economy and the livelihoods of a large part of the population. Various serious problems like changes in monsoon patterns, drought, forest fires, torrential rain, and animal behavior have been observed. Changing the monsoon pattern has led to unbelievable situations such as decreased food security, increased poverty, unbalanced agriculture production systems, and many environmental issues in the country. Climate change includes physical characteristics that cause consequences in association with multidimensional effects (Visschers, 2018). The influence of climate change on livestock production and productivity is a serious concern. This anxiety includes livestock diseases, infertility, repeat breeding, sterility fertility behavior, breeding-related risks, the incidence of livestock vector-borne diseases, parasites, and changes in the availability of feed, livestock class, and range-land forage types. Livestock holders apply different types of knowledge to mitigate climate change using their inherent knowledge and skills (MOPE, 2004; Devkota et al., 2011; Pokhrel, 2011; Jan Petzold et al., 2020).

The investment in livestock development should be given priority by those countries based on this projected evidence. In Nepal, 66% of the population is involved in the agricultural sector. The livestock sector plays an important role in the national economy and household food security in Nepal. The sector contributes about 27.1% to the agricultural gross domestic product (AGDP) in the country (MoF, 2019), while the contribution from the sector to the total gross domestic product (GDP) is 11.5% (MoALD, 2020). At the global level, the livestock sector contributes 18% of dietary energy supplies and 34% of protein intake (Thornton, 2011; Ripple et al., 2013). Due to the rapid upsurge in the population, the demand for livestock products will rise by more than 50% by 2050 in Asia and Africa (FAO, 2012).

The livestock sector provides meat, milk, and hides for industries, and draught power and manure for crop production. For food and nutritional security, the livestock sector contributes to farm households and higher segments of communities (MoALD, 2019).

The risk of climate change has forced many issues to be addressed promptly by making policy-level decisions regarding the adoption of adverse effects. In regards to climate change, Nepal lies on the southern limit of the dispersed boundary of the Indian plate under thrusts and the northern limit with the Eurasian Chinese plate oppressing the Himalayan arc, Hinduku-shparbat. Geographically, Nepal is subdivided into five ecological regions: Terai Plain, Sivalik Range, and Mahabharat Range with mid valleys, high hills, and high mountains. By intersecting two types of plates, Nepal is ranked 11th in the high-risk geological area for earthquake hitting in the world. The development will be linked to a technology combination of nutrition, feed, genetics and breeding, health, and environment management combination of the different systems. Climate change in the physical and biological system significantly occurs on all continents, and in most oceans, and most of the change is in the warming of temperature. Developing countries are changing to urbanization, and in the coming decades, the demand for livestock products will increase significantly (Rosenzweig et al., 2008; Delgado et al., 2020).

Agriculture accounts for nearly 69% of greenhouse gas (GHG) emissions in Nepal (Nepal Second National Communication, UNFCCC, 2014). Within agriculture, enteric fermentation of ruminants and emissions from animal manure represent about 68% of the total or 46% of total emissions at the national level. Climate change poses a significant risk to humans and the natural system of global and regional climate patterns. Weather is a set of all the phenomena occurring in a given atmosphere at a given time (Stephenson et al., 2008; Pathak et al., 2011). Climate Smart Agriculture (CSA) is an approach for transforming and reorienting agricultural systems to support food security under changing environmental conditions. CSA is promoted only by collaborative actions of smallholder farmers, marginalized groups, policymakers, and other related sectors for adaptation and practices, programs, policies, and projects (Swodesh & Abin, 2019). The implementation of measures on livestock production at the local and national levels is also equally important for a sustainable environment; therefore, major concerns and policy issues on these matters were reviewed in this paper.

The research aims to present a basic systematic review of climate change's implications on Nepal's livestock industry and how Nepal's government is handling CC's negative consequences on the livestock sector. This work takes the form of reviews of peer-reviewed journal papers and other essential programs and policy documents available in Nepal.

13.1.1 RESEARCH QUESTION GENERATING

The Population, Interest, and Contest (PICO) guidelines were adapted to generate the research question applied by several methodical papers (Shaffril et al., 2020). Here, people refer to resident farmers who are connected to livestock farming. These local rural farmers' climate change adaptation intervention on milk production in Nepal. The developed research question is: What are the major local adaptation techniques of rural farmers to cope with climate change on livestock production?

This study will help policymakers, program implementers, researchers, and rural farmers in receiving local techniques and knowledge to adapt to climate change and unpredictability around the globe. The local techniques and knowledge are primary inputs for the design of sustainable environment and temperature change-associated plans, policies, strategies, and programs for rural livestock farmers. Such types of local techniques and knowledge are very useful to recognize the current gaps in climate change effects to cope with and adaptation implementation.

13.2 MATERIAL AND METHODOLOGY

13.2.1 REVIEW GUIDELINE-PRISMA

The analysis is based on the Preferred Reporting Items for Systematic Reviews and Meta-Analyses (PRISMA). For environmental management, the important tool is the review protocol (Moher et al., 2009). This systematic review approach aims to reduce the risk of bias and increase transparency at every stage of the review process by relying on explicit, systematic methods to reduce bias in the selection and inclusion of studies, appraise the quality of the included studies, and summarize them objectively (Liberati et al., 2009; Petticrew, 2001). The protocol of identification in a systematic review process comprises screening (defining inclusion and exclusion criteria) and eligibility evaluation. It is followed by an assessment of excellence, data acquisition, concept, and analysis (Ishtiaque et al., 2020).

13.2.2 SYSTEMATIC LITERATURE REVIEW

An organized examination study was implemented for the attainment of information which comprises the research question, the strategic steps of the invention, identification, screening, eligibility, quality appraisal, and data removal which are readymade PRISMA protocols as shown in the diagram below. This is presented in detail in the sections below (Table 13.1).

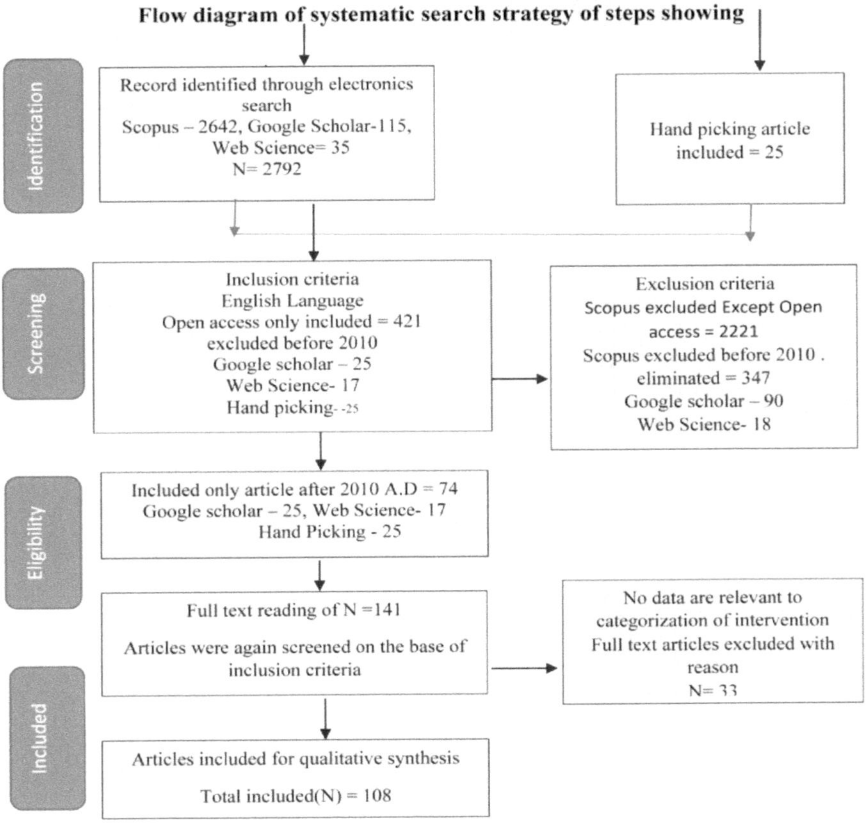

TABLE 13.1 Using PRISMA Model for Article Screening (Moher et al., 2009)

2.3.1 Identification of Article		**2.3.2 Screening of Article Results**		**2.3.3 Eligibility of Article Results**		**2.3.4 Included Article for Quality Appraisal**
Data is interrelated (Okoli, 2015) to the search process (Zhu & Liu, 2020). Suitable for systematic process (Falagas et al., 2008) multiple disciplines included (Gusenbauer & Haddaway, 2020; Martín et al., 2018).		• Included only English • Included open access only • Subject area livestock and climate change only • Abs-key words only' • Source – journal, book, conference proceeding • Document type – climate change impact and livestock production		• Excluded before 2010 A.D. • Full-text articles only included • Duplication was excluded		• No data are relevant to the categorization of the intervention • Full-text articles excluded with reason
Article sources						Articles included for qualitative synthesis
Scopus	2642	Scopus	421	Scopus	74	
Google Scholar	115	Google Scholar	17	Google Scholar	17	Total included (N) = 108
Science web Direct	35	Science Web Direct	25	Science Web Direct	25	
Other sources Hand picking		Another source – hand-picking	25	Another source – hand-picking	25	

13.2.3 DATA ANALYSIS

The developed data is used to analyze the consolidative review method. This is a method to create a comprehensive understanding of the earlier works and the research question (Whittemore & Knafl, 2005). In integrative review works, thematically synthesizing (Flemming et al., 2019). The collected literature shows the consolidative review into themes and sub-themes. The livestock producer adaptation strategies and action theme and sub-theme were developed by Gentle (2013). These developed themes and sub-themes focus on factors of inclusion and consistency. The adaptation strategies were summarized through an integrative review approach.

13.3 RESULTS AND DISCUSSION

The investigations have shown how climate change conditions have unpleasantly impacted livestock production. These impacts on livestock production are due to thermal heat stress, irregular precipitation, extreme climate events, and an increase in natural risks causing pasture degradation, deficiencies in feed production, flooding, and land degradation. Significant results were found on the degradation of natural resources, decrease in milk production, increase in anestrus in animals, and increases in disease incidence and risk events in livestock production. The livelihood of poor livestock farmers is directly impacted. The corrective measures include practicing diversification of livelihoods, and income sources, and the adoption of local and scientific practices for livestock production (Table 13.2).

These energies of local traditional and scientific knowledge adoption practices are thematically presented and studied in the selected literature as below:

Thematic categorization: The adoption strategies identified thematically (Below Annex II) were characterized into 7 major themes and 41 subsequent sub-themes. These explored literatures are practices for adaptation are presented below in Annex II (table 2). In the selected articles, the adaptation strategies are identified and dissected.

13.3.1 TEMPERATURE STRESS CONTROL TECHNIQUE

In livestock production, the first theme is Temperature Stress Control Techniques, which are considered excellent means of adapting to climate change. To minimize the decline in milk production and anestrus condition of animals,

TABLE 13.2 Summary of Local Adaptation Measures Explored from the Systematic Review of the Selected Articles

	Theme	Adoption Measure
3.1	Temperature Stress Control Technique	Housing improvement, construction of shed, concrete floor, asbestos sheet roofing material,
		Cold water during hot and humid climates, Sprinkle water on the roof of the shed
		Provision of shed to reduce hot and cold stress. Provision of bedding (cow mat) in the shed
		Provide fresh air/fan/cooler during extreme hot conditions.
		Kept animals outside during the hot summer.
		Loose housing system.
		Provision of fire in cold season.
3.2	Risk management	The government provided subsidies for boosting production to smallholders and marginal farmers.
		Insurance system: Government gives the 75% subsidies on premiums. Dairy cooperatives are doing insurance systems to minimize the risk of mortality. The cooperative model is unique and beneficial to farmers.
		Awareness of diseases and vaccination by dairy cooperatives and provision of free vaccination to control disease
		Credit of bank/loan/cooperative from Bank for animal purchase and housing management.
3.3	Technology for adoption	Conservation of local indigenous breeds and genetic climate tolerance cross breeds types.
		Improved animal health by controlling vector-borne, parasitic, and protozoal diseases.
		Water harvesting at wells in the mountains and pounds in terai for animal drinking purposes and recharge of soil management by improving alternative irrigation,
		Improving pasture and forage cultivation in alpine and tropics use of high-yielding variety.
		Securing water access for livestock in drought conditions by harvesting wells and ponds.
		Controlling animal health science-based practice using vaccines and medicines preventing infectious and contagious diseases.
3.4	Animal Management system adopted by farmers	Integration of pasture management.
		Stall feeding of animals and collection of urine.
		Introducing a mixed farming system (stall-fed and pasture grazing)
		Changing the plantation time of fodder and pasture in fodder cultivated and grazing land

TABLE 13.2 (Continued)

	Theme	Adoption Measure
		Reduction in unproductive and developing high-yielding livestock numbers.
		Replacement of exotic breed to local breed of pasture grass.
		Provide bedding for extreme cold.
3.5	Breeding strategies	Breed changing introduces the local breed resistant to climate change. Exchange of breeding bulls in communities (control of inbreeding) to control inbreeding system.
		Artificial insemination with proven breed semen and proper recording.
		Promotion of local breeds through selection methods
		Improving genetics through pedigree performance Recording system in important dairy cooperative pocket area
		Infertility management, supply of mineral mixture, vitamins, and minerals tablets
3.6	Animal feeds, pasture, and fodder Management	Deforestation and tree plantation on pasture land. Rotational plantation
		Replacing roughage with locally balanced concentrate.
		Feeding leguminous forage than grass-based diet.
		Improving pasture management with climate resistance seed and breed.
		Feeding of locally available ingredients of balanced feed
		Feeding of the additive mineral mixture (Wheat soaked-germination–orally feed)
		Rotational looping of vegetative biomass of fodder, trees, shrubs, herbs, and grasses.
		Deforestation and tree protection on pasture land.
3.7	Manure Management	Storage of manure in the pit and aerobic treatment.
		Practice with shortening of storage duration.
		Improving timing in pit and application of manure in pasture land.
		Covering manure with a solid separator reduces the methanation of GHG gas to minimize the effect of climate on livestock production.

provisions should include shed improvement, providing fresh air/fans/coolers, keeping animals outside, using a movable housing system during hot and humid climates, providing cold water to reduce heat stress, and providing heat sources in cold weather. Climate change-induced heat stress on animals will decrease the rate of feed consumption, leading to a decline in milk production, growth performance, increased transmission of vector-borne diseases, the appearance of new diseases, and the risk of local breeds becoming extinct. Climate change poses a real danger in this regard. Therefore, Nepal needs to take concrete actions at both local and global levels to address this issue. To mitigate the risks of climate change in livestock production, priority should be given to implementing Nepal's National Adaptation Program of Action (NAPA) and Local Adaptation Plan of Action (LAPA). In hotter climates, the majority of losses occur during the summer months. The temperature has increased three times since the 1970s, and the effects of climate change are unexpectedly sudden or irreversible (IPCC, 2007).

The dairy cow exposed to heat stress during late gestation had lower birth weight and less milk production (Collier, 1982). The male breeds of buffalo exposed to hot summer had decreased scrotal circumference, testicular consistency, tone, size, and weight compared to the same breed in temperate environmental conditions (Collier, 1982). Under thermal stress in lactating and non-lactating dairy cattle, the preovulatory estradiol in blood plasma at 32°C was lower, and the duration of estrus was shorter (Gwazdauskas, 1981). Heat stress increases the plasma growth hormone concentration, and breeds that can reduce plasma thyroid hormone concentration are more adaptable to heat stress (Pragna et al., 2018).

Adaptation strategies of livestock producers to various climate vagaries vary from farm to farm and from one area to another based on the existing support system and their native knowledge. High-milk-producing animals are more susceptible than low-producing animals. Livestock exhibits standard adaptive technology, which includes morphological, behavioral, physiological, neuroendocrine, blood-biochemical, and cellular responses that act in coordination to promote their welfare and favor their survival in a specific environment (Angel et al., 2018).

13.3.2 RISK MANAGEMENT

Nowadays, the government of Nepal is giving importance to self-employment youth-oriented programs, insurance programs, subsidized interest on credit soft loans (low interest 5%), awareness of disease vaccination, and grants/

subsidies for income generation. They are emphasizing training and infrastructure from both the government and non-government sectors, resulting in increased livestock productivity (MoLD, 2011; Pradhanang et al., 2015). The policies at local, national, and international levels aim to increase production but do not focus seriously on essential measures to adapt to climate change (Thornton et al., 2011). The livestock sector has contributed to development through its various policies. These policies have highlighted the importance of commercializing breed improvement and promoting private and public partnerships in the livestock sector (Pathak, 2011; Pradhanang et al., 2015). Nepal's contribution to climate change is minor, but geographically it poses risks worldwide to the livestock sector. To minimize greenhouse gas emissions, an integrated sector approach with government, national, and international nongovernmental organization support is essential (Pathak, 2011). To enhance governance delivery services and increase accessibility for deprived families, prerequisite social security schemes such as access to loan money for labor, crop, and livestock insurance, and improved ways to ensure nutritional security enhance adaptive capacity (Gental et al., 2012). Selecting appropriate breeds, better greenery forage and fodder, and refining feeding management help build social capital through community awareness of the livestock insurance system and prevention of vector-borne diseases (Karn, 2011; Sushila et al., 2021).

The dairy cooperatives are practicing the cooperative insurance system to minimize the risk to livestock producers. They developed the insurance guidelines and approved them by the general assembly. Like this, they become legal documents. They are providing 60–80% compensation decided by the cooperative committee. The benefit of such an insurance system is that the premiums deposited by farmers will be collected in cooperatives. Again, they can utilize the premiums based on the decision of the cooperatives. There will not be a dishonest situation in cooperatives. The farmers receive fast compensation for their animal losses. The cooperative can have provisions for compensation for infertility, cirrhosis of the udder, mastitis, and acute disease conditions. The local and provincial governments can also support the premium costs of insured animals that need to be paid by the farmers.

The livestock producers are small-scale marginalized and deprived by the economic situation. Sometimes the livestock face difficulties in getting nutrition for survival. In critical situations of household expenditure or the spreading of infectious diseases in neighboring places, livestock producers resort to selling livestock to buy food. The government is providing a 75% subsidy on insurance premiums. Livestock producers, by any means of loans, credits, or borrowing money, find that the best practice to minimize risk is insuring their livestock. This can help avoid the effects of climate change on production loss, anestrus, infertility, and disease incidence and mortality of animals.

13.3.3 TECHNOLOGY FOR ADAPTATION

In Nepal, the effect of climate change is adapting local communities through their local knowledge and plans (Budhathoki, 2020; Nuorteva, 2010). This technology includes the development of new breeds, replacement of exotic breeds with local breeds, providing cow mats or bedding in extreme cold or heat, new techniques to control vector-borne diseases, managing alternate irrigation systems, developing high-yielding varieties of pasture and forage, and the use of vaccines and medicines that help with climate tolerance (Singh, 2012; Dhakal et al., 2016). The knowledge of native residents to acclimatize to global climate change and improve its effects will also depend on their socio-economic and environmental conditions, as well as on the accessible resources (IFAD, 2009; Gental et al., 2012). The difficulties of torrents, dry periods, landslides, and erosion can be reduced by applying climate-smart technology using locally accessible assets and ideas. As temperature alteration has a significant consequence on agricultural production, in the long run, the government should prepare suitable plans and policies and execute them effectively (Poudel, 2020). The effects of climate change are more noticeable at a local level, and to respond to local movements, a combined method should be constructed at the national and district levels for pro-poor plan strategies and policies to ensure fair distribution (Gental et al., 2012; Poudel, 2020). Indigenous knowledge needs to be combined with recent technical information as a unified method to better formulate methodological, economic, and adaptation plans within the government (Gental et al., 2012; Sushila et al., 2021).

13.3.4 ANIMAL MANAGEMENT SYSTEM ADOPTED BY FARMERS

The livestock holders apply different types of knowledge to mitigate climate change using their inherent skills (Devkota et al., 2011). The majority of farmers use different techniques to cope with climate effects by providing bedding, seasonal migration, destocking, replacement of local breeds, and fodder tree lines intended for animals during temperature and emotional stress to avert them from pressure (Singh, 2012). Climate change affects adaptation accessibility and the quality of water is a major challenge in the future (Nardone, 2021). For concrete livestock administration systems, competent and reasonable adaptation preserves forage and grass for the dry season, provides livestock with a nutritious diet, deworms, and treats livestock regularly, and executes existing high charges of liveliness related to air conditioning. Providing natural shade is a low-cost solution that is additionally applied for rural poor producers (Singh, 2012; Chase, 2014; Koirala & Shrestha, 2019). Adaptation approaches rely on socioeconomic and physical issues (Hussain, 2018; Sushila et al., 2021). The impact of heat stress on animal production and profitability in dairy cattle can be reduced by a number of management and housing alterations, challenging with balanced investment to develop new facilities (Chase, 2014).

13.3.5 BREEDING STRATEGIES

The local and rare breeds are at risk of being lost in localized disasters due to their geographically restricted distribution (Heffernan et al., 2006; Carson, 2009). The effects of climate change and disease outbreaks on native and rare breeds have led to a loss of genetic diversity with global implications (IFAD, 2009). Native breeds, resilient to disease and requiring less water, can thrive in contrasting climatic conditions (Tailor, 2005). Improved fertility and management of anestrus in dairy cattle, resulting in a smaller number of breeding and replacement animals, help reduce methane and nitrous oxide emissions (Giampiero et al., 2019; Thakur, 2011). It has been stated that adopting climate mitigation measures includes promoting native breeds of livestock and poultry, cultivating drought-tolerant fodder and forage, and raising awareness among rural communities. This requires coordinated efforts with the flow of information and interaction among local, national, and international agencies through active involvement at all levels. Native animal breeds are more adaptable to temperature changes compared to exotic and hybridized animals (Thornton, 2007).

Homegrown and nationwide-level strategies and approaches to a variety of adoption and maintainable and practicable (Thakur, 2011; Singh, 2012; Sushila et al., 2021). Genetic biodiversity, organisms, and ecosystems found within a specific environment that is at risk and decreasing to climate change appropriate breeding is the option to tolerant cold and heat, to alleviate the negative effects and reduce the disease prevalence of climate change in livestock (Devkota et al., 2011; UNEP, 2012; Koirala & Bhandari, 2019).

13.3.6 ANIMAL FEED, PASTURE, AND FODDER MANAGEMENT

The production of livestock is dependent on a protected or controlled environment and quality feed to mitigate the direct and indirect effects of climate change (Adams, 1998; Freitas, 2006). Changes in temperature and CO_2 levels will affect the arrangement of grasslands by shifting the species competition dynamics due to variations in growth charges (Thornton, 2007; Thornton, 2009; IFAD, 2010; Thornton, 2015). Plant competition is influenced by seasonal shifts in water availability (Polley et al., 2013). Primary throughput in pastures may be improved due to changes in species alignment if temperature, precipitation, and synchronized nitrogen deposition rise (IPCC, 2007). By increasing concentrated feeding, diet digestibility can be improved per unit of protein fat corrected, reducing methane emissions by 15% (Knapp, 2014). Livestock products supply protein and essential nutrients, impacting climate change due to the expansion of pasturelands and croplands contributing 9.2% of total livestock GHG emissions (Gerber et al., 2013). As temperature differences become more adjustable, animal food and the capability of smallholders to succeed in feedstuff deficits for different types of changes (IFAD, 2009).

Appropriate administration of feed and fodder, along with a reliable water source, is an option to tolerate cold and heat and alleviate the negative effects. Indigenous breeds have higher feed conversion ability on low-quality pasture, greater adaptability, and a reduced capacity for disease prevalence originating in tropical and subtropical regions (Wilkes et al., 2012; Koirala & Bhandari, 2019). The impact of climate change affects the dairy industry directly and indirectly. The direct effects include milk production, reproduction, and disease resistance, while the indirect effects involve a reduction in concentrated feedstuff (Silanikovea, 2015). The portion of total nutritional energy contributed by dairy products is around 14% in advanced countries and 4% in emerging countries (Stefano & Skoet, 2012). Appropriate breeding, administration of feed and fodder, and a reliable water source are options to tolerate cold and heat, alleviate negative effects, and reduce the disease prevalence of climate change in livestock (Devkota et al., 2011; UNEP, 2012; Koirala & Bhandari, 2019). The feed should contain higher quality forages, ration fiber (ADF, NDF) with high digestibility, yeast or yeast cultures to balance ration protein levels, and ration potassium levels to counteract the higher potassium losses in heat-stressed animals (Chase, 2014).

13.3.7 MANURE MANAGEMENT

The guidelines for reinstating water quality and reducing manure irritation have influenced agronomic practices related to the use and treatment of animal manure, as well as the implementation of new policies for the joint management of livestock manure (Kupper, 2015; Menzi, 2015). Livestock manure is mainly applied on farmers' land, either raw or treated, usually in close proximity to the farm (Quideau, 2010). Manure slurry or solid produces varying concentrations of dry matter content (DM) (Degueurce, 2016).

This paper does not discuss biogas production from livestock but emphasizes the importance of manure management in livestock operations. Good practices include composting and covering manure heaps, while bad practices involve leaving manure exposed in open fields without soil incorporation for months. Therefore, manure management is a key aspect of this discussion. Manure can be composted in pits, covered with solid separators, stored, undergo anaerobic treatment, and have storage duration shortened, with improved timing and application. This technology is cost-effective and can be easily implemented by small-scale and marginalized farmers. Proper waste/manure management reduces methane emissions and helps livestock producers adapt local knowledge to minimize the impact of climate change on livestock production.

13.4 DISCUSSION

Developing countries do not have sufficient capacities to cope with the challenges posed by climate change (UNFCCC, 2021). The Paris Agreement calls for capacity building in developing countries to support action for all developing countries (UNFCCC, 2021) through the use of practical tools that can help participants better recognize the problem and take steps to address it. Climate change directly and indirectly affects livestock health and production parameters, influencing growth, milk and meat yield and quality, reproductive performance, metabolic and health status, and carcass traits (Nardone, 2010). Improving livestock efficiency has a significant impact on reducing the strength per unit of resources, which is a priority for both developing and developed countries. However, the outcomes will also establish a foundation such as the genetic scope of the livestock and the adoption of monitoring technology (Giampiero, 2018).

Livestock producers are using local knowledge to mitigate the effects of climate change on livestock production. To reduce temperature stress, they provide shade, keep their animals outside at night, use loose housing for cooling, and use blow fans during hot summer months. The animals are free to move in search of feed and a safe place in adverse climatic conditions. Farmers' adoption strategies to address various climate effects vary from farm to farm and region to region based on current conditions and their local knowledge.

Although most reports have emphasized the detrimental effects of ambient temperature and humidity on reproductive processes, cattle exhibit estrus throughout the year, but buffaloes show sexual activity during the cooler part of the year (Oct-Feb) (Upadhaya, 2009). Due to increased temperature and hotter days, fertility behavior, incidence of livestock vector-borne diseases, and parasites, decrease in fodder and forage production, as well as a decline in the recharge of natural water (Pokhrel, 2011) are observed.

For Nepalese dairy cooperatives, facilities are provided to cooperative farmers for lending money to purchase and insured their animals. For Nepalese dairy cooperatives, facilities are provided to cooperative farmers for lending money to purchase their animals. There will not be a dishonest situation in cooperatives. The farmers receive fast compensation for their animal losses as well as compensation for infertility, cirrhosis of the udder, mastitis, and acute disease conditions. The local and Provincial governments appear to be ready to support the premium cost of insured animals for the farmers' benefit. The cooperative self-managed insurance system is becoming popular for the commercialization of livestock farming. The farmers are developing confidence in livestock farming by overcoming losses due to mortality. They can utilize the cooperative-managed premium for the benefit of rural farmers. Such a practice in risk management of the impact of climate change is unique and can be a good model to explore for its suitability in other developing countries.

The local adaptation techniques for providing communal ponds near the villages in Terai are a major source of water for animals. Wallowing in ponds and rivers is a traditional practice to cope with high temperatures during summer months in both hills and Terai. Rainwater harvesting is also practiced in the hills to recharge wallowing ponds. Similarly, in high mountains, seasonal migration from high mountains to lower plains during winter and to high mountains during summer was an adaptation strategy adopted for a long time by our ancestors. It was also a strategy to utilize the feed resources available at different altitudes. This local traditional knowledge and these techniques are very helpful in dealing with situations arising due to climate change. These techniques are specific to the local area and require no external help. Documenting such knowledge and techniques is necessary to contribute to academic institutions and set priorities for research. Developing adaptation and mitigation strategies against climate change in coordination and forming alliances with suitable planning is effective (Rijal & Babin, 2019).

Adapting concrete livestock administration care and managing different classes of livestock is mandatory to reduce the impact of climate change. Ensuring nutrition availability in the lean season and making hay and silage for the dry season are essential parts of management. Controlling the prevalence of infections can be achieved through vaccination, parasitic control, and timely treatment. Improving animal shelters by providing natural shade, rather than using coolers or air conditioners, is a low-cost solution that is widely applied by rural small-scale producers. Addressing the socioeconomic and physical issues impacting production and profitability in dairy cattle can be achieved by managing the number of livestock and improving housing, balancing investments to develop new facilities.

Farmers adopted plantations of fodder trees near animal sheds or shelter farms to reduce the effects of heat waves. Seasonal migration along with livestock and destocking their large animals, keeping small animals that require less water and feed, and ensuring those healthy animals survive during adverse conditions are good practices. Nepal has a large number of low-productive animals per unit of cultivated land. It is important to reduce the number of low-productive animals to minimize the impact on the environment (Gorkhali et al., 2011). Farmers have adopted local breeds that are resistant to many diseases, require less water, and can survive well in adverse climatic conditions.

Nepal has well-adapted breeds in different species of livestock, and it is necessary to breed them in suitable bioclimatic zones to allow them to express their full genetic potential. Farmers' preferences should be considered when prioritizing breeds. Some transformations are already taking place in this direction. Farmers in Sunsari, Saptari, Siraha, Dhanusa, Rautahat, Bara, Sarhali, Kavre, and Lalitpur have predominantly opted for buffalo rearing. There is also an increasing preference for buffalo in parts of Dhangadi, Bardia, and Nepaljung instead of cattle. In Chitwan, Rupandehi, Jhapa, Morang, and Ilam, farmers initially opted for crossbred cattle and are now eager to rear pure Holstein Friesian. Appropriate breeding is necessary to adapt to hazardous climatic conditions. The native breeds have better adaptive characteristics, which need to be selected for rearing and breeding to minimize the effects of climate change.

In adverse climatic conditions, farmers preserve fodder crops in the form of hay (mainly in high mountain areas). They also store agricultural by-product remains to feed their livestock in the lean period. They change the planting dates, provide bedding to prevent them from cold stress and grow for livestock during extreme winter/cold (Thakur, 2011; Singh et al., 2012; Guillaume, 2015). The climate change adaptation measures need to implement improved pasture grass species breeding, water management, and removal of weeds to improve the pasture land. The animal pasture and fodder management in climate change include deforestation and tree protection on pasture land, replacing roughage with concentrate, feeding leguminous forage rather than a grass-based diet, improving pasture management, feeding locally balanced feed, feeding additive mineral mixture (Wheat soaked-germination-orally feed), rotational looping of vegetative biomass of fodder, trees, herbs, shrubs, and grasses. These measures are applied to minimize the effect of climate change and enhance the productivity of livestock.

Livestock manure application is mainly done on farmer land close to application of livestock manure (whether raw or treated) is mainly done on farmers' land. The manure will be composted in a pit and covering manure solid separator, storage, and anaerobic treatment, practicing shortening storage duration and improving timing and application of manure are some important practices being used. These technologies are cost-effective and can easily be used by small-scale and marginalized farmers. Proper management of waste/manure reduces the methanation of GHG gas and helps to adopt local knowledge in livestock producers to minimize the effect of climate on livestock production.

This study appears to have some limitations on the impact of livestock from climate change. It discusses the impact of climate change on livestock production. Following the study, based on a comprehensive literature review on the subject, specific areas' characteristics will be discussed at great length. Specific adaptation measures for different thematic areas will be highlighted, showing their implications. More investigation work is required to find out the implementation of climate-smart technology by livestock producers, from heat stress, feeding, technology adoption, husbandry practices, and risk management systems that can alter dairy cattle homeostasis mechanisms. Province-specific adaptation strategies are necessary to find out in the live-stock producers. The livestock farmers' perception of adopting climate-smart technology practices in livestock production needs to be further studied in different zones/provinces of Nepal. The implementation of climate-smart technology practices in livestock production at the local and national levels is equally important for a sustainable environment.

ACKNOWLEDGMENTS

The profound acknowledgment goes to the Prince of Songkla University, Hat Yai, Thailand, for allowing the first author to pursue a PhD and providing a scholarship. He is indebted to Drs. Nuttaya Yuangyai and Kuaanan Techato. A special acknowledgment goes to Dr. Saroj Gyawali for inspiring him to enroll in a PhD. An honorable mention goes to Associate Professor Megh Raj Dangal, Adjunct Professor Dr. Sreeram Neupane, and Mr. Shyam Prasad Poudel for guiding the publication and providing valuable comments on the manuscript as professional reviewers.

KEYWORDS

- **agricultural gross domestic product**
- **biodiversity**
- **climate change**
- **farm management**
- **greenhouse gas**
- **gross domestic product**
- **sustainable development**

REFERENCES

Adams, R. H. (1998). Effects of global climate change on agriculture: An interpretative review. *Climate Research, 11*, 19–30.

Ananta Koirala & Shrestha, K. B. (2019). Effects of climate change on the livestock. *Asian Journal of Agriculture and Development, 14*(1), 37–49.

Angel, S. P., Amitha, J. P., Rashmol, V. P., Bandana, G. T., Savitha, S. T., Afsal, A., Bagath, M., Krishnan, G., & Sejian, V. (2018). Climate change and cattle production: Impact and adaptation. *Journal of Veterinary Medicine and Research, 5*(4), 1134.

Berman, A. (2005). Estimates of heat stress relief needs for Holstein dairy cows. *Journal of Animal Science, 83*(6), 1377–1384.

Brooks, D. R. (2007). How will global climate change affect parasite-host assemblages? *Trends in Parasitology, 23*, 571–574.

Brown, R., & Wilhite, H. W. (2013). Climate change and North American rangelands: Trends, projections, and implications. *Elsevier*, 493–511.

Budhathoki, N. K. (2020). Nepalese farmers' climate change perceptions, reality and farming strategies. *Climate and Development, 12*(3), 204–215. https://doi.org/10.1080/17565529.2019.1612317.

Carson, A., Elliott, M., Groom, J., Winter, A., & Bowles, D. (2009). Geographical isolation of native sheep breeds in the UK—Evidence of endemism as a risk factor to genetic resources. *Livestock Science, 123*(2–3), 288–299.

Central Bureau of Statistics, Nepal. (2013). National Planning Commission Secretariat: Kathmandu, Nepal. *National Report.*

Change, U. (2007). Convention on Climate Change. (2007). Investment and financial flows to address climate change. Bonn: UNFCCC.

Chase, E. (2014). Climate Change Impacts on Dairy Cattle. Climate Change and Agriculture: Promoting Practical and Profitable Responses. Department of Animal Science, Cornell University.

Climate change and livestock: Impacts, adaptation, and mitigation. (2008). *Climate Change and Livestock: Impacts, Adaptation, and Mitigation.*

Collier, R. J. (1982). Influences of environment and its modification on dairy animal health and production. *Journal of Dairy Science, 65*, 2213–2227.

Dahal, H., & Khanal, D. R. G. (2010). Food security and climate change adaptation framework: Issues and challenges. *Impacts of Climate Change: Voices of the People, A Collision of Crises and the Impact of Sustained Food Security Deterioration in Nepal.* United Nations World Food Programme, Nepal. Paper presented in the second stakeholders' workshop on NAPA in the agriculture sector held on 23 February 2010 in Kathmandu. (pp. 19–23). Nepal Agriculture Research Council.

Degueurce, A. C. (2016). Fumiers de Bovins, une Ressource à Fort Potentiel Pour la Filière de Méthanisation en France. *Sciences Eaux &Territoires Hors-série numéro, 24*, 9.

Delgado, C. R. (2020). Livestock to 2020: The Next Food Revolution, Food, Agriculture, and the Environment VL – 30, DO - 10.5367/000000001101293427.

Devkota, N. R., Kolachhapati, M. R., Bhattarai, N., & Shrestha, R. (2011). Climate change and its possible consequences to the livestock species from adaptation perspectives. In *Climate Change: Livestock Sector Vulnerability & Adaptation in Nepal* (pp. 51–63). Nepal Agriculture Research Council.

Dhakal, C. K., Regmi, P. P., Dhakal, I. P., Khanal, B., Bhatta, U. K., Barsila, S. R., & Acharya, B. (2013). Perception, impact, and adaptation to climate change: An analysis of livestock system in Nepal. *Journal of Animal Science Advances, 3*(9), 462–471.

Dhakal, S. Poudel, Uddab, Umesh Dahal, Nabin Upadhyaya, Saroj Chaudhari, & Santosh Dhakal (2020). "Livestock and Poultry Production in Nepal and Current Status of Vaccine Development" *Vaccines, 8*(2), 322. https://doi.org/10.3390/vaccines8020322.

Dikshit, A. D. (2010). FAO. (2009). Grasslands: Enabling their potential to contribute to greenhouse gas mitigation. Submission by the UN Food and Agriculture Organization to the intergovernmental panel on climate change. Wheat flour. Agribusiness handbook. Investment center division.

Doungnapa Promket & Wiwanitkit, K. (2020). Effects of climate change on milk yield and milk composition in Thai crossbred Holstein cows. *International Journal of GEOMATE, 18*(67), 108–113.
Everett, R. A. (1982). Environmental influences on sperm output. *Journal of Dairy Science, 65*, 1303–1310.
Falagas, M. E. (2008). Comparison of PubMed, Scopus, Web of Science, and Google Scholar: Strengths and weaknesses. *The FASEB Journal, 22*(2), 338–342. https://doi.org/10.1096/fj.07-9492LSF.
FAO (Food and Agriculture Organization). (2012). World agriculture towards 2030/2050: The 2012 revision. ESA Working Paper No. 12-03. Rome, FAO.
FAO. (2011). Statistical Database, FAOSTAT. http://faostat.fao.org/site/567/default.aspx (accessed on 5th June 2024).
FAO. (2014). Poultry sector Nepal. FAO animal production and health livestock country reviews; No. 8; FAO Poultry Sector Nepal: Rome, Italy.
FAO. (2020). Sustainable small-scale livestock production. Available online: http://www.fao.org/3/ca7660en/CA7660EN.pdf (accessed on 5th June 2024).
FAO. (2021). Climate-smart livestock production: A practical guide for Asia and the Pacific region (Publication No. 50). https://doi.org/10.4060/cb3170en (accessed on 5th June 2024).
Flemming, K. B. (2019). Qualitative evidence synthesis for complex interventions and guideline development: Clarification of the purpose, designs, and relevant methods. *BMJ Global Health, 4*(1), e000882.
Freitas, M. M. (2006). Utility of on- and off-farm weather records for studies in genetics of heat tolerance. *Livestock Science, 105*, 223–228.
Gauly, M., Hölt, M., & Erhardt, G. (2013). Future consequences and challenges for dairy cow production systems arising from climate change in central Europe: A review. *Animal, 7*, 843–859.
Gentle, P., & Maraseni, T. N. (2012). Climate change, poverty and livelihoods: Adaptation practices by rural mountain communities in Nepal. *Environmental Science & Policy, 21*, 24–34. https://doi.org/10.1016/j.envsci.2012.03.007.
Gentle, P., Thwaites, R., Race, D., & Alexander, K. (2013). The changing role of local institutions to enable individual and collective actions for adapting to climate change.
Gerber, P. (2013). Tackling climate change through livestock – A global assessment of emissions and mitigation. FAO, Rome. http://www.fao.org/3/a-i3437e.pdf (accessed on 5th June 2024).
Gerber, P. J., Steinfeld, H., Henderson, B., Mottet, A., Opio, C., Dijkman, J., Falcucci, A., & Tempio, G. (2013). Tackling climate change through livestock: A global assessment of emissions and mitigation opportunities. Rome: Food and Agriculture Organization.
Gezu Tadesse & Moges Dereje (2018). Impact of Climate Change on Smallholder Dairy Production and Coping Mechanism in Sub- Saharan Africa: Review. *Agri Res & Tech :Open Access J. 16*(4), 556000. DOI: 10.19080/ARTOAJ.2018.16.556000.
Ghimire, Y. N. (2011). Chapter 3. Assessment of the vulnerability of livestock farmers to climate change-related drought in the hills of NEPAL. In *Climate Change: Livestock Sector Vulnerability & Adaptation in Nepal*, 24–33.
Ghosh, D. S. (2014). Impact of climate change on livestock production: A review. *Journal of Animal Research, 4*(2), 223–239.
Grossi, G., Goglio, P., Vitali, A., & Williams, A. G. (2019). Livestock and climate change: Impact of livestock on climate and mitigation strategies. *Animal Frontiers, 9*(1), 70–74.

Guillaume Mauger, Yoram Bauman, Tamilee Nennich & Eric Salathé (2015). Impacts of Climate Change on Milk Production in the United States, *The Professional Geographer, 67*(1), 121–131, DOI: 10.1080/00330124.2014.921017. http://dx.doi.org/10.1080/00330124.2014.921017.

Gurung T. B., Pokharel, P. K., & Wright, I. (2011). Climate change: Livestock and Vulnerability in Nepal. *Proceedings of Consultative Technical Workshop on Kathmandu,* Nepal, 156 pp.

Gurung T. B., Pokharel, P. K., & Wright, I. (2011). Climate change: Livestock and Vulnerability in Nepal. *Proceedings of consultative technical workshop on Kathmandu,* Nepal, 156 pp.

Gusenbauer, M., & Haddaway, N. R. (2015). The role of Google Scholar in evidence reviews and its applicability to grey literature searching. *PLoS One, 10*(9), e0138237.

Hussain, A. R. (2018). A case study on climate change impacts on livestock farming. *Natural Hazards, 91*(3), 1365–1383. https://doi.org/10.1007/s11069-018-3187-1.

IFAD (International Fund for Agricultural Development). (2010). Livestock and climate change. Climate Change and Livestock: Impacts, Adaptation, and Mitigation. Retrieved from: www.ifad.org/lrkm/index.htm.

IPCC (Intergovernmental Panel on Climate Change) Summary for policymakers. (2013). In T. F. Stocker et al. (Eds.), *Climate Change 2013: The Physical Science Basis*. Cambridge, UK: Cambridge University Press.

Ishtiaque, A. M. (2020). Remote sensing-based research for monitoring progress towards SDG 15 in Bangladesh: A review. *Remote Sensing, 12*(14), 2231.

Karn, P. (2011). Climate change and livestock sector in NEPAL. In *Climate Change: Livestock Sector Vulnerability & Adaptation in Nepal, Chapter 12,* 87–92.

Knapp, J. R., Laur, G. L., & Vadas, P. A. (2014). Enteric methane in dairy cattle production: Quantifying the opportunities and impact of reducing emissions. *Journal of Dairy Science, 97*(6), 3231–3281. doi: 10.3168/jds.2013-7234.

Koirala, A., & Bhandari, P. (2018). Impact of climate change on livestock production. *Nepalese Veterinary Journal, 35,* 178–183.

Koubkova, M., Knizkova, I., & Hartl, J. (2002). Influence of high environmental temperatures and evaporative cooling on some physiological, hematological, and biochemical parameters in high-yielding dairy cows. *Czech Journal of Animal Science, 47*(7), 309–318.

Kupper, T. B., Gygax, L., & Hillmann, E. (2015). Evolution of farm and manure management and their influence on ammonia emissions from agriculture in Switzerland between 1990 and 2010. *Atmospheric Environment, 103,* 215–221.

Liberati, A., Altman, D. G., Tetzlaff, J., Mulrow, C., Gøtzsche, P. C., Ioannidis, J. P. A., & Moher, D. (2009). The PRISMA statement for reporting systematic reviews and meta–Analyses of studies that evaluate health care interventions: Explanation and elaboration. *PLoS Medicine, 6*(7), e1000100. https://doi.org/10.1371/journal.pmed.1000100.

Llonch, P., Haskell, M. J., & Dewhurst, R. J. (2017). Current available strategies to mitigate greenhouse gas emissions in livestock systems: An animal welfare perspective. *Animal Frontiers, 7*(4), 69–76.

Malla, G. (2008). Climate change and its impact on Nepalese agriculture. *The Journal of Agriculture and Environment, 9,* 62–71.

Martín-Martín, A., Orduna-Malea, E., & Delgado-López-Cózar, E. (2018). Google Scholar, Web of Science, and Scopus: A systematic comparison of citations in 252 subject categories. *Journal of Informetrics, 12*(4), 1160–1177. https://doi.org/10.1016/j.joi.2018.09.002.

Maust, L. M., Kirschten, D. P., & Lloyd, J. W. (1972). Effect of summer weather on performance of Holstein cows in three stages of lactation. *Journal of Dairy Science, 55*(8), 1133–1139.

Menzi, H., Staub, R., & Muggli, M. (2015). Country Manure Management Profile.

MoALD (Ministry of Agriculture and Livestock Development). (2011). Nepal agricultural statistics 2074/75. Ministry of Agriculture and Livestock Development, Singhdurbar, Kathmandu, Nepal.

MoALD (Ministry of Agriculture and Livestock Development). (2019). Nepal agricultural statistics 2074/75. Ministry of Agriculture and Livestock Development, Singhdurbar, Kathmandu, Nepal.

MoF (Ministry of Finance). (2019). Economic survey of Nepal (2018/19). Ministry of Finance, Government of Nepal, Kathmandu.

Mohamed Shaffril, H. A., Samsuddin, S. F. M., & Shaharudin, M. R. (2020). Systematic literature review on adaptation towards climate change impacts among indigenous people in the Asia Pacific regions. *Journal of Cleaner Production, 258*, 120595. https://doi.org/10.1016/j.jclepro.2020.120595.

Moher, D., Liberati, A., Tetzlaff, J., Altman, D. G., & The PRISMA Group. (2009). Preferred reporting items for systematic reviews and meta-analyses: The PRISMA statement. *PLoS Medicine, 6*(7), e1000097. https://doi.org/10.1371/journal.pmed.1000097.

Molee, A. B. (2011). Suitable percentage of Holstein in crossbred dairy cattle in climate change situation. *Journal of Animal and Veterinary Advances, 10*(7), 828–831.

MOPE. (2004). Initial National Communication to the Conference of the Parties of the United Nations framework convention on climate change, Government of Nepal. In *Climate Change: Livestock Sector Vulnerability & Adaptation in Nepal* (pp. 42–47). Ministry of Population and Environment.

Murphy, S., & Allen, L. N. (2003). Nutritional importance of animal source foods. *The Journal of Nutrition, 133*, 3932S–3935S.

Naqvi, V. S. (2012). Concept of multiple stresses and their significance on livestock productivity. In Environmental Stress and Amelioration in Livestock Production, Nature, Springer and Palgrave Macmillan, (pp. 129–150).SBN : 978–3-642–29204–0.

Nardone, A., Ronchi, B., Lacetera, N., Ranieri, M. S., & Bernabucci, U. (2010). Effects of climate changes on animal production and sustainability of livestock systems. *Livestock Science, 130*, 57–69.

Neena, A., Gorkhali, R. Shrestha, Bhola S. Shrestha & Prakash K Pokharel (2011). A review on impact of climate change on livestock breeds and breeding system. In *Climate Change: Livestock Sector Vulnerability & Adaptation in Nepal,* Chapter 14, 112–117.

Neopane, S. P. (2010). Effect of climate change on livestock, vulnerability, and adaptation plans in Nepal. Paper presented at a workshop on climate change adaptation and disaster risk management organized by the Ministry of Agriculture and Cooperatives & Food and Agriculture Organization (FAO) on 23rd February 2010, Kathmandu, Nepal. Nepal Agriculture Research Council.

Neopane, S. P., & Tuladhar, K. (2011). Livestock: An asset for enhancing the adaptive capacity of climate-vulnerable communities in Nepal. In *Climate Change: Livestock Sector Vulnerability & Adaptation in Nepal*, 19–23. Nepal Agricultural Research Council.

Nepal, G. O. (2004). Nepal's Initial national communication to the conference of the parties of the United Nations framework. Convention on climate change (UNFCCC), July 2004, Kathmandu, Ministry of Population and Environment.

Nissim Silanikovea, N. K. (2015). Impact of climate change on the dairy industry in the temperate zone: Predictions on the overall negative impact and the. *Small Ruminant Research, 123*, 27–34.

NSCoA. (2013). National sample census of agriculture-2011/12. Central bureau of statistics, national planning commission secretariat, Government of Nepal: Kathmandu, Nepal.

Nuorteva, P. K. (2010). Water, livelihoods and climate change adaptation in the Tonle Sap Lake area Cambodia. *Journal of Water and Climate Change, 1*(1), 87–101, https://doi.org/10.2166/wcc.2010.010.

Okoli, C. (2015). A guide to conducting a standalone systematic literature review. *Communications of the Association for Information Science, 37*, 43. https://doi.org/10.17705/1CAIS.03743.

Orduña-Malea, E., Ayllón, J. M., Martín-Martín, A., & López-Cózar, E. D. (2017). The lost academic home: Institutional affiliation links in Google Scholar Citations. ArXiv abs/1804.07268.

Pathak, P., Thakur, U. C., & Shrestha, D. (2011). Climate change and livestock sector: Vulnerability and adaptation in Nepal. *Change: Livestock Sector Vulnerability & Adaptation in Nepal*, 42.

Petticrew, M. (2001). "Systematic reviews from astronomy to zoology: myths and misconceptions," *BMJ: The BMJ, 322*(7278), 98–101. http://doi.org/10.1136/bmj.322.7278.98.

Philip Thornton, Mario Herrero, and Polly Ericksen. (2011). Livestock and climate change. *Livestock Exchange Issue Brief 3*, 1–1.

Pokhrel, D. M. (2011). The agriculture sector's strategy for climate change adaptation focused on livestock vulnerability. In *Climate Change: Livestock Sector Vulnerability & Adaptation in Nepal* (pp. 34–41). Nepal Agriculture Research Council.

Polley, H. W., Kolodziejski, C. A., Jones, K. A., Smith, (2013). Climate change and North American rangelands: Trends, projections, and implications. *Rangeland Ecology & Management, 66*(5), 493–511.

Poudel, S., & Subedi, N. (2020). Effects of climate change on agriculture and its mitigation. Tropical Agrobiodiversity (TRAB),.*01.2020.* 47–51.

Prabhakar Pathak*, Uday Chandra Thakur, and Damayanti Shrestha. (2011). Climate change and livestock sector: Vulnerability and adaptation in NEPAL. In *Climate Change: Livestock Sector Vulnerability & Adaptation in Nepal* (pp. 42–47). Nepal Agriculture Research Council.

Pradhanang, U., Pradhanang, S., Sthapit, A., Krakauer, N., Jha, A., & Lakhankar, T. (2015). National livestock policy of Nepal: Needs and opportunities. *Agriculture 2015, 5*, 103–131.

Pragna, P., Sejian, V., Bagath, M., Krishnan, G., Archana, P., Soren, N., & Bhatta, R. (2018). Comparative assessment of growth performance of three different indigenous goat breeds exposed to summer heat stress. *Journal of Animal Physiology and Animal Nutrition, 102*, 825–836.

Purwanto, B. Y. (2014). Diurnal patterns of heat production and heart rate under thermo neutral conditions in Holstein Friesian cows differing in milk production. *Journal of Agricultural Science, 114*, 139–142.

Quideau, P. (2010). Les effluents d'élevage, les coproduits de traitement et leurs. Versailles: Les effluents d'élevage, les coproduits de traitement et leurs, 260.DOI: 10.3917/edagri.espag.2010.01.0119, In: *Livestock Farming and the Environment* (pp.119–185).

Ravagnolo, O., & Misztal, I. (2007). The genetic component of heat stress in dairy cattle, parameter estimation. *Journal of Dairy Science, 83*, 2126–2130.

Rijal, S. R. (2019). Climate-smart agriculture concept and adaptation. *International Journal of Research & Review, 47*, 47–56.

Rijal, S., & Rijal, B. R. (2019). Climate Smart Agriculture Concept and Adaptation in Nepal: An Overview. *International Journal of Research & Review, 6*, 47–56.

Rijal, S., Gentle, P., Khanal, U., & Rimale, B. (2021). A systematic review of Nepalese farmers' climate change adaptation. *Climate Policy, 22*(1), 1–15. DOI:10.1080/14693062.2021.1977600.

Ripple, W., Smith, P., Haberl, H., Montzka, S., McAlpine, C., & Boucher, D. (2013). Ruminants, climate change, and climate policy. *Nature Climate Change*, *4*, 2–5.

Rosenzweig, C., Karoly, D., Vicarelli, M., Neofotis, P., Wu, Q., Casassa, G., Menzel, A., Root, T. L., Estrella, N., Seguin, B., Tryjanowski, P., Liu, C., Rawlins, S., & Imeson, A. (2008). Attributing physical and biological impacts to anthropogenic climate change. *Nature*. *453*(7193), 353–357. doi: 10.1038/nature06937. PMID: 18480817.

Rust, J. R. (2013). Climate change and livestock production: A review with emphasis on Africa. *South African Journal of Animal Science*, *43*(3), pp.256–267. ISSN 2221–4062.

Singh, K. B., Nauriyal, D. C., Oberoi, M. S., & Baxi, K. K. (1996). Studies on the occurrence of clinical mastitis in relation to climatic factors. *Indian Journal of Dairy Science*, *49*(8), 534–536.

Singh, S. K., Meena, H. R., Kolekar, D. V., & Singh, Y. P. (2012). Climate change impacts on livestock and adaptation. *Journal of Veterinary Advances*, *2*(7), 407–412.

Skoet, S. G. (2012). Milk availability trends in production and demand and medium-term. Food and Agriculture Organization of the United Nations.

Slenning, B. D. (2010). Global climate change and implications for disease emergence. *Veterinary Pathology*, *47*, 28–33.

Steinfeld, H., Gerber, P., Wassenaar, T., Castel, V., Rosales, M., & de Haan, C. (2006). Livestock's long shadow: Environmental issues and options. *Food and Agriculture Organization of the United Nations, 24*, SN – 978–92–5-105571–7.

Stephenson, D. B., & Cavazos, T. (2008). Definition, diagnosis, and origin of extreme weather and climate events. *Climate Extremes and Society*, *340*, 11–23.

Tailor, S. P., & Rijal, N. R. (2005). Conception rate in buffaloes maintained under a sub-humid climate of Rajasthan. *Indian Journal of Dairy Science*, *58*(1), 69–70.

Taylor, J. B. (1985). Genetic and environmental components of semen production traits of artificial insemination Holstein bulls. *Journal of Dairy Science*, *68*, 2703–2722.

Team, C. N. (2020). Dairy Sector Strategy – Nepal. Commercial Agriculture for Smallholders and Agribusiness.

Thakur, R. P. (2011). Climate change: Livestock health and production in Nepal. In *Climate Change: Livestock Sector Vulnerability & Adaptation in Nepal* (pp. 76–80). Nepal Agriculture Research Council.

Thornton, P. K. (2021). Potential for reduced methane and nitrous oxide emissions from livestock. *Tropical Agrobiodiversity1(1): 47–51.*

Thornton, P. K., Boone, R. B., & Ramirez-Villegas, J. (2015). Climate change impacts on livestock. CCAFS Working Paper no. 120. CGIAR Research Program on Climate Change, Agriculture and Food Security (CCAFS). Copenhagen, Denmark. Available online at: www.ccafs.cgiar.org.

Thornton, P., Herrero, M., & Freeman, H. A. (2007). Vulnerability, climate change, and livestock: Research opportunities and challenges for poverty alleviation. *An Open Access Journal published by ICRISAT*, *4*(1), 1–23.

Thornton, P., van de Steeg, J., Notenbaert, A., & Herrero, M. (2009). The impacts of climate change on livestock and livestock systems in developing countries: A review of what we know and what we need to know. *Agricultural Systems*, *101*, 113–127.

Toma, A. P. (2012). A typology of dairy farmer perceptions towards climate change. *Climatic Change*, *112*, 507–522.

United Nations Framework Convention on Climate Change. (2007). Investment and financial flows to address climate change. Bonn: UNFCCC.

United Nations Framework Convention on Climate Change. (2019). Definitions of climate change. Retrieved from: http://unfccc.int/files/essential_background/background_publications_htmlpdf/application/pdf/conveng.pdf (accessed on 5th June 2024).

United Nations Framework Convention on Climate Change. (2021). The Paris Agreement. Retrieved from: https://unfccc.int/process-and-meetings/the-paris-agreement/ (accessed on 5th June 2024).

Upadhaya, R. C. (2007). Impact of climate change on milk production of Murrah buffaloes. *Italian Journal of Animal Science*, *6*(2), 1329–1332.

Upadhaya, R. C. (2009). Inventory of methane emission from livestock in India. In P. K. Aggarwal (Ed.), *Global Climate Change and Indian Agriculture: Case Studies from the ICAR Network Project* (pp. 117–122). ICAR, New Delhi.

Upadhyay, R. C., & Singh, S. S. (2016). Impact of Climate change on Milk production of Murrah buffaloes. *Italian Journal of Animal Science*, *6*(sup2), 1329–1332. DOI: 10.4081/ijas. 2007.s2.1329.

Upadhyay, R. S. (2008). Impact of climate change on livestock. *Indian Dairyman, 60*(3), 98–102.

Upreti, C. R. (2011). Greenhouse Gas (GHG) production from ruminant animals – Management strategies for reduction in Nepal, In *Climate Change: Livestock Sector Vulnerability & Adaptation in Nepal*. (pp. 61–69). Publisher, Nepal Agricultural Research Council (NARC).

Upreti, D. (1999). Rising Atmospheric CO and crop response. *SASCOM Scientific Report*, 1–8. World Climate News.

Upreti, D. (2008). Rising Atmospheric CO and crop response 2006, SASCOM Scientific Report, *World Climate News, The Journal of Agriculture and Environment Vol*.9. (pp 1–8). Science and Education Publishing.

Whittemore, R., & Knafl, K. (2005). The integrative review: Updated methodology. *Journal of Advanced Nursing, 52*(5), 546–553.

Wilkes, M., Hynd, P., & Pitchford, W. (2012). Damara sheep have higher digestible energy intake than merino sheep. *Animal Production Science, 52*, 30–34.

Williams, C. M., & Davis, J. C. (2011). Management and utilization of poultry wastes. *Reviews of Environmental Contamination and Toxicology, 214*, 81–86.

Zhu, J., & Liu, W. (2020). A tale of two databases: The use of Web of Science and Scopus in academic papers. *Scient Metrics, 123*(1), 321–335. doi: 10.1007/s11192-020-03387-8.

ANNEX I

Boolean Database Search from Web of Science and Scopus, and Advance Search in Google Scholar

Scopus	Search String
	(impact AND of AND climate AND change AND on AND livestock AND production AND its AND local AND adaptation) (climate AND change AND livestock AND production)) AND ((related AND to AND livestock AND production)) AND (climate AND change AND related AND to AND adoption AND on AND livestock AND production) AND (EXCLUDE (PUBYEAR, 2010)) AND (LIMIT TO (OA, “all”)) (climate change AND livestock AND production)) AND (climate AND change AND livestock AND production AND adoption) climate AND change AND livestock AND production)) AND ((related AND to AND livestock AND production)) AND (climate AND change AND related AND to AND adoption AND on AND livestock AND production) AND (EXCLUDE (PUBYEAR, 2010)) AND (LIMITTO (OA, “all”)) (climate AND change AND adaptation; AND livestock AND producers)
Web of Science	(impact AND of AND climate AND change AND on AND livestock AND production AND its AND local AND adaptation) (climate AND change AND livestock AND production)) AND ((related AND to AND livestock AND production)) AND (climate AND change AND related AND to AND adoption AND on AND livestock AND production) AND (EXCLUDE (PUBYEAR, 2010)) AND (LIMITTO (OA, “all”)) (climate change AND livestock AND production)) AND (climate AND change AND livestock AND production AND adoption) climate AND change AND livestock AND production)) AND ((related AND to AND livestock AND production)) AND (climate AND change AND related AND to AND adoption AND on AND livestock AND production) AND (EXCLUDE (PUBYEAR, 2010)) AND (LIMITTO (OA, “all”)) (climate AND change AND adaptation; AND livestock AND producers)
Google Scholar	‘Climate change livestock production adaptation in Nepal’

ANNEX II

Summary of the Local Adaptation Approaches Implemented by the Livestock Producers Discovered from the Systematic Review of the Selected Articles

Theme	Sub-Theme	Number of Articles
Temperature stress control technique	1. Housing improvement	7
	2. Coldwater during hot and humid climate	5
	3. Provision of shed to reduce hot and cold stress.	5
	4. Provide fresh air/fan/cooler during extreme hot conditions.	4
	5. Kept animals outside during hot summer.	3
	6. Loose housing system.	4
	7. Provision of fire during the cold season.	4
Risk management	8. Government subsidies.	2
	9. Insurance system.	3
	10. Awareness of diseases and vaccination.	5
	11. Credit of bank/loan from Bank	6
Technology for adoption	12. Development of new breeds and genetic climate tolerance types.	7
	13. Improved animal health by controlling vector-borne diseases.	7
	14. Water and soil resources management by improving alternative irrigation	9
	15. Improving pasture and forage use of high-yielding variety.	5
	16. Securing water access for livestock in drought conditions.	7
	17. Controlling animal health science-based practice using vaccines and medicines.	9
Animal management system adopted by farmers	18. Integration of pasture management.	7
	19. Stall feeding of animals	13
	20. Introducing a mixed farming system (stall-fed and pasture grazing)	15

ANNEX II *(Continued)*

Theme	Sub-Theme	Number of Articles
	21. Changing the plantation time	6
	22. Reduction in livestock number.	8
	23. Replacement of exotic breed to the local breed of pasture grass.	9
	24. Provide bedding for extreme cold.	7
Breeding strategies	25. Breed changing (control of inbreeding)	9
	26. Artificial insemination with proven breed semen.	7
	27. Promotion of local breeds through selection	5
	28. Improving genetics through a pedigree performance recording system	4
	29. Infertility management	7
Animal feeds, pasture, and fodder management	30. Deforestation and tree protection on pastureland.	5
	31. Replacing roughage with concentrate.	4
	32. Feeding leguminous forage than a grass-based diet.	5
	33. Improving pasture management	7
	34. Feeding of locally balanced feed	5
	35. Feeding of the additive mineral mixture (Wheat soaked-germination – orally feed)	7
	36. Rotational looping of vegetative biomass of fodder, trees, shrubs, herbs, and grasses.	8
	37. Deforestation and tree protection on pasture land.	6
Manure management	38. Storage and anaerobic treatment.	8
	39. Practice shortening storage duration.	4
	40. Improving timing and application of manure.	5
	41. Covering manure solid separator.	4

ANNEX III

Results

Summary Findings of Local Adoption and Mitigation Measure

	Theme	Adoption Measure
3.1	Temperature stress control technique	• Housing improvement, construction of shed, concrete floor, asbestos sheet roofing material. • Cold water during hot and humid climates, Sprinkle water on the roof of the shed • Provision of shed to reduce hot and cold stress. Provision of bedding (cow mat) in the shed • Provide fresh air/fan/cooler during extreme hot conditions. • Kept animals outside during the hot summer. • Loose housing system. • Provision of fire in cold season.
3.2	Risk management	• The government provided subsidies for boosting production to smallholders and marginal farmers. • Insurance system: Government gives the 75% subsidies on premiums. Dairy cooperatives are doing insurance systems to minimize the risk of mortality. The cooperative model is unique and beneficial to farmers. • Awareness of diseases and vaccination by dairy cooperatives and provision of free vaccination to control disease • Credit of bank/loan/cooperative from Bank for animal purchase and housing management.
3.3	Technology for adoption	• Conservation of local indigenous breeds and genetic climate tolerance cross-breed types. • Improved animal health by controlling vector-borne, parasitic, and protozoal diseases. • Water harvesting at wells in mountains and pounds in terai for animal drinking purposes and recharge of soil management by improving alternative irrigation, • Improving pasture and forage cultivation in alpine and tropics use of high-yielding variety. • Securing water access for livestock in drought conditions by harvesting wells and ponds. • Controlling animal health science-based practice using vaccines and medicines preventing infectious and contagious diseases.

ANNEX III *(Continued)*

	Theme	Adoption Measure
3.4	Animal management system adopted by farmers	• Integration of pasture management. • Stall feeding of animals and collection of urine. • Introducing a mixed farming system (stall-fed and pasture grazing) • Changing the plantation time of fodder and pasture in fodder cultivated and grazing land • Reduction in unproductive and developing high-yielding livestock numbers. • Replacement of exotic breed to local breed of pasture grass. • Provide bedding for extreme cold.
3.5	Breeding strategies	• Breed changing introducing the local breed resistant to climate change. Exchange of breeding bulls in communities (control of inbreeding) to control inbreeding system. • Artificial insemination with proven breed semen and proper recording. • Promotion of local breeds through selection methods • Improving genetics through pedigree performance Recording system in the important dairy cooperative pocket area • Infertility management, supply of mineral mixture, vitamins and minerals tablets
3.6	Animal feeds, pasture, and fodder management	• Deforestation and tree plantation on pasture land. Rotational plantation • Replacing roughage with locally balanced concentrate. • Feeding leguminous forage than a grass-based diet. • Improving pasture management with climate resistance seed and breed. • Feeding of locally available ingredients of balanced feed • Feeding of the additive mineral mixture (Wheat soaked-germination – orally feed) • Rotational looping of vegetative biomass of fodder, trees, shrubs, herbs, and grasses. • Deforestation and tree protection on pasture land.

ANNEX III *(Continued)*

	Theme	Adoption Measure
3.7	Manure management	• Storage of manure in the pit and aerobic treatment. • Practice with shortening of storage duration. • Improving timing in pit and application of manure in pasture land. • Covering manure with a solid separator reduces the methanation of GHG gas to minimize the effect of climate on livestock production.

ANNEX IV

Summary Findings of Local Adoption and Mitigation Measure in Terai

	Theme	Adoption Measure	
3.1	Temperature Stress Control Technique	• Housing improvement, construction of shed, concrete floor, asbestos sheet roofing material,	• Jhapa, Morang, Sunsari, Saptari, siraha, Mohatari, Dhanusha, Sarlahi, Rautahat, Bara, Parsa
		• Cold water during hot and humid climates, Sprinkle water on the roof of the shed	• Chitwan, Nawalparasi, Rupandehi, Dqng, Bardia, Nepal Jung Kailali, Kanchanpur
		• Provision of shed to reduce cold stress. Provision of bedding (cow mat) in the shed	
		• Provide fresh air/fan/cooler during extreme hot conditions.	
		• Kept animals outside during the hot summer.	
		• Loose housing system.	
3.2	Risk management	• The government provided subsidies for boosting production to smallholders and marginal farmers.	
		• Insurance system: Government gives the 75% subsidies on premiums. Dairy cooperatives are doing insurance systems to minimize the risk of mortality. The cooperative model is unique and beneficial to farmers.	
		• Awareness of diseases and vaccination by dairy cooperatives and provision of free vaccination to control disease	
		• Credit of bank/loan/cooperative from Bank for animal purchase and housing management.	
3.3	Technology for adoption	• Conservation of local indigenous breeds and genetic climate tolerance cross-breed types.	
		• Improved animal health by controlling vector-borne, parasitic, and protozoal diseases.	

ANNEX IV *(Continued)*

Theme	Adoption Measure
	• Securing water access for livestock in drought conditions by harvesting wells and ponds.
	• Controlling animal health science-based practice using vaccines and medicines preventing infectious and contagious diseases.
3.4	• Stall feeding of animals and collection of urine.
	• Introducing a mixed farming system (stall-fed and pasture grazing)
	• Changing the plantation time of fodder cultivated land
	• Reduction in unproductive and developing high-yielding livestock numbers.
	• Replacement of exotic breed to local breed of pasture grass.
3.5 Breeding strategies	• Breed changing introducing the local breed resistant to climate change. Exchange of breeding bulls in communities (control of inbreeding) to control inbreeding system.
	• Artificial insemination with proven breed semen and proper recording.
	• Promotion of local breeds through selection methods
	• Improving genetics through pedigree performance Recording system in the important dairy cooperative pocket area
	• Infertility management, supply of mineral mixture, vitamins and minerals tablets
3.6	• Replacing roughage with locally balanced concentrate.
	• Feeding leguminous forage than a grass-based diet.
	• Feeding of locally available ingredients of balanced feed
	• Feeding of the additive mineral mixture (Wheat soaked-germination – orally feed)

ANNEX IV *(Continued)*

Theme	Adoption Measure
	• Rotational looping of vegetative biomass of fodder, trees, shrubs, herbs, and grasses.
3.7 Manure Management	• Storage of manure in the pit and aerobic treatment. • Practice with shortening of storage duration. • Improving timing in pit and application of manure in pasture land. Covering manure with a solid separator reduces the methanation of GHG gas to minimize the effect of climate on livestock production.

ANNEX V

Summary Findings of Local Adoption and Mitigation Measures in Hill and Mountain

	Theme	Adoption Measure	Districts
3.1	Temperature Stress Control Technique	• Housing improvement, construction of shed, concrete floor, asbestos sheet roofing material, • Provision of shed to reduce cold stress. Provision of bedding (cow mat) in the shed. • Provision of fire in cold season.	• Ilam • Terathum • Kavre
3.2	Risk management	• The government provided subsidies for boosting production to smallholders and marginal farmers. • Insurance system: Government gives the 75% subsidies on premiums. Private companies are doing and cooperative system yet to develop an insurance system. • Awareness of diseases and free vaccination to control the disease. • Credit of bank/loan from Bank for animal purchase and housing management.	• Bhaktapur • Kathmandu, Kaski Palpa, Syanja • Dadheldhura • Surkhet, Gorkha, Solukhumbu
3.3	Technology for adoption	• Conservation of new local indigenous breeds. • Improved animal health by controlling vector-borne, parasitic, and protozoal diseases. • Water harvesting at wells in hills and mountains for animal drinking purposes and recharge of soil management by improving alternative irrigation. • Improving pasture and forage cultivation in alpine and tropics use of high-yielding variety. • Securing water access for livestock in drought conditions by harvesting wells and ponds. • Controlling animal health science-based practice using vaccines and medicines preventing infectious and contagious diseases.	• Jumla • Mustang
3.4	Animal Management system adopted by farmers	• Integration of pasture management. • Introducing a mixed farming system (stall-fed and pasture grazing) • Changing the plantation time of fodder and pasture in grazing land	

ANNEX V *(Continued)*

	Theme	Adoption Measure	Districts
		• Replacement of exotic breed to local breed of pasture grass.	
		• Provide bedding for extreme cold.	
3.5	Breeding strategies	• Breed changing introducing the local breed resistant to climate change. Exchange of breeding bulls in communities (control of inbreeding).	
		• Promotion of local breeds through selection methods	
		• Infertility management, supply of mineral mixture, vitamins and minerals tablets	
3.6	Animal feeds, pasture, and fodder Management	• Deforestation and tree plantation on pasture land. Rotational plantation	
		• Replacing roughage with locally balanced concentrate.	
		• Feeding leguminous forage than a grass-based diet.	
		• Improving pasture management with climate resistance seed and breed.	
		• Feeding of locally available ingredients of balanced feed	
		• Feeding of the additive mineral mixture (Wheat soaked-germination – orally feed)	
		• Rotational looping of vegetative biomass of fodder, trees, shrubs, herbs, and grasses.	
		• Deforestation and tree protection on pasture land.	
3.7	Manure Management	• Storage of manure in the pit and aerobic treatment.	
		• Practice with shortening of storage duration.	
		• Improving timing in pit and application of manure in pasture land.	
		• Covering manure with a solid separator reduces the methanation of GHG gas and helps to adopt the local knowledge in livestock producers to minimize the effect of climate on livestock production.	

CHAPTER 14

New Insights on the Characteristics of Equatorial Plasma Bubbles: A Case Study

IDAHWATI SARUDIN,[1] NURUL SHAZANA ABDUL HAMID,[2,3] MARDINA ABDULLAH,[3,4] and SUHAILA M. BUHARI[5]

[1]*School of Physics, Universiti Sains Malaysia, USM, Pulau Pinang, Malaysia*

[2]*Department of Applied Physics, Faculty of Science and Technology, Universiti Kebangsaan Malaysia, UKM Bangi, Selangor, Malaysia*

[3]*Space Science Center (ANGKASA), Institute of Climate Change, Universiti Kebangsaan Malaysia, UKM Bangi, Selangor, Malaysia*

[4]*Department of Electrical, Electronic, and Systems Engineering, Faculty of Engineering and Built Environment, Universiti Kebangsaan Malaysia, Bangi, Malaysia*

[5]*Department of Physics, Faculty of Science, Universiti Teknologi Malaysia, UTM, Skudai, Johor, Malaysia*

ABSTRACT

The purpose of the presented work is to investigate the trends of both Equatorial Plasma Bubble (EPB) zonal drift velocity (V_E) and eastward winds which applied two different ground-based instruments. Malaysia's real-time kinematics GNSS Network (MyRTKnet), consisting of more than 78 GPS receivers was used to observe EPB along 96° East to 120° East longitudes. The eastward wind data were measured using a Fabry-Perot interferometer (FPI) located at the station of Kototabang, Indonesia. In general, those temporal variations of EPB V_E are consistent with the eastward winds. The results illustrate similar trends for both V_E of EPB and eastward winds, even though the EPB V_E tend to be slightly higher than the eastward wind velocity.

Engineering Design and Technical Applications of Physical Science: An Integrated Approach for Sustainable Development. Chin Hua Chia, Sarani Zakaria, Ruey Shan Chen, & Md. Shahariar Chowdhury (Eds.)

14.1 INTRODUCTION

Equatorial plasma bubbles (EPB) are an ionospheric phenomenon near the Earth's geomagnetic equator at nighttime. It is defined as the depletion of total electron content (TEC) in the ionosphere. EPBs are able to interrupt communication and navigation systems between satellites and receivers (Bhattacharyya, 2022). The investigation of EPBs has been done before by applying different techniques such as optical imaging and radar measurements. However, most of the techniques are not able to monitor the EPB occurrences in a large-scale area as well as have difficulty in estimating the EPB zonal drift velocity. This V_E is one of the characteristics of EPBs which have been commonly reported in the past years using various techniques.

The study of this characteristic is important in maintaining stable communication between satellites and the ground (Sarudin et al., 2020). Most of the previous studies determine the V_E of EPB using ground-based instruments such as optical techniques (Gurav et al., 2018). They reported that the V_E of EPB appears between 21:00 and 22:00 LT and decreases toward the remaining course of the night. Apart from that, past studies believed that the V_E of EPB was related to the eastward winds. There are numerous discussions regarding the relationship between the V_E of EPB and the eastward winds. For instance, a study by Chapagain (2016) indicated that the eastward wind was higher in the evening hours and decreased as time progressed. He, in particular, focused on the relationship between EPB V_E and eastward wind velocity. The present study focuses on the EPB V_E estimated using Rate of TEC change index (ROTI) measurements from approximately 130 stations received from the Global Positioning System (GPS) networks in Southeast Asia (SEA) during the equinox season in the period of 2011–2013 and compared it with the eastward wind velocity from the Fabry-Perot interferometer (FPI).

14.2 DATA OBSERVATION

The EPBs V_E were calculated using the rate of TEC change index, ROTI, obtained using high-density GPS receivers in SEA (Buhari et al., 2014). The Malaysia Real-Time Kinematics GNSS Networks (MyRTKnet) has more than 78 GPS receiver stations across Malaysia. The GPS data from Singapore, Indonesia, and Thailand were collected from IGS and SuGAr networks through the Scripps Orbit and Permanent Array Center website (sopac.ucsd.edu/dataArchive). Total electron content (TEC) was obtained along the sightline between the receiver and satellite, considering 1 TECU = 10^{16} electron/m^2 for the two signals conveyed by the satellites (Sarudin et al., 2020).

The structure of EPBs extends in the north-to-south directions and moves eastward with time (Otsuka et al., 2004). Data processing to obtain a ROTI keogram can be referred to in a previous study by Sarudin et al., 2020. According to past studies, the onset of EPB structure began at 104°E longitude at 1230 UT and developed faster over time (Yokoyama & Fukao, 2006). The EPBs are believed to be moving in an eastward direction (Paulino et al., 2011). In this keogram, we can see that every EPB moves in an eastward direction and vanishes almost after 1800 UT. The present work focused on three years of ROTI keogram acquired from GPS networks to investigate the V_E EPB. The occurrence of EPBs was confirmed with a ROTI threshold between 0.04 and 0.07 TECU/min in the keogram. This range of values is beneficial for recognizing the existence of EPBs in the observation area (Buhari et al., 2017). The V_E EPB can be estimated from the change of longitude with time (Sarudin et al., 2017).

The eastward winds were studied using data obtained from FPI. The FPI (FP03) in operation at Kototabang, Indonesia is under the Optical Mesosphere Thermosphere Imagers (OMTIs) (Shiokawa et al., 2012) operated by the Institute for Space-Earth Environmental Research (ISEE), Nagoya University. Details of these data-processing procedures are described by Shiokawa et al., (2012).

14.3 RESULTS AND DISCUSSION

We simultaneously observed EPBs using GPS ROTI measurements from large-density GPS networks in SEA as well as eastward winds using FPI, obtained from the KTB station. Based on the data analysis from 2011–2013 during the equinox season, there were 12 days of data that can be compared for EPB V_E and eastward wind velocity. This is based on the availability of data for both measurements. The data on the equinox season was selected because the occurrence of EPBs at that time was the most frequent compared to other seasons (Buhari et al., 2014).

Figure 14.1 and Figure 14.2 depict the temporal distribution, starting from 12 to 23 UT for both V_E EPB and the zonal neutral winds. The main focus of this study is the area in the gray column that shows the onset period of EPB occurrences. In that particular area, both EPB V_E and eastward winds were compared in terms of their pattern of changes. The results illustrate similar trends for both EPB V_E and eastward wind, even though the EPB

V_E tends to be slightly higher than the eastward wind velocity. In Figure 14.1, the EPB V_E obtained the highest peak of about ~200 m/s with eastward winds of ~150 m/s at the onset period on 5 April 2011. This result is slightly higher compared to a previous study carried out by Mukherjee (2003) in India which focused on monthly data during the equinox. They reported the V_E EPB peak to be about ~175 m/s.

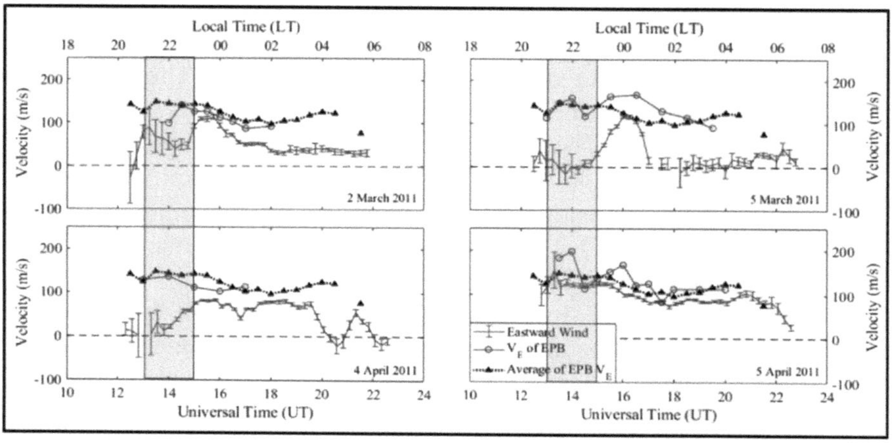

FIGURE 14.1 Temporal variation of both EPB V_E and eastward winds measured from GPS ROTI and FPI during the equinox in the year 2011.

Figure 14.2 reveals similar trends to the previous figure but differs in the peak values of EPB. The highest peak value (~200 m/s) of EPB V_E was on 8 March 2013, where the eastward wind velocity was also higher at ~190 m/s. This shows good agreement between both measurements, which is assumed to be in line with past studies that reported the contribution of eastward winds to the motion of EPB. From the results, 3 out of 8 days do not have data for the eastward wind velocity during some periods of time (in the gray column). This might be due to the sky conditions during those particular times. In that case, a comparison between the two measurements at the onset period could not be observed.

This current work is the preliminary analysis to observe the trends of both EPB V_E using GPS ROTI and eastward winds from FPI which cover the SEA sector. This study is in contrast with previous studies that conducted the investigation in other sectors using different approaches. For future study, we plan to consider and report the details about the number of bubbles in representing the days which contribute to the average V_E of EPB.

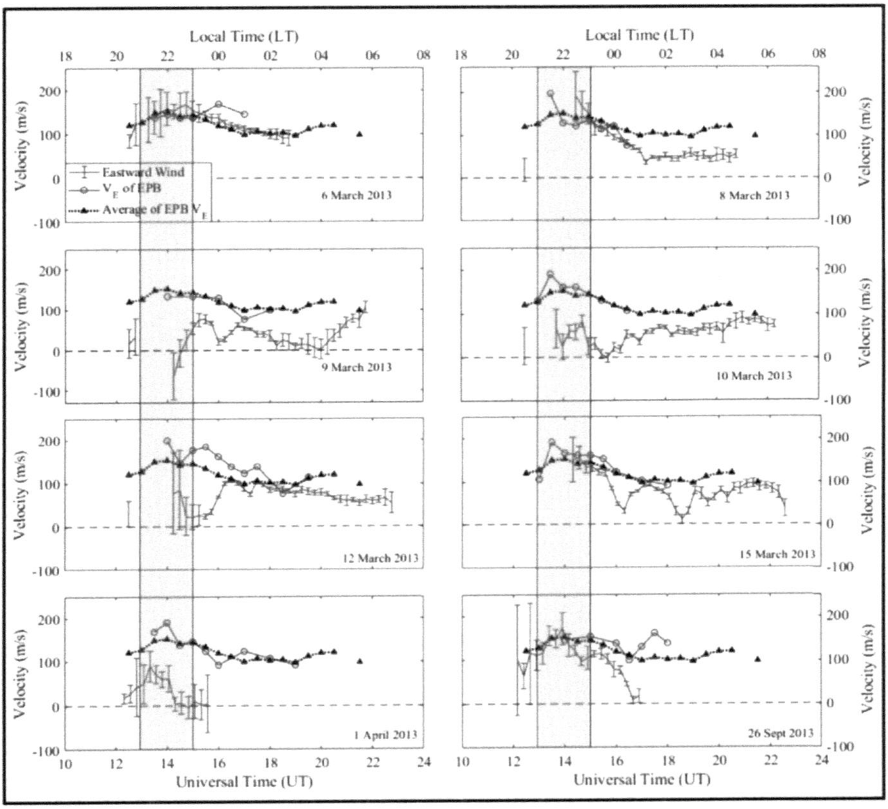

FIGURE 14.2 Temporal variation of both EPB V_E and eastward winds measured from GPS ROTI and FPI during the equinox in the year 2013.

14.4 CONCLUSION

In this current work, the EPB that was estimated from GPS ROTI keogram in SEA and eastward winds from FPI during the equinox in the period of 2011–2013 was investigated. The results obtained show similar trends for both V_E EPB and eastward wind, even though the EPB V_E tends to be slightly higher than the eastward wind velocity. The highest peak value reported from three years of data observation is about ~200 m/s with the highest eastward winds of ~150 to ~190 m/s. This work agrees with past studies that the eastward winds play an important role in drifting the EPB.

ACKNOWLEDGMENT

We would like to extend our appreciation to MyRTKnet through the link (http://www.rtknet.gov.my/) as a source of request for GPS data due to the data policy of the Department of Survey and Mapping Malaysia, JUPEM. Other supported GPS data (SuGAr and IGS networks) were obtained from http://sopac.ucsd.edu/. This work was supported by Universiti Sains Malaysia, Short-Term Grant with Project No: 304/PFIZIK/6315730. The availability of FPI data was checked from http://stdb2.isee.Nagoya.u.ac.jp/omti/index.html, and gratitude is also extended to the Institute for Space-Earth Environmental Research (ISEE), Nagoya University, for providing the FPI data at Kototabang.

KEYWORDS

- **eastward winds**
- **equatorial plasma bubble**
- **equatorial plasma bubbles**
- **optical mesosphere thermosphere imagers**
- **Southeast Asia**
- **total electron content**
- **zonal drift velocity**

REFERENCES

Bhattacharyya, A. (2022). Equatorial plasma bubbles: A review. *Journal of Atmosphere MDPI, 13*, 1637.

Buhari, S. M., Abdullah, M., Hasbi, A. M., Otsuka, Y., Tokoyama, T., Nishioka, M., & Tsugawa, T. (2014). Continuous generation and two-dimensional structure of equatorial plasma bubbles observed by high-density GPS receivers in Southeast Asia. *Journal of Geophysical Research, 119*, 10569–10580.

Buhari, S. M., Abdullah, M., Yokoyama, T., Otsuka, Y., Nishioka, M., & Hasbi, A. M. (2017). Climatology of successive equatorial plasma bubbles observed by GPS ROTI over Malaysia. *Journal of Geophysical Research: Space Physics, 122*, A023202.

Chapagain, N. P. (2016). Ionospheric plasma drift and neutral winds modeling. *Research Journal of Physical Sciences, 4*(7), 5–10.

Chapagain, N. P., Taylor, M. J., Nielsen, K., & Jarvis, M. J. (2011). Airglow observations and modeling of F region depletion zonal velocities over Christmas Island. *Journal of Geophysical Research, 116*, A02301.

Gurav, O. B., Sharma, A. K., Ghodpage, R. N., Nade, D. P., Chavan, G. A., Gaikwad, H. P., & Patil, P. T. (2018). Zonal drift velocity of equatorial plasma bubbles during the ascending phase of 24th solar cycle using all-sky imager over Kolhapur, India. *Journal of Geophysical Research: Space Physics, 123*, 10260–10282.

Mukherjee, G. K. (2003). Studies of equatorial-F region depletions and dynamics using multiple-wavelength nightglow imaging. *Journal of Atmospheric and Solar-Terrestrial Physics, 65*, 379–390.

Otsuka, Y., Shiokawa, K., Ogawa, T., Yokoyama, T., Yamamoto, M., & Fukao, S. (2014). The spatial relationship of equatorial plasma bubbles and field-aligned irregularities was observed with an all-sky airglow imager and the Equatorial Atmosphere Radar. *Geophysical Research Letters, 31*, L20802.

Paulino, I., Medeiros, A. F. D., Buriti, R. A., Takahashi, H., Humberto, J., Sobral, A., & Gobbi, D. (2011). Plasma bubble zonal drift characteristics observed by airglow images over Brazilian tropical region. *Brazilian Journal of Geophysics, 29*(2), 239–246.

Sarudin, I., Hamid, N. S. A., Abdullah, M., & Buhari, S. M. (2017). Investigation of zonal velocity of equatorial plasma bubbles (EPBs) by using GPS data. *Journal of Physics: Conference Series, 852*(1), 1–5. UK: IOP Publishing.

Sarudin, I., Hamid, N. S. A., Abdullah, M., Buhari, S. M., Shiokawa, K., Otsuka, Y., & Yatini, C. Y. (2020). Equatorial plasma bubble zonal drift velocity variations in response to season, local time, and solar activity across Southeast Asia. *Journal of Geophysical Research: Space Physics, 125*, e2019JA027521.

Shiokawa, K., Otsuka, Y., Oyama, S., Nozawa, S., Satoh, M., & Katoh, Y., et al. (2012). Development of low-cost sky-scanning Fabry-Perot interferometer for airglow and auroral studies. *Earth, Planets and Space, 64*, 1033–1046.

Yokoyama, T., & Fukao, S. (2006). Upwelling backscatter plumes in the growth phase of equatorial spread F observed with the Equatorial Atmosphere Radar. *Journal of Geophysical Research, 33*, L08104.

PART IV

Environment, Health, and Sustainable Development

CHAPTER 15

Indoor Air Quality (IAQ) and Particle Dispersion in Healthcare Facilities: A Case Study and Systematic Review

ERMIE ABDULL HAMID, SUBKI AHMAD, ANUAR ITHNIN, and KHAIRUNNISA BAHAMAN

Environmental Health and Industrial Safety Program, Faculty of Health Sciences, Universiti Kebangsaan Malaysia, Jalan Raja Muda Abdul Aziz, Kuala Lumpur, Malaysia

ABSTRACT

Indoor air is becoming an increasingly concerning health hazard, especially in hospitals that provide a therapeutic environment. This study sought to determine the IAQ of selected medical wards in the University Hospital at Klang Valley, Malaysia. We measured physical parameters (temperature, relative humidity, and air movement), chemical components (CH_2O, CO_2, CO, TVOC, and PM_{10}), and biological indicators (bacterial and fungal counts) following the Industry Code of Practice on Indoor Air Quality 2010 (ICOP IAQ 2010) in the morning and afternoon using relevant equipment. Data were analyzed with One-Way ANOVA using Statistical Package for The Social Sciences Version 25 (SPSS25). The study revealed significant differences ($p<0.05$) in temperature, RH, CO, CO_2, PM_{10}, CH_2O, TVOC, air movement (morning), and total bacterial count (morning) between selected medical wards, while no differences were found in fungal concentrations. All selected medical wards are in compliance with the recommended limits by ICOP IAQ (2010) except for temperature, RH, air movement, CO_2, and CH_2O. Therefore, a good ventilation system and improvements in safe work

Engineering Design and Technical Applications of Physical Science: An Integrated Approach for Sustainable Development. Chin Hua Chia, Sarani Zakaria, Ruey Shan Chen, & Md. Shahariar Chowdhury (Eds.)

practices need to be implemented to lower the exposure risk of indoor air contaminants among hospital workers.

15.1 INTRODUCTION

The complex hospital environment requires special attention to ensure healthy indoor air quality (IAQ). The diversity of facilities and occupants makes the complex hospital environment (Hiwar, Marco-Felipe, Farag, Louise, Stephanie, & Catherine, 2021; Ghanizadeh, & Godini, 2018) unlike that of any other residential and industrial buildings. Achieving good hospital IAQ is a challenge as hospital air quality is compromised by pollutants such as physical factors (gas), particulate matter, and microbiological contaminants, which can lead to patient infections and negatively affect hospital employees (Huiyi, Keng, Othman, Hong et al., 2022; Ibrahim, Samsudin, Ishak, & Sathasivam, 2022).

It is common knowledge that occupants are one of the factors that interfere with air quality in indoor environments (Huiyi, Keng, Othman, Hong et al., 2022; Vardoulakis, Giagloglou, Steinle, Davis et al., 2020; Slezakova, Peixoto, Pereira, & Morais, 2018). An increasing number of patients admitted to the hospital can affect the transmission and distribution of infectious agents. Previously, in 2020, Tran et al. (2020) stated that humans can create favorable conditions for the development of millions of molds, fungi, pollen, spores, bacteria, viruses, and insects, such as dust mites and cockroaches. This microbial spread occurs mostly via large droplets, direct contact with infectious material, or through contact with inanimate objects contaminated by infectious materials, making the indoor environment a great concern for public health (Tungjai & Kubaha, 2017). Furthermore, activities in the hospital take place around the clock, 7 days a week, causing chemical and biological pollutants and pathogenic agents to be continuously present, which may cause problems in the indoor environment.

Meanwhile, according to the World Health Organization (2021), indoor air pollution (IAP) is responsible for the deaths of 3.8 million people annually. It is well known that the existence of pollutants in the air of indoor environments in healthcare facilities, such as a variety of particles, chemicals, and gaseous contaminants like carbon dioxide (CO_2), carbon monoxide (CO), formaldehyde (CH_2O), total volatile organic compounds (VOCs), respirable suspended particulates, radon, glutaraldehyde ($C_5H_8O_2$), nitrous oxide (N_2O), latex allergens, and total bacterial count can highly influence the air quality, which has negative impacts on the human body (Tran, Duckshin, & Lee, 2020; Marco, Gaetano, & Stefano, 2019; Basinska, Michatkiewicz, & Ratajczak, 2019). For example, in hospitals, cleaning solutions and detergents are used to reduce the risk of infection, but they are associated with high levels of exposure to total volatile organic compounds (TVOCs) in healthcare settings (Fend-Chico, Friesen, Stefaniak, Henneberger et al., 2018). As pointed out by other studies, formaldehyde exposure may also increase the risk of cancer (Vardoulakis, Giagloglou, Steinbeck's, Davis et al., 2020; Heaton, Hurst, Amiri, Lungu, & Oh, 2019), as well as TVOC (Chuanqi & Qianqian, 2020). Additionally, high indoor CO_2 levels can cause allergies, serious medical conditions, and death (Azuma, Kagi, Yanagi, & Osawa, 2018; Zhang, Wargocki, Lian, & Thyregod, 2017). Baures et al. (2018) have identified that the concentration of chemical compounds in hospital air is determined by healthcare activities and materials used in the hospital environment (Baurès, Blanchard, Mercier, Surget et al., 2018). Even Tang et al. (2020) also reviewed that the distributions of IAQ can vary throughout a hospital since medical activities and materials vary across different work areas.

At this point, ventilation in healthcare facilities is so important as it provides protection from variability of the indoor air contaminants: physical, chemical, and microbiological (Basinska, Michalkiewicz, & Ratajczak, 2019). This statement is also agreed by Khan et al. (2020), which stated that effective ventilation is at the core of the wellness standards of a healthcare facility. Besides, a few studies have also pointed out there was an association between ventilation rates with occupant health and performance (Huiyi, Keng, Othman, Hong et al., 2022; Zhong, Yuan, & Fleck, 2019). Ikhtiar et al. (2017) from his article stated that an inadequate ventilation system can lead to Sick Building Syndrome, which leads to eye and nose irritation, cough, dizziness, sore throat, and chest tightness. Therefore, improving ventilation systems in buildings is the key issue for providing better IAQ to the occupants and minimizing the possibility of health problems as a consequence (Huiyi, Keng, Othman, Hong et al., 2022; Chenari, Carrilho, & Silva, 2016).

As mentioned above, IAQ has caused increasing concern due to the adverse effects that it may have on human health. Many studies on this topic have shown both qualitative and quantitative IAQ variations through the years, underlining an increase in pollutants and their levels. To this aim, IAQ-related standards, regulations, and monitoring plans have been developed in several countries, including Malaysia, which is under ICOP IAQ (2010). Thus, the purpose of this study is to investigate the distributions and sources of biological and non-biological pollutants in four selected medical wards of the University Hospital in Klang Valley, Malaysia. The results of this study will be compared to the recommended limit of the Industry Code of Practice on Indoor Air Quality 2010 (ICOP IAQ, 2010).

15.2 METHOD

15.2.1 WALKTHROUGH INSPECTION

This study was conducted at Klang Valley University Hospital, involving medical wards, namely Medical Ward F, G, H, and J. In this study, we used three phases of investigations to perform IAQ monitoring. Phase 1 involved baseline information through walkthrough inspections of the wards and collecting information from literature reviews and previous studies. The advantage of the preliminary survey is to identify the sampling area and sources of poor IAQ in the hospital building. Therefore, four sampling points were selected, with each ward assigned one sampling point at the center of the ward following the ICOP IAQ 2010 (Brągoszewska, 2019).

15.2.2 MEASUREMENT OF PARAMETERS

Phase II involved measuring IAQ at the selected medical wards according to ICOP IAQ 2010 (Brągoszewska, 2019). The equipment used in this study was maintained and calibrated according to the Standard Operating Procedure and Quality Control before use. Following the ICOP IAQ 2010, the measurement of IAQ parameters was carried out for 8 hours with 4-time slots in the morning and afternoon only, as no permission was obtained for evening and night-time measurements. The IAQ physical parameters assessed are temperature, relative humidity (RH), and air movement by using EVM 3 (Shawcity Ltd., Oxfordshire, UK). The chemical parameters were also measured by the same method using Aeroqual Series 500 (Aeroqual Ltd., Auckland, New Zealand), and Tetra 3 Crowcon (Crowcon Detection Instruments Ltd, Abingdon, UK). Besides, Formaldemeter (ENMET, An-Nur Arbor, USA) was also used in this study to measure O_3, TVOC, CO, and CH_2O. Regarding the biological risk factors related to pollution, Colony Counter Galaxy 230 (WIGGENS GmbH, Straubenhardt, Germany) using the Trypticase Soy Agar (TSA) and Sabouraud Dextrose Agar (SDA) media exposure method is used to assess total bacterial and fungal counts. All microbial samples were incubated for 1 day at 37°C, while fungi were incubated for 3 days at 25°C. The growth of bacterial and fungal colonies was counted and calculated then converted as colony-forming unit/m³ (cfu/m³).

15.2.3 STATISTICAL ANALYSIS

Phase III is a statistical analysis involving data analysis by using Statistical Package for Social Sciences (SPSS) Version 25. The level of $p<0.05$ was set as significance in the study. Based on the data obtained, the statistical analysis, One-Way ANOVA was carried out under the stated specific objectives. These include determining the physical parameters (temperature, RH, and air movement), chemical parameters (CO, CO_2, CH_2O, PM_{10}, and TVOC), and biological parameters (total bacterial count and total fungal count) in the four selected medical wards of University Hospital at Klang Valley Malaysia. Then, all the data collected will be compared between the wards.

15.3 RESULTS AND DISCUSSION

15.3.1 RESULT

- **Overall IAQ in Selected Medical Wards:** The mean with standard deviation for physical (temperature, RH, air movement), chemical (CH_2O, CO_2, CO, TVOC, PM_{10}), and biological (Total bacterial count, TBC, and Total fungal count, TBF) parameters of all medical wards involved in this study were given in Table 15.1. The evaluation of air quality in all medical wards showed compliance with the requirements set by the ICOP 2010, except for the physical parameter which was temperature in Ward G (morning – 22.96 ± 1.0°C), and the chemical parameters which were RH reading in Ward F (morning–70.91 ± 3.50%), and both the values of CO_2 and formaldehyde in Ward J (afternoon–1127.75 ± 51.30 ppm) and (morning–0.17 ± 0.1 ppm). Although some parameters were above the ICOP standard limit, there was no large variation in biological parameters observed among the medical wards.
- **Comparison Data of IAQ Parameters between Medical Wards (F, G, H & J) in the Morning and Afternoon:** Tables 15.2, 15.3, and 15.4 summarize the mean differences by the Independent T-test of all physical (temperature, RH, air movement), chemical (CH_2O,

CO_2, CO, TVOC, PM_{10}), and biological parameters (TBC and TBF) in the morning and afternoon in medical wards. Overall, a total of three physical parameters were studied. The air movement reading in the afternoon ($p = 0.290$) was reported as a statistically significant mean difference, $p>0.05$. Temperature (morning, $p = 0.000$ & afternoon, $p = 0.000$), RH (morning, $p = 0.000$ & afternoon, $p = 0.000$), and air movement (morning, $p = 0.002$) were found to be statistically significant mean differences ($p<0.05$) respectively, between Wards F, G, H, and J (Table 15.2). Ward F recorded the highest RH reading in the morning ($67.358 \pm 1.801\%$) among the medical wards, exceeding the standards limit specified by the ICOP IAQ (2010). Meanwhile, the air movement reading in the morning for all medical wards except Ward J (0.18 ± 0.1 ms^{-1}) showed lower readings than the acceptable limit, which was set between 0.15 and 0.50 ms^{-1}. Air movement in Ward F (0.10 ± 0.1 ms^{-1}), Ward G (0.12 ± 0.1 ms^{-1}), and Ward H (0.10 ± 0.1 ms^{-1}). Table 15.3 showed there were statistically significant mean differences ($p<0.05$) in the morning and afternoon for CO (morning, $p = 0.009$ & afternoon, $p = 0.000$), CO_2 (morning, $p = 0.000$ & afternoon, $p = 0.000$), CH_2O (morning, $p = 0.000$ & afternoon, $p = 0.000$), PM_{10} (morning, $p = 0.000$ & afternoon, $p = 0.000$), and TVOC (morning, $p = 0.000$ & afternoon, $p = 0.031$). The concentration of CO_2 in Ward J ($1,127.75 \pm 51.3$ ppm) was much higher in the afternoon compared to other medical wards, and it also recorded as the highest level of CH_2O (morning, 0.17 ± 0.1 ppm) in the indoor environment. As for the biological parameters (TBC & TFC), TBC reading (morning, $p = 0.005$) between medical wards was reported as a statistically significant mean difference, $p<0.05$ (Table 15.4). Meanwhile, TFC reading (morning, $p = 0.927$ & afternoon, $p = 0.156$) and TBC (afternoon, $p = 0.108$) were found to be statistically not significant mean differences, $p>0.05$. In this study, Ward F showed the highest concentrations of bacteria in the morning (233 ± 38.2 cfu/m^3).

- **Comparison Data of IAQ Parameters Between Medical Wards (F, G, H & J) in the Morning and Afternoon with ICOP IAQ (2010) Standard:** To compare data of IAQ parameters (physical, chemical & biological) between medical wards (F, G, H, & J) to the ICOP standard, ICOP IAQ (2010), mean difference by the One-Sample T-Test was used and results were recorded in Table 15.5, 15.6, and 15.7. For physical parameters, RH reading showed a significant mean difference ($p<0.05$) with the standard values of ICOP IAQ (2010)

TABLE 15.1 Air Quality Profiles of Medical Wards (F, G, H & J) In the Studied University Hospital

Sites	Temperature (°C)	Rh (%)	AM (ms^{-1})	CO (ppm)	CO_2 (ppm)	CH_2O (ppm)	PM10 (mg/m^3)	TVOC (ppm)	Total Bacterial Counts (CFU/m^3)	Total Fungal Counts (CFU/m^3)
	mean ± SD	mean ± SD	mean ± SD	mean ± SD	mean ± SD	mean ± SD	mean ± SD	mean ± SD	mean ± SD	mean ± SD
Morning reading										
F	23.39 ± 0.4	70.91 ± 3.7	0.11 ± 0.0	0.02 ± 0.0	766.00 ± 61.3	0.05 ± 0.0	0.03 ± 0.0	0.25 ± 0.5	233 ± 38.2	23 ± 4.2
G	22.96 ± 1.0	65.32 ± 4.7	0.12 ± 0.0	0.02 ± 0.0	749.42 ± 66.8	0.03 ± 0.0	0.010 ± 0.0	0.00 ± 0.0	116 ± 59.4	23 ± 1.4
H	25.50 ± 0.5	65.53 ± 1.9	0.11 ± 0.0	0.29 ± 0.4	825.17 ± 43.0	0.08 ± 0.1	0.029 ± 0.0	0.75 ± 0.5	187 ± 26.9	24 ± 2.8
J	24.83 ± 0.6	63.86 ± 2.0	0.18 ± 0.1	0.19 ± 0.2	953.50 ± 12.2	0.17 ± 0.1	0.023 ± 0.0	1.01 ± 0.0	156 ± 14.1	25 ± 7.1
Afternoon reading										
F	23.68 ± 0.6	67.44 ± 1.3	0.14 ± 0.1	0.00 ± 0.0	781.92 ± 107.3	0.01 ± 0.0	0.020 ± 0.0	0.33 ± 0.5	169 ± 106.1	22 ± 2.8
G	22.87 ± 1.5	60.31 ± 2.1	0.18 ± 0.1	0.02 ± 0.0	812.50 ± 82.6	0.002 ± 0.005	0.008 ± 0.001	0.50 ± 0.5	086 ± 056.6	24 ± 2.8
H	25.47 ± 0.5	64.19 ± 1.0	0.15 ± 0.1	0.25 ± 0.3	965.58 ± 29.7	0.06 ± 0.0	0.022 ± 0.001	0.50 ± 0.5	193 ± 026.9	25 ± 1.4
J	25.33 ± 0.7	58.76 ± 0.9	0.17 ± 0.1	0.03 ± 0.1	1,127.75 ± 51.3	0.04 ± 0.0	0.015 ± 0.0	0.92 ± 0.3	185 ± 012.7	26 ± 2.8
Standard Value (ICOP IAQ 2010)	**23.0–26.0°C**	**40–70%**	**0.15–0.50 ms^{-1}**	**10 ppm**	**1000 ppm**	**0.1 ppm**	**0.15 mg/m^3**	**3.0 ppm**	**500 CFU/m^3**	**1000 CFU/m^3**

Temp, temperature; RH, relative humidity; AM, air movement; CO, carbon monoxide; CO_2, carbon dioxide; CH_2O, formaldehyde; PM_{10}, particulate matter; TVOC, total volatile organic compounds.

TABLE 15.2 Comparison of Temperature, Relative Humidity and Air Movement Between the Selected Medical Wards

Parameter	Session	Medical Ward	Mean ± STD	F	p*
Temperature (°C)	Morning	F	23.39 ± 0.4	39.528	0.000*
		G	22.96 ± 1.0		
		H	25.50 ± 0.5		
		J	24.83 ± 0.6		
	Afternoon	F	23.68 ± 0.6	23.885	0.000*
		G	22.87 ± 1.5		
		H	25.47 ± 0.5		
		J	25.33 ± 0.7		
Relative Humidity (%)	Morning	F	70.91 ± 3.5	11.041	0.000*
		G	65.32 ± 4.7		
		H	65.53 ± 1.9		
		J	63.86 ± 2.0		
	Afternoon	F	67.44 ± 1.3	97.844	0.000*
		G	60.31 ± 2.1		
		H	64.19 ± 1.0		
		J	58.76 ± 0.9		
Air Movement (m/s)	Morning	F	0.11 ± 0.0	14.586	0.002*
		G	0.12 ± 0.0		
		H	0.11 ± 0.0		
		J	0.18 ± 0.1		
	Afternoon	F	0.14 ± 0.1	3.750	0.290
		G	0.18 ± 0.1		
		H	0.15 ± 0.1		
		J	0.17 ± 0.1		

Significant difference ($p<0.05$)

TABLE 15.3 Comparison of Carbon Monoxide, Carbon Dioxide, Formaldehyde, Particulate Matter (PM_{10}) and Total Volatile Organic Compounds (TVOC) Between the Selected Medical Wards

Parameter	Session	Medical Ward	Mean ± SD	F	p*
Carbon Monoxide (ppm)	Morning	F	0.02 ± 0.0	4.636	0.009*
		G	0.02 ± 0.0		
		H	0.29 ± 0.4		
		J	0.19 ± 0.2		
	Afternoon	F	0.00 ± 0.0	19.531	0.000*
		G	0.02 ± 0.0		
		H	0.25 ± 0.3		
		J	0.03 ± 0.1		
Carbon Dioxide (ppm)	Morning	F	766.00 ± 61.3	44.067	0.000*
		G	749.42 ± 66.8		
		H	825.17 ± 43.0		
		J	953.50 ± 12.2		
	Afternoon	F	781.92 ± 107.3	55.533	0.000*
		G	812.50 ± 82.6		
		H	965.58 ± 29.7		
		J	1,127.75 ± 51.3		
Formaldehyde (ppm)	Morning	F	0.05 ± 0.0	17.856	0.000*
		G	0.03 ± 0.0		
		H	0.08 ± 0.1		
		J	0.17 ± 0.1		

TABLE 15.3 *(Continued)*

Parameter	Session	Medical Ward	Mean ± SD	F	p*
	Afternoon	F	0.01 ± 0.0	27.816	0.000*
		G	0.002 ± 0.005		
		H	0.06 ± 0.0		
		J	0.04 ± 0.0		
		F	0.030 ± 0.0		
	Morning	G	0.010 ± 0.0	26.121	0.000*
		H	0.029 ± 0.0		
PM_{10} (mg/m^3)		J	0.023 ± 0.0		
		F	0.020 ± 0.0		
	Afternoon	G	0.008 ± 0.0	156.132	0.000*
		H	0.022 ± 0.001		
		J	0.015 ± 0.0		
		F	0.25 ± 0.5		
	Morning	G	0.00 ± 0.0	29.375	0.000*
		H	0.75 ± 0.5		
TVOC		J	1.00 ± 0.0		
(ppm)		F	0.33 ± 0.5		
	Afternoon	G	0.50 ± 0.5	8.869	0.031*
		H	0.50 ± 0.5		
		J	0.92 ± 0.3		

*Significant difference ($p<0.05$).

TABLE 15.4 Comparison of the Total Bacterial Count and Total Fungal Count Between the Selected Medical Wards

Parameter	Session	Medical Ward	Mean ± SD	F	p^*
Total Bacterial Count (CFU/m^3)	Morning	F	233 ± 38.2	12.741	0.005*
		G	116 ± 59.4		
		H	187 ± 26.9		
		J	156 ± 14.1		
	Afternoon	F	169 ± 106.1	6.071	0.108
		G	86 ± 56.6		
		H	193 ± 26.9		
		J	185 ± 12.7		
Total Fungi Count (CFU/m^3)	Morning	F	23 ± 4.2	0.463	0.927
		G	23 ± 1.4		
		H	24 ± 2.8		
		J	25 ± 7.1		
	Afternoon	F	22 ± 2.8	5.222	0.156
		G	24 ± 2.8		
		H	25 ± 1.4		
		J	26 ± 2.8		

Significant difference ($p<0.05$)

TABLE 15.5 Comparison of Medical Wards (F, G, H, and J) Physical Parameters in the Morning and Afternoon with the Standard Value of ICOP IAQ (2010)

Location	Parameter	Mean ± SD		t		p	
		Morning	Afternoon	Morning	Afternoon	Morning	Afternoon
Ward F	Temperature (oC)	23.391 ± 0.405	23.683 ± 0.590	–5.462	–2.767	0.012*	0.070
	Relative Humidity (%)	70.908 ± 3.756	67.441 ± 1.069	8.470	23.272	0.003*	0.000*
	Air Movement (ms^{-1})	0.108 ± 0.002	0.141 ± 0.050	–186.568	–7.533	0.000*	0.005*
Ward G	Temperature (oC)	22.958 ± 1.092	22.866 ± 1.555	–2.821	–2.100	0.067	0.127
	Relative Humidity (%)	65.316 ± 5.025	60.308 ± 1.799	4.106	5.901	0.026*	0.010*
	Air Movement (ms^{-1})	0.116 ± 0.033	0.183 ± 0.057	–12.800	–5.081	0.001*	0.015*
Ward H	Temperature (oC)	25.500 ± 0.503	25.466 ± 0.579	3.974	3.338	0.028*	0.044*
	Relative Humidity (%)	65.525 ± 1.957	64.191 ± 0.675	10.754	27.228	0.002*	0.000*
	Air Movement (ms^{-1})	0.105 ± 0.001	0.150 ± 0.043	–233.827	–8.366	0.000*	0.004*
Ward J	Temperature (oC)	24.825 ± 0.602	25.333 ± 0.771	1.080	2.160	0.359	0.120
	Relative Humidity (%)	63.858 ± 2.024	58.758 ± 0.753	8.750	9.976	0.003*	0.002*
	Air Movement (ms^{-1})	0.183 ± 0.103	0.166 ± 0.060	–2.830	–5.368	0.066	0.013*

**Significant value is at $p<0.05$*

Standard Value ICOP IAQ (2010): Temperature, 23.0–26.0°C; Relative humidity, 40–70%; Air movement, 0.15–0.50 ms^{-1}.

TABLE 15.6 Comparison of Medical Wards (F, G, H, and J) Chemical Parameters in The Morning and Afternoon with The Standard Value of ICOP IAQ (2010)

Location	Parameter	Mean ± SD		t		p	
		Morning	Afternoon	Morning	Afternoon	Morning	Afternoon
Ward F	Formaldehyde (ppm)	0.050 ± 0.033	0.011 ± 0.006	–2.990	–27.678	0.058	0.000*
	Carbon Dioxide (ppm)	766.000 ± 65.277	781.916 ± 118.329	–7.169	–3.686	0.006*	0.035*
	Carbon Monoxide (ppm)	0.016 ± 0.033	0.00 ± 0.0	–599.000	0	0.000*	0
	PM_{10} (mg/m^3)	0.029 ± 0.007	0.019 ± 0.002	–30.189	–129.900	0.000*	0.000*
	TVOC (ppm)	0.250 ± 0.500	0.333 ± 0.471	–11.000	–11.314	0.002*	0.001*
Ward G	Formaldehyde (ppm)	0.027 ± 0.031	0.002 ± 0.005	–4.562	–39.000	0.020*	0.000*
	Carbon Dioxide (ppm)	749.416 ± 37.574	811.500 ± 91.266	–13.338	–4.131	0.001*	0.026*
	Carbon Monoxide (ppm)	0.016 ± 0.033	0.016 ± 0.019	–599.000	–1037.49	0.000*	0.000*
	PM_{10} (mg/m^3)	0.010 ± 0.002	0.008 ± 0.001	–134.765	–188.723	0.000*	0.000*
	TVOC (ppm)	0.00 ± 0.0	0.500 ± 0.577	0	–8.660	0	0.003*
Ward H	Formaldehyde (ppm)	0.075 ± 0.072	0.055 ± 0.025	–.663	–3.576	0.555	0.037*
	Carbon Dioxide (ppm)	825.166 ± 33.597	965.583 ± 21.129	–10.408	–3.258	0.002*	0.047*
	Carbon Monoxide (ppm)	0.291 ± 0.395	0.250 ± 0.261	–49.070	–74.496	0.000*	0.000*
	PM_{10} (mg/m^3)	0.029 ± 0.006	0.022 ± 0.001	–37.961	–143.836	0.000*	0.000*
	TVOC (ppm)	0.750 ± 0.500	0.500 ± 0.577	–9.000	–8.660	0.003*	0.003*
Ward J	Formaldehyde (ppm)	0.165 ± 0.060	0.038 ± 0.012	2.135	–9.773	0.122	0.002*
	Carbon Dioxide (ppm)	953.500 ± 11.213	1127.750 ± 52.353	–8.294	4.880	0.004*	0.016*
	Carbon Monoxide (ppm)	0.191 ± 0.191	0.033 ± 0.047	–102.574	–422.850	0.000*	0.000*
	PM_{10} (mg/m^3)	0.023 ± 0.008	0.015 ± 0.001	–29.419	–145.406	0.000*	0.000*
	TVOC (ppm)	1.083 ± 0.019	0.916 ± 0.166	–199.186	–25.000	0.000*	0.000*

**Significant value is at $p<0.05$*

Standard Value ICOP IAQ (2010): Formaldehyde, 0.1 ppm; Carbon Dioxide, 1000 ppm; Carbon Monoxide, 10 ppm; PM_{10}, 0.15 mg/m^3.

TABLE 15.7 Comparison of Medical Wards (F, G, H, and J) Biological Parameters in the Morning and Afternoon with the Standard Value of ICOP IAQ (2010)

Location		Mean ± SD		t		p	
		Morning	Afternoon	Morning	Afternoon	Morning	Afternoon
Ward F	Total bacterial count (cfu/m^3)	233.000 ± 31.176	169.000 ± 86.602	–17.128	–7.644	0.000*	0.005*
	Total fungal count (cfu/m^3)	23.000 ± 3.464	22.000 ± 2.309	–564.071	–846.973	0.000*	0.000*
Ward G	Total bacterial count (cfu/m^3)	116.000 ± 48.497	86.000 ± 46.188	–15.836	–17.927	0.001*	0.000*
	Total fungal count (cfu/m^3)	23.000 ± 1.154	24.000 ± 2.309	–1692.214	–845.241	0.000*	0.000*
Ward H	Total bacterial count (cfu/m^3)	187.000 ± 21.939	193.000 ± 21.939	–28.533	–27.986	0.000*	0.000*
	Total fungal count (cfu/m^3)	24.000 ± 2.309	25.000 ± 1.154	–845.241	–1688.750	0.000*	0.000*
Ward J	Total bacterial count (cfu/m^3)	156.000 ± 11.547	185.000 ± 10.392	–59.583	–60.622	0.000*	0.000*
	Total fungal count (cfu/m^3)	25.000 ± 5.773	26.000 ± 2.309	–337.750	–843.509	0.000*	0.000*

**Significant value is at $p<0.05$*

Standard Value ICOP IAQ (2010): Total bacterial count, 500 cfu/m^3; Total fungal count, 1000 cfu/m^3.

in the morning and afternoon for all the medical wards (Ward F, G, H & J) (Table 15.5). All the relative humidity readings in Ward F (morning, 70.908 ± 3.756% & afternoon, 67.441 ± 1.069%), Ward G (morning, 65.316 ± 5.025% & afternoon, 60.308 ± 1.799%), Ward H (morning, 65.525 ± 1.957% & afternoon, 64.191 ± 0.675%), and Ward J (morning, 63.858 ± 2.024% & afternoon, 58.758 ± 0.753%), were found to be within the ICOP standard range, respectively except Ward F. Meanwhile, PM_{10} and CO_2 readings were the chemical parameters which recorded to have a significant mean difference ($p<0.05$) with the standard values of ICOP IAQ (2010) in the morning and afternoon for all medical wards (Table 15.6). The PM_{10} and CO_2 readings in medical wards were within the ICOP standard range in the morning and afternoon, respectively; Ward F (morning, 0.029 ± 0.007 mg/m^3 and afternoon, 0.019 ± 0.002 mg/m^3), Ward G (morning, 0.010 ± 0.002 mg/m^3 and afternoon, 0.008 ± 0.001 mg/m^3), Ward H (morning, 0.029 ± 0.006 mg/m^3 and afternoon, 0.022 ± 0.001 mg/m^3), and Ward J (morning, 0.023 ± 0.008 mg/m^3 and afternoon, 0.015 ± 0.001 mg/m^3). For CO_2, surprisingly Ward J (evening, 1127.750 ± 52.353 ppm) has the highest reading and even exceeded the standard of ICOP IAQ (2010) range, compared with other medical wards, Ward F (morning, 766.000 ± 65.277 ppm and afternoon, 781.916 ± 118.329 ppm), Ward G (morning, 749.416 ± 37.574 ppm and afternoon, 811.500 ± 91.266 ppm), Ward H (morning 825.166 ± 33.597 ppm and afternoon, 965.583 ± 21.129 ppm) which remained CO_2 reading below the ICOP IAQ (2010) standard range. For biological parameters (TBC & TFC), TFC was found to have a significant mean difference ($p<0.05$) with the standard values of ICOP IAQ (2010) in the morning and afternoon for all the medical wards (Ward F, G, H & J) (Table 15.7). All readings for TFC were recorded under the ICOP IAQ (2010) standard range; Ward F (morning, 23.000 ± 3.464 cfu/m^3, evening, 22.000 ± 2.309 cfu/m^3), Ward G (morning, 23.000 ± 1.154 cfu/m^3, evening, 24.000 ± 2.309 CFU/m^3), Ward H (morning, 24.000 ± 2.309 cfu/m^3, evening, 25.000 ± 1.154 cfu/m^3), Ward J (morning, 25.000 ± 5.773 cfu/m^3, evening, 26.000 ± 2.309 cfu/m^3) respectively.

15.4 DISCUSSION

Indoor air quality parameters were measured from four different medical wards in this study. This study revealed that temperature (morning &

afternoon) and air movement (morning) readings were lower in Ward G than the acceptable range of ICOP IAQ (2010). But the temperature condition in Ward F was just on the zone border (23.39 ± 0.4°C; 23.68 ± 0.6°C), whereas air movement readings both morning and afternoon were just below the recommended standard (Table 15.1), agreeing in this sense with a prior study finding, which showed an effect of air movement on temperature (Liu, Ma, Cao, Meng, & He, 2018; Brągoszewska, 2019). Central air-conditioning that has been used in the hospital to maintain the temperature might have contributed to the lower temperature in Ward G compared to other wards in this hospital. This was because, in Ward G, they were using auto-locked doors and had a limited number of visitors, with two visitors per patient at one time. The presence of humans can increase the heat, as well as the frequency of opening the doors and the total floor area, which can contribute to the differences in temperature between the wards (Ikhtiar, Alzad, & Paramita, 2017; Aguilera, Andersen, & Toftum, 2019; Onmek, Kongcharoen, Singtong, Penjumrus, & Junnoo, 2020). Furthermore, this can also be seen in Table 15.1 where CO_2 levels were recorded as the lowest in Ward G (morning), proving fewer human activities in Ward G. In this study, we also noted low occupancy and temperature in Ward G, but both were high in Ward J, whereas in Ward H, there was high temperature but low occupancy compared to Ward J. This indicated that the temperature variations at the sampling sites were influenced by the occupancy factor (Aguilera, Andersen, & Toftum, 2019), but this variation was not considered if the air movement reading is not within the standard limit specified by the ICOP IAQ (2010) in Malaysia.

For humidity levels, Ward F was found to be slightly higher (70.91±3.5%) in the morning compared to other selected medical wards and the acceptable range of ICOP IAQ (2010). Remediation of this condition should be given priority as the humidity factor can influence indoor air quality (Bozic, Llic, & Llic, 2019) by promoting the growth of microbes (Ikhtiar, Alzad, & Paramita, 2017; Bayani, Mirhoseini, Koolivand, Sarlak et al., 2021; Larrey, Laryea, Kpordze, & Saba, 2020). This finding aligns with our results, which showed that the total bacterial count in Ward F was the highest among the selected medical wards in the morning (Figure 15.1). The temperature in Ward F was also observed to be on the border of the ICOP standard zone. This is consistent with previous studies where a temperature drop affected humidity levels to increase (Valsson, & Bharat, 2011), as shown in Table 15.1. Additionally, the temperature is known to be one of the environmental factors that affect bacterial growth, as indicated by Amaliyah (2017).

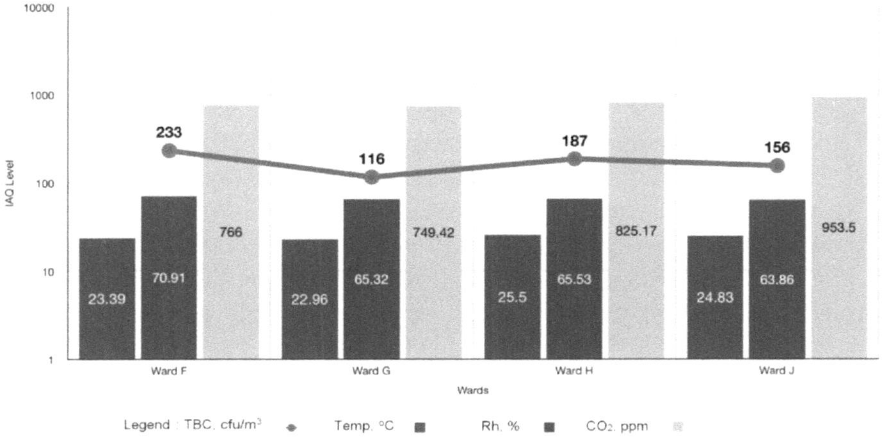

FIGURE 15.1 The influence of temperature, RH, and CO_2 reading for bacterial growth at selected Medical Ward (F, G, H & J) in the morning.

Based on One-Way ANOVA, there were significant mean differences in temperature, relative humidity, air movement (morning), carbon dioxide, and bacterial count (morning) between all selected medical wards ($p<0.05$), while the trends were the opposite (not significant) for air movement and bacterial count in the afternoon (Tables 15.2–15.4). Due to this fact, there was a correlation between temperature readings with relative humidity, carbon dioxide, and bacterial growth but not with air movement. Therefore, there is a need to control the air speeds as it can affect the user's thermal comfort. If air movement increases, a person starts to get cold because of the evaporation process of the water on the body's surface (Ozdamar, & Umarogullari, 2018). The ambient air flow rate should be kept within the acceptable range of ICOP IAQ. This is what happened in Ward J, where the ceiling fan was switched on during sampling and the result analysis showed that the air movement of Ward J complies with the standard of ICOP 2010 either in the morning or in the evening.

Gas and chemical pollutants are air pollutants that can be found in indoor spaces. Table 15.1 shows the comparison of concentration levels for chemical pollutants (CO, CO_2, CH_2O, PM_{10}, and TVOC) between Wards F, G, H, and J in the morning and afternoon. The data released by this study shows a higher level of CO_2 in the afternoon compared to the morning session for all selected medical wards (Table 15.1). The highest CO_2 level was observed in Ward J (1,127.75 ± 51.3 ppm), which exceeded the legislated maximum. The results suggest that high occupancy might be the cause of the high

CO_2 concentration in Ward J, and this is related to the visiting hours that open in the afternoon. This finding is also supported by the higher bacterial counts (Figure 15.2) in the afternoon (185 ± 12.7 cfu/m^3) compared to the morning (156 ± 14.1 cfu/m^3), as people are the main source of CO_2 (Pipat, Netchanok, Noppanun, Aranya, Pornpimol, & Susan, 2019; Fu Shaw, Chen, Chen, Wu et al., 2018; Bragoszewska, Mainka, Pastuszka, Lizonczyk, & Desta, 2018) and bacteria (Ikhtiar, Alzad, & Paramita, 2017; Bragoszewska, Mainka, Pastuszka, Lizonczyk, & Desta, 2018; D'Amico, Montagna, & Caggiano, 2019; Stanaszek-Tomal, 2020). Therefore, the statistical results from the One-Way ANOVA mentioned above for CO_2 and bacterial count have confirmed a positive relationship between CO_2 readings and bacterial growth (Ana, Isabel, Maria, Cristina et al., 2019).

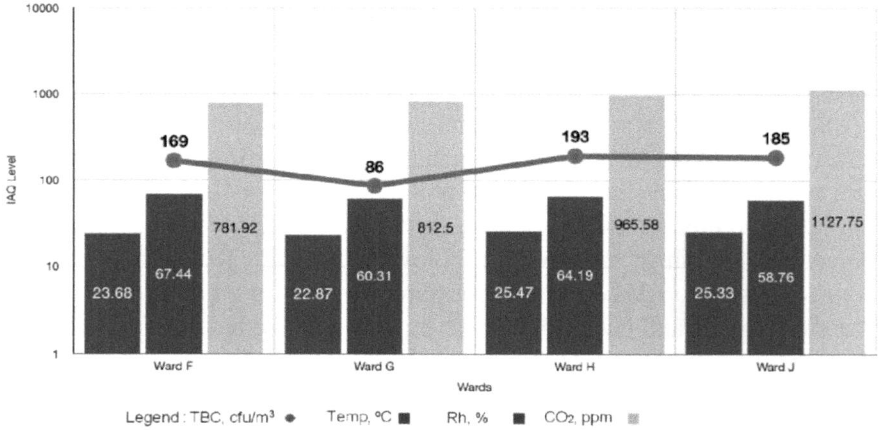

FIGURE 15.2 The influence of temperature, RH, and CO_2 reading for bacterial growth at selected medical WARD (F, G, H & J) in the afternoon.

On the other hand, Ward J was also presented as the highest concentration of CH_2O among all selected medical wards especially during working hours in the morning with 0.17±0.1 ppm, exceeding the ICOP IAQ (2010) standard of 0.1 ppm. This condition may be related to the nephrology services provided by Ward J, where hospital workers were exposed to CH_2O during their preparation of formalin solution, processing tissue specimens, and disinfecting hemodialysis machines (Adamovic, Cepic, Adamovic, Stosic et al., 2021). This finding was consistent with the One-Way ANOVA analysis, where all selected medical wards were found to have a significant mean difference ($p<0.05$) for CH_2O in the morning and afternoon (Table 15.3). We also discovered that the total bacterial count was lower in the morning (156 ± 14.1 cfu/m^3) compared to the afternoon (185 ± 12.7 cfu/m^3) in Ward J. Therefore, based on the observed positive correlation between CH_2O concentration and total bacterial count, it is possible that CH_2O data can be used as an indicator of decreased concentration of microorganism growth (Figure 15.1 & 15.2). This may be due to disinfecting procedures being done before the samples were taken, as we know that CH_2O is one of the sources detected in cleaning products, as mentioned by the American Cancer Society (2022), in order to maintain a hygienic hospital environment.

The concentration of CO as a source of occupants' paths containing dust (Ahmad, Abdull Hamid, Ithnin, & Mohamad Pain, 2022) showed a contrasting result, where the level in the morning was higher than in the afternoon (Table 15.1) for all medical wards. Furthermore, the CO level was recorded ranging from 0.02 ± 0.0 to 0.29 ± 0.4 ppm in the morning and 0.00 ± 0.0 to 0.25 ± 0.3 ppm in the afternoon and it was still under the indoor air quality standards limit by ICOP IAQ (2010) (25) which is 10 ppm. The location of this hospital in an urban area and nearby heavy traffic can also contribute to high CO pollution levels (Ahmad Mohtar, Latif, Baharudin, Ahmad et al., 2018) in the morning when so many automobiles pass by as people head for work. Referring to One-Way ANOVA, significant differences for CO were found among all selected medical wards either in the morning or in the afternoon (Table 15.3).

In a hospital environment, PM_{10} was considered an important source of indoor air pollution. According to the studies conducted by Vardoulakis et al. (2020), people's movements indoors may be able to influence the level of PM_{10} in the air. The measurements of PM_{10} levels, which were conducted across the medical wards, are shown in Table 15.1. PM_{10} levels were found to be lower (ranging from 0.010 ± 0.0 mg/m^3 to 0.030 ± 0.0 mg/m^3 in, the morning & 0.008 ± 0.0 mg/m^3 to 0.023 ± 0.0 mg/m^3 in, afternoon) than the acceptable limit of ICOP IAQ (2010). One-way ANOVA showed a significant mean difference ($p<0.05$) of PM10 both in the morning and afternoon sessions between all medical wards, yet no significant impact on bacterial growth was seen because of the low level of PM_{10} concentration.

On the other hand, there was a positive correlation between TVOC levels and bacterial growth, where higher levels of TVOC will inhibit the growth of bacteria (Table 15.1). This finding is agreeable to a previous study by Fend-Chico et al. (2018) that there will be an increase in TVOC levels in many hospitals as they use cleaning solutions and detergents to reduce the risk of infection. However, in this study, TVOC concentrations in all selected medical ward areas were found to be low and did not exceed the ICOP IAQ (2010) standard. This situation was in line with the One-Way ANOVA analysis, which presented a significant mean difference ($p<0.05$) in bacterial count only in the morning between all selected medical wards (Table 15.4).

According to Table 15.1, the bacterial and fungal counts in all wards showed different concentrations. Moreover, the statistical analysis showed that the concentrations of bacteria measured in all medical wards were significantly different ($p<0.05$) only in the morning session. This can be mainly explained by the variation of environmental factors as well as the variation in the density of occupants and activities during sampling time (Ahmad, Abdull Hamid, Ithnin, & Mohamad Pain, 2022). When we looked at the relative humidity parameter, which is known to be crucial for micro-organism growth even at low temperatures (23°C, 30°C–32°C), we found that bacterial concentrations were higher in Ward F (233 ± 38.2 cfu/m^3) in the morning compared to other medical wards, even though its temperature was lower than Ward H (Figure 15.1). This may be explained by the relative humidity level in Ward F ($70.91\pm3.5\%$) being higher compared to other selected medical wards and within the acceptable range of ICOP IAQ (2010). Our understanding is that high-level relative humidity can also be influenced by factors such as the number of people, their activities like regularly opening the entrance door, and the size of medical wards (Ahmad, Abdull Hamid, Ithnin, & Mohamad Pain, 2022). In contrast to a previous study (Onmek, Kongcharoen, Singtong, Penjumrus, & Junnoo, 2020), the low air movement may also affect the relative humidity reading, as Ward F, which was investigated, was reported to have low air movement in the morning. This finding is supported by the One-Way ANOVA analysis, which showed a significant mean difference ($p<0.05$) in air movement in the morning among the selected medical wards. When examining the distribution of microorganisms in the investigated medical wards, it was found that the concentrations of airborne bacteria were higher than those of fungi. This result is consistent with a previous study by Ahmad et al. (2022). However, the concentrations of bacteria and fungi in this study do not exceed the maximum levels set by ICOP IAQ (2010). The concentrations of fungi that were measured in all selected medical wards were not significantly different ($p<0.05$), suggesting no contamination and significant input by fungi in selected medical wards. However, all wards in our study had daily cleaning operations, but during the measurements were conducted, ceiling fans were switched on, there was a frequency of opening the wards' doors, visiting hours happened to be in the afternoon, and disinfecting procedures were done before sampling in the morning, which limits our ability to distinguish between the effects of these exposures to the microbial growth and influencing the IAQ. Besides, there is a need to control the airflow through the wards as it may lead to moisture problems, thus increasing the growth of microorganisms. The prevention and control of microorganism growth can reduce the human health risk.

15.5 CONCLUSION

Based on our measurements in four selected medical wards in the University Hospital at Klang Valley, Malaysia, there were significant mean differences ($p<0.05$) for physical, chemical, and biological parameters except for air movement (afternoon), bacterial concentration (afternoon), and fungal concentration, respectively between Ward F, Ward G, Ward H, and Ward J. To some degree, the airborne bacterial concentrations were significantly higher than the fungal concentrations in this study. The general air quality characteristics indicated that bacteria can grow and are more prevalent than mold. From our assessment, the concentrations of airborne bacterial and fungal concentrations in all selected medical wards still showed satisfying levels when compared to the standard limits set by DOSH Malaysia. Besides, we have noted that fungi growth is not influenced by all physical and chemical parameters. This study also revealed that the occurrence of high humidity, temperature, and occupancy were the factors associated with bacterial concentrations. Interestingly, although we identified that the level of CO_2 in Ward J exceeded the standard limits while the level of air movement in Ward F was lower than the standard limits, whereby both factors can affect airborne concentrations, but the airborne concentrations still showed satisfying levels in both wards. Therefore, we presume that all the selected medical wards in the University Hospital at Klang Valley, Malaysia, have intensive ventilation, and more attention is paid to the cleanliness of the medical ward areas as shown by our data that formaldehyde impacts inhibit microbial growth. We come to the conclusion that this is valuable data for developing interventions to improve indoor air quality, which may affect the patient's health and the quality of working life of healthcare personnel in hospitals.

KEYWORDS

- **biological parameters**
- **chemical parameters**
- **ICOP IAQ (2010)**
- **Indoor air quality**
- **physical parameters**
- **sabouraud dextrose agar**
- **statistical package for social sciences**
- **trypticase soy agar**

REFERENCES

Adamović, D., Čepić, Z., Adamović, S., Stošić, M., Obrovski, B., Morača, S., & Vojinović Miloradov, M. (2021). Occupational exposure to formaldehyde and cancer risk assessment in an anatomy laboratory. *International Journal of Environmental Research and Public Health, 18*, 11198. https://doi.org/10.3390/ijerph182111198.

Aguilera, J. J., Andersen, R. K., & Toftum, J. (2019). Prediction of indoor air temperature using weather data and simple building descriptors. *International Journal of Environmental Research and Public Health, 16*(22), 4349. https://doi.org/10.3390/ijerph16224349.

Ahmad Mohtar, A. A., Latif, M. T., Baharudin, N. H., Ahmad, F., Chang, J. X., Othman, M., & Juneng, L. (2018). Variation of major air pollutants in different seasonal conditions in an urban environment in Malaysia. *Geoscience Letters, 5*, 21.

Ahmad, S., Abdull Hamid, E., Ithnin, A., & Mohamad Pain, I. A. (2022). Indoor air quality level at Medical Clinic, University Hospital in Klang Valley, Malaysia. *Malaysian Journal of Medicine and Health Sciences, 18*(6), 158–165.

Amaliyah, N. (2017). *Penyehatan Makanan dan Minuman-A*. Yogyakarta: Deepublish.

American Cancer Society. (2022). Formaldehyde and cancer risk. Retrieved from: https://www.cancer.org/cancer/cancer-causes/formaldehyde.html (accessed on 5th June 2024).

Ana, F., Isabel, A., Maria, J. G., Cristina, A., Ricardo, S., & Nelson, B. (2019). Indoor air quality & sustainability management – A case study in three Portuguese healthcare units. *Sustainability, 11*, 101. https://doi.org/10.3390/su11010101.

Azuma, K., Kagi, N., Yanagi, U., & Osawa, H. (2018). Effects of low-level inhalation exposure to carbon dioxide in indoor environments: A short review on human health and psychomotor performance. *Environment International, 121*(Pt 1), 51–56. https://doi.org/10.1016/j.envint.2018.08.059.

Basinska, M., Michalkiewicz, M., & Ratajczak, K. (2019). Impact of physical and microbiological parameters on proper indoor air quality in the nursery. *Environment International, 132*, 105098.

Basinska, M., Michatkiewicz, M., & Ratajczak, K. (2019). Impact of physical and microbiological parameters on proper indoor air quality in the nursery. *Environment International, 132*, 105098. https://doi.org/10.1016/j.envint.2019.105098.

Baurès, E., Blanchard, O., Mercier, F., Surget, E., Le Cann, P., Rivier, A., Gangneux, J., & Florentin, A. (2018). Indoor air quality in two French hospitals: Measurement of chemical and microbiological contaminants. *Science of the Total Environment, 642*, 168–179.

Bayani, M., Mirhoseini, S. H., Koolivand, A., Sarlak, H., Moradzadeh, R., Ghamari, F., Sheykhan, A., & Taheri, A. (2021). Indoor air quality inwards of dental school clinic: Variations of VOCs, CO_2, and airborne particle levels. *Journal of Air Pollution and Health, 6*(3), 171–180. https://doi.org/10.18502/japh.v6i3.8230.

Bozić, J., Ilić, P., & Ilić, S. (2019). Indoor air quality in the hospital: The influence of heating, ventilating, and conditioning system. *Brazilian Archives of Biology and Technology, 62*, 1–11.

Brągoszewska, E. (2019). Exposure to bacterial and fungal aerosols: Microorganism indices in a waste-sorting plant in Poland. *International Journal of Environmental Research and Public Health, 16*(18), 3308. https://doi.org/10.3390/ijerph16183308.

Brągoszewska, E., Mainka, A., Pastuszka, J. S., Lizonczyk, K., & Desta, Y. G. (2018). Assessment of bacterial aerosol in a preschool, primary school, and high school in Poland. *Atmosphere, 9*, 87. https://doi.org/10.3390/atmos9030087.

Chenari, B., Carrilho, J. D., & Silva, M. G. (2016). Towards sustainable, energy-efficient and healthy ventilation strategies in buildings: A review. *Renewable and Sustainable Energy Reviews, 59*, 1426–1447.

Chuanqi, L., & Qianqian, L. (2020). Environmental impact and health risk assessment of volatile organic compound emissions during different seasons in Beijing. *Journal of Environmental Sciences, 93*, 1–12.

D'Amico, A., Montagna, M. T., & Caggiano, G. (2019). Observational study on hospital building heritage and microbiological air quality in the orthopedic operating theater: The IMPACT project. *Annali di Igiene, 31*, 482–495. https://doi.org/10.7416/ai.2019.2309.

Department of Occupational Safety and Health (DOSH). (2010). Industry Code of Practice on Indoor Air Quality 2010. Ministry of Human Resources, Malaysia. Retrieved from: https://www.dosh.gov.my/ (accessed on 5th June 2024).

Fend-Chico, S., Friesen, M. C., Stefaniak, A., Henneberger, P. K., LeBouf, R. F., Stanton, M. L., Liang, X., Humann, M., & Virginia, M. A. (2018). Exposure to volatile organic compounds among healthcare workers: Modelling the effects of cleaning tasks and product use. *Annals of Work Exposures and Health, 62*(7), 852–870.

Fu Shaw, L., Chen, I. H., Chen, C. S., Wu, H. H., Lai, L. S., Chen, Y. Y., & Wang, F. D. (2018). Factors influencing microbial colonies in the air of operating rooms. *BMC Infectious Diseases, 18*(1), 4. https://doi.org/10.1186/s12879-017-2928-1.

Ghanizadeh, F., & Godini, H. A. (2018). A review of the chemical and biological pollutants in indoor air in hospitals and assessing their effects on the health of patients, staff, and visitors. *Reviews on Environmental Health, 33*(3), 231–245.

Heaton, T., Hurst, L. K., Amiri, A., Lungu, C. T., & Oh, J. (2019). Laboratory estimation of occupational exposures to volatile organic compounds during nail polish application. *Workplace Health & Safety, 67*, 288–293.

Hiwar, W., Marco-Felipe, K., Farag, S., Louise, A. F., Stephanie, J. D., & Catherine, J. N. (2021). What is the relationship between indoor air quality parameters and airborne microorganisms in hospital environments? A systematic review and meta-analysis. *Indoor Air, 31*(5), 1308–1322.

Huiyi, T., Keng, Y. W., Othman, M. H. D., Hong, Y. K., Abdul Wahab, R., Garry Kuan, P. E., Wen, T. C., & Kee, Q. L. (2022). Current and potential approaches on assessing airflow and particle dispersion in healthcare facilities: A systematic review. *Environmental Science and Pollution Research, 29*, 80137–80160.

Ibrahim, F., Samsudin, E. Z., Ishak, A. R., & Sathasivam, J. (2022). Hospital indoor air quality and its relationships with building design, building operation, and occupant-related factors: A mini-review. *Frontiers in Public Health, 10*, 1067764. https://doi.org/10.3389/fpubh.2022.1067764.

Ikhtiar, M., Alzad, H., & Paramita, S. (2017). Microbiological assessment of indoor air of Takalar County Hospital wards in South Sulawesi, Indonesia. *Science Journal of Public Health, 5*(3), 172–177.

Khan, M., Thaheem, M. J., Khan, M., Maqsoom, A., & Zeeshan, M. (2020). Thermal comfort and ventilation conditions in healthcare facilities–Part 1: An assessment of indoor environmental quality (IEQ). *Environmental Engineering and Management Journal, 6*, 917–933.

Larrey, E. K., Laryea, J. N. A., Kpordze, S. W., & Saba, C. K. S. (2020). Microbial load of indoor airborne bacteria and fungi in a teaching hospital in Ghana. *African Journal of Microbiology Research, 14*(3), 100–105. https://doi.org/10.5897/AJMR2020.9297.

Liu, Z., Ma, S., Cao, G., Meng, C., & He, B. J. (2018). Distribution characteristics, growth, reproduction and transmission modes and control strategies for microbial contamination in HVAC systems: A literature review. *Energy and Buildings, 177*, 77–95. https://doi.org/10.1016/j.enbuild.2018.07.050.

Marco, G., Gaetano, S., & Stefano, C. (2019). Indoor air quality in inpatient environments: A systematic review on factors that influence chemical pollution in inpatient wards. *Journal of Healthcare Engineering, 2019*, 8358306. https://doi.org/10.1155/2019/8358306.

Onmek, N., Kongcharoen, J., Singtong, A., Penjumrus, A., & Junnoo, S. (2020). Environmental factors and ventilation affect concentrations of microorganisms in hospital wards of Southern Thailand. *Journal of Environmental and Public Health, 2020*, Article ID 7292198. https://doi.org/10.1155/2020/7292198.

Ozdamar, M., & Umarogullari, F. (2018). Thermal comfort & indoor air quality. *International Journal of Scientific Research and Innovative Technology, 5*(3), 2313–3759.

Pipat, L., Netchanok, S., Noppanun, N., Aranya, P., Pornpimol, K., & Susan, W. (2019). Indoor air quality at different sites of a governmental hospital, in Thailand. *Nursing and Palliative Care, 4*, 1–5. https://doi.org/10.15761/NPC.1000206.

Slezakova, K., Peixoto, C., Pereira, M. C., & Morais, S. (2018). Indoor air quality in health clubs: Impact of occupancy and type of performed activities on exposure level. *Journal of Hazardous Materials, 235*, 56–66.

Stanaszek-Tomal, E. (2020). Environmental factors causing the development of microorganisms on the surfaces of national cultural monuments made of mineral building materials–Review. *Coatings, 10*, 1203. https://doi.org/10.3390/coatings10121203.

Tang, H., Ding, J., & Lin, Z. (2020). On-site measurement of indoor environment quality in a Chinese healthcare facility with a semi-closed hospital street. *Building and Environment, 173*, 106637.

Tran, V. V., Duckshin, P., & Lee, Y. C. (2020). Indoor air pollution, related human diseases, and recent trends in the control and improvement of indoor air quality. *Journal of Environmental Research and Public Health, 17*(8), 2927.

Tungjai, A., & Kubaha, K. (2017). Indoor air quality evaluation of isolation room for a hospital in Thailand. *Energy Procedia, 138*, 858–863.

Valsson, S., & Bharat, A. (2011). Impact of air temperature on relative humidity–A study. *Journal of Architecture–Time, Space and People, 11*(2), 38–41

Vardoulakis, S., Giagloglou, E., Steinbeck, S., Davis, A., Sleeuwenhoek, A., Galea, K. S., Dixon, D., & Crawford, J. O. (2020). Indoor exposure to selected air pollutants in the home environment: A systematic review. *International Journal of Environmental Research and Public Health, 17*, 8972.

Vardoulakis, S., Giagloglou, E., Steinle, S., Davis, A., Sleeuwenhoek, A., Galea, K. S., Dixon, K., & Crawford, J. O. (2020). Indoor exposure to selected air pollutants in the home environment: A systematic review. *International Journal of Environmental Research and Public Health, 17*(23), 8972.

World Health Organization (WHO). (2021). Household air pollution and health. Retrieved from: https://www.who.int/en/news-room/fact-sheets/detail/household-air-pollution-and-health (accessed on 5th June 2024).

Zhang, X., Wargocki, P., Lian, Z., & Thyregod, C. (2017). Effects of exposure to carbon dioxide and bioeffluents on perceived air quality, self-assessed acute health symptoms, and cognitive performance. *Indoor Air, 27*, 47–64. https://doi.org/10.1111/ina.12284.

Zhong, L., Yuan, J., & Fleck, B. (2019). Indoor environmental quality evaluation of lecture classrooms in an institutional building in a cold climate. *Sustainability, 11*, 6591. https://doi.org/10.3390/su11236591.

CHAPTER 16

A Case Study on the Impacts of Global Warming on Climate Change

TOFAN AGUNG EKA PRASETYA,[1] CENDANA FITRAHANJANI,[2]
SILMI MUNA,[2] ABDULLAH AL MAMUN,[3,4]
RATNANINGTYAS WAHYU KUSUMA WARDANI,[1] and
MD. SHAHARIAR CHOWDHURY[5]

[1]*Health Department, Faculty of Vocational Studies, Universitas Airlangga, Campus B, Surabaya, East Java, Indonesia*

[2]*Biostatistics and Demography Department, Faculty of Public Health, Universitas Airlangga, Surabaya, Indonesia*

[3]*Doctorate Degree Program in Public Health, Faculty of Public Health, Universitas Airlangga, Campus C, Surabaya, Indonesia*

[4]*Center for Multidisciplinary and Contemporary Research, Rangpur, Bangladesh*

[5]*Faculty of Environmental Management, Prince of Songkhla University, Hat Yai, Thailand*

ABSTRACT

Land Surface Temperature (LST) is a terrestrial essential climate variable (ECV) whose changes could have a significant impact on global climate change and sea level rise. The Tiwi Islands, which are situated in Australia's Northern Territory, are an important part of the country's flora and fauna conservation efforts. Most of the area on the Tiwi Islands is covered by vegetation, and only a small area is inhabited. Small island states like Tiwi are considered one of the ten most vulnerable places to climate change, yet with

Engineering Design and Technical Applications of Physical Science: An Integrated Approach for Sustainable Development. Chin Hua Chia, Sarani Zakaria, Ruey Shan Chen, & Md. Shahariar Chowdhury (Eds.)

their biodiversity capacities, they can play a significant role in mitigating the effects of climate change. This study focuses on determining whether LST tends to rise or decrease in the Tiwi Islands. We used autoregression (AR) and multivariate regression with cubic spline to assess LST changes in the Tiwi Islands for a 19-year period, starting from February 2000 to November 2019. Our study found that LST in the Tiwi Islands is stable because of the highly vegetated area and few built-up areas. LST in the Tiwi Islands decreased by 0.057°C per decade (Z: –0.718; CI: –0.217–0.101), and in one of the selected sub-regions, a location of a dense forestry plantation, LST greatly decreased by 0.593°C per decade.

16.1 INTRODUCTION

The world has encountered global warming and climate change issues, both caused by humans or natural processes. These issues affect Earth's ecosystems including terrestrial ecosystems, which correspond to approximately one-third of the total surface area of the Earth. Terrestrial ecosystems are characterized by greater temperature fluctuations due to the imbalance of energy transferred under the Earth's energy budget scheme. Land use and land cover (LULC), especially urban cover, is a driving factor in energy budget balance, determining the warming or cooling trend of the Earth's temperature (Duveiller et al., 2018). Hereby, multiple terrestrial ecosystem climate parameters are used to understand how severe global warming has impacted the Earth as well as to strengthen adaptation and mitigation under global climate change (Freudenberger et al., 2012). One of them is Land Surface Temperature (LST).

LST is known as one of the main parameters of energy balance on the surface. It is a climate parameter that is used to assess the environmental impacts of regional warming. It is also the main climatological variable controlling the atmospheric long-wave energy flux of the atmosphere. In Australia, warming LST is apparent as land cover changes from forest to urban (MacLachlan et al., 2017). Population growth and rapid changes in LULC into an urban area are predictors of LST and cause the LST warming trend (Fonseka et al., 2019; Mustafa et al., 2020; Wang et al., 2018a). However, in some regions of the world, especially in agricultural regions and areas with high vegetation cover, LST decreases (Al Kafy et al., 2020; Alshaikh, 2015; Song et al., 2018a).

The Tiwi Islands are part of the Northern Territory of Australia, which is known for its prosperous biodiversity. It is home to thousands of endemic plants and animals. The Tiwi Islands consist of two main islands, namely

Melville and Bathurst Islands. The Tiwi Islands are inhabited by approximately 2000 people who are concentrated in three cities: Wurrumiyanga (Bathurst Island), Milikapiti, and Pirlangimpi (Melville Island). Based on the islands' characteristics–high vegetation and small township area–this study assessed whether LST on the Tiwi Islands had also increased regionally as in Australia.

The Terra satellite carries a Moderate Resolution Imaging Spectroradiometer (MODIS) sensor. It is an environmental observation satellite used to extract LST data regionally. LST retrieval with remote sensors covers a larger area over time, making the data highly spatially correlated, autocorrelated, and vulnerable to missing data. However, in previous research, no fixed statistic has been used in analyzing LST (Firoozi et al., 2020; Khandelwal et al., 2018; Muro et al., 2018; Retamales-Muñoz et al., 2019; Suwanwong & Kongchouy, 2016; J. Wang et al., 2018b). To assess LST changes in the Tiwi Islands, this study used simple and well-known methods: autoregression (AR) time-series analysis model and multivariate regression with cubic spline. This study aimed to assess LST changes in the Tiwi Islands from 2000 to 2019.

16.2 METHOD

16.2.1 STUDY AREA

The Tiwi Islands are part of the Northern Territory of Australia, located at 11°36′S 130°49′E (Figure 16.1). The Tiwi Islands consist of two inhabited main islands separated by a sea barrier, namely Melville (the largest main island with an area of 5,786 km^2) and Bathurst Islands (the smaller main island with an area of 2,600 km^2). The area also includes nine smaller uninhabited islands (Buchanan, Harris, Seagull, Karslake, Irritutu, Clift, Turiturina, Matingalia, and Nodlaw) which are not the areas of concern in this study. The Tiwi Islands' combined area is 8,320 km^2 with a population of approximately 2,435. The population is concentrated in Wurrumiyanga of Bathurst Island, Milikapiti, and Pirlangimpi of Melville Island (accounting for 25.7% of the Tiwi Islands' total area).

The Tiwi Islands are recognized as a key area for Australia's flora and fauna conservation. The islands are known for their biodiversity values and high vegetation cover (around 81.1% of the Tiwi Islands' total area). Therefore, as shown in Figure 16.1, the vast majority of the Tiwi Islands' land is covered by forest. Vegetation types on the Tiwi Islands consist of mixed-species closed forest, open forest, woodland, and open shrubland.

Most of the areas are covered with eucalypt open forests and woodlands (Woinarski et al., 2003). Thousands of endemic species and subspecies of plants and animals have been found there.

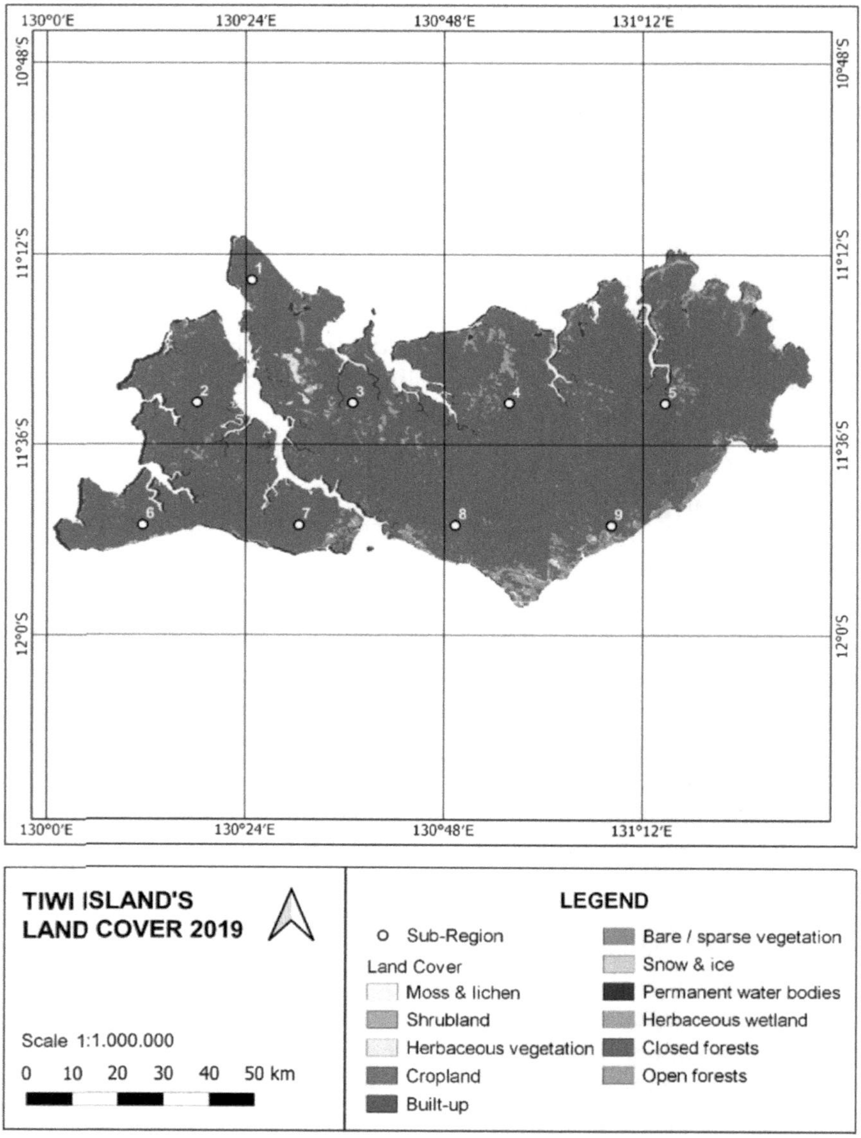

FIGURE 16.1 Map of the Tiwi islands and land cover map 2019. The data was adapted from Copernicus global land service by Buchhorn et al., (2020).

16.2.2 DATASET

This study used data obtained from NASA's MODIS Terra satellite. The LST data used in this study were the 8-day LST on the Tiwi Islands. The data were retrieved every eight days. The data used in this study were the average of 8-day LST of clear sky per pixel (0.859 km^2). The data used were from February 2000 to November 2019. The dataset was downloaded from https://modis.ornl.gov by inputting the coordinates of the study area. To get a precise coordinate, latitude and longitude were calculated or converted with Modland tile calculator tool (https://landweb.modaps.eosdis.nasa.gov/cgi-bin/developer/tilemap.cgi). The LST dataset was sent by email and had to be downloaded before being analyzed. The data subset was in kelvin conversion. In this study, we converted it into centigrade.

16.2.3 SAMPLE SELECTION

We selected nine subregions to represent the entire Tiwi Islands. They were samples that we used to hinder spatial autocorrelation that possibly occurred in the LST data retrieval. In this study, each subregion consists of 49 pixels in a 7×7 array. Terra Satellites made an area of 1 km × 1 km as the focal point for building a 3 km × 3 km of 7 km × 7 km dimension. The dimension used in this study is also the data subset recommendation provided by MODIS.

TABLE 16.1 Land Cover Percentages

Land Cover Type (UN-FAO's Land Cover Classification System)	Percentages (%)
Moss & lichen	0%
Shrubland	0.36%
Herbaceous vegetation	1%
Cropland	0.05%
Built-up	0.01%
Bare/sparse vegetation	1%
Snow & ice	0%
Permanent water bodies	3%
Herbaceous wetland	1%
Closed forests	90%
Open forests	4%

16.2.4 STATISTICAL METHOD

16.2.4.1 SEASONAL PATTERNS

We assumed that each subregion had constant seasonal variations and seasonal patterns. To create seasonal patterns in the Tiwi Islands, we plotted the average of 8-day LST data from February 2000 to November 2019 (Figure 16.3). Natural cubic spline was used in this study to ensure a smooth periodicity. A natural cubic spline was shown by selecting a set of knots. Choosing the right knots is essential because too few knots will result in underfitting while too many knots will result in overfitting. There were eight knots used throughout the data points in this study. They were in the Julian Days (10, 35, 60, 90, 115, 310, 335, and 355). These predetermined numbers of knots and locations were the numbers and locations that had been deemed as the most appropriate to the LST data (Wongsai et al., 2017). These knots will provide the smallest *p*-value and the largest R squared.

16.2.4.2 TIME SERIES CORRELATION MODEL

Spline-smoothed seasonal patterns were then subtracted to create a season-adjusted time series. The 8-day LST data were filtered with second-order autoregression (AR (2)) because AR(2) was equivalent to the dataset (it was decided after performing the first-order autoregression). Then the data were fitted with multivariate regression. Multivariate regression is used to accommodate temporal and spatial variations in the effects of the covariate, including the intercept (Gamerman & Moreira, 2004). Variance-covariance matrix in multivariate regression was used to create confidence intervals of the LST data.

16.2.4.3 ADJUSTING SPATIAL CORRELATION

In the tropics, which have a lot of cloud cover and rainy season, LST data sets become uncertain and create gaps in the time series (Shiff et al., 2021). It is difficult and imprecise, in many circumstances, to estimate values that are lost in succession in the LST time series because surface temperatures in the given time vary greatly for many reasons such as surface topography (Peng et al., 2020), altitude (Phan et al., 2018), land cover (Jiang & Tian, 2010; Bharath et al., 2013; Tan et al., 2020), climate conditions, and surface solidity

(Khandelwal et al., 2018). In addition, it often requires a large amount of historical data and complicated methods to interpolate the values that are lost in succession in the LST time series data. With the lack of available seasonal extraction methods suitable for time series with successive missing values, it is very important but challenging to detect seasonality in the MODIS LST time series due to gaps and series of noise. Thus, a simple method is needed to detect seasonality in long-term LST data analysis. In this study, we fitted a cubic spline to handle the occurrence of missing data.

The cubic spline function is widely used to refine data in various fields of study such as interactive computer graphics, real-time digital signal processing, and satellite-based time series data. The main advantage of spline fit is that the first and second derivatives are continuous, including simple calculations, good stability, high precision, and smoothness. Because MODIS LST data are collected from time to time, they fluctuate throughout the seasons. We assumed that the seasonal patterns were the same each year. Sudden natural disasters – such as forest fires, landslides, or tsunamis that can cause changes in data behavior instantly – were not included in this consideration model. Ideally, the model should provide continuous seasonal patterns every day. This consideration shows that the most appropriate model is a cubic spline with certain boundary conditions (Wongsai et al., 2017).

To predict acceleration in the square of degree Celsius ($°C^2$) per decade and cyclic pulse of LST, respectively, a second-derivative spline in 4 knots and 7 knots was carried out. Four knots were chosen and set at Julian days 1, 276, 552, and 828, while seven knots were chosen and placed at Julian days 1, 138, 276, 414, 552, 690, and 828. Acceleration is presented in a positive or negative value, which describes how rapidly the LST changed over the 19-year study period.

16.3 RESULTS AND DISCUSSION

The two issues of biodiversity and climate change are inextricably linked. Biodiversity is a vital feature of the planet that has the capacity to supply a wide spectrum of life, where we rely on it to provide food, and water, and preserve climate stability. With this potential, biodiversity loss caused by changes in land use such as rapid urbanization would have an impact on global climate change. A study of global land change from 1982–2016 revealed that 60% of all land changes are associated with human land use (Song et al., 2018b). Urbanization is predicted to be the cause of agricultural area loss which is 1.6 to 3.3 Mha per year (Lambin & Meyfroidt, 2011).

Biodiversity, on the other hand, is extremely vulnerable to climate change. Therefore, the efforts to maintain its sustainability will help in mitigating climate change.

The Tiwi Islands are a biodiversity conservation and home to thousands of endemic plants and animals. The Tiwi Islands have an integral role as the Flora and Fauna Conservation in Australia. Approximately 94% of the Tiwi Islands are plantation areas while only 0.01% are built-up areas (Table 16.1). Built-up areas are characterized by impervious land which controls LST (Pal & Ziaul, 2017). Built-up areas in the Tiwi Islands function as the support for viable economic opportunities for the Tiwi people. Only the Wurrumiyanga, Milikapiti, and Pirlangimpi areas are occupied by residents. These areas are quite small and mean a lot for sustainable conservation.

Figure 16.2 displays each of the sub-region seasonal patterns in the Tiwi Islands. The annual LST average range was between 30.39°C and 32.72°C. The temperature is warm due to the tropical monsoon climate in the Tiwi Islands. Tropical monsoon climate is characterized by a humid wet season and a hot dry season. The local climate is divided into three major seasons: Jamutakari or the wet season (December-February), Kumunupunari or the dry season (March-August), and Tiyari (September-November), or the build-up season. The LST in the Tiwi Islands was dynamic but started to warm up in mid-June when the dry season occurred. It was at its highest at the beginning of September. September is the season of hot weather and high humidity. Colder days occur between April and early June.

Figure 16.3 shows a seasonally adjusted time series for the Tiwi Islands. Our study found that the LST in the Tiwi Islands decreased by 0.057°C per decade (Z: –0.718; CI: –0.217–0.101). The rightmost panel shows that during 19 years, a drop in the LST trend occurred in 2010. A similar study also found that in 2010, Australia experienced a very sharp drop in the LST trend and a massive increase in the following year (Rasul et al., 2020). Compared to our findings, in some parts of Australia where the majority of the land cover is urban, e.g., Brisbane, Cairns, Mareeba, Melbourne, Newcastle, Perth, Sydney, LST is warming. The findings show that LST warming is a serious issue, ranging from 5–11°C (30). A study found that impervious land is exponentially correlated with LST (Xu et al., 2013), which means that urban expansion will accelerate LST warming. Warming also appears over the arid and semi-arid zones in Australia (Haynes et al., 2018). However, an additional space of green areas and water bodies is predicted to lower the LST (Xu et al., 2013). Urban expansion should be parallel with a greening trend. A continuous greening trend may stabilize local temperatures (Rasul et al., 2020).

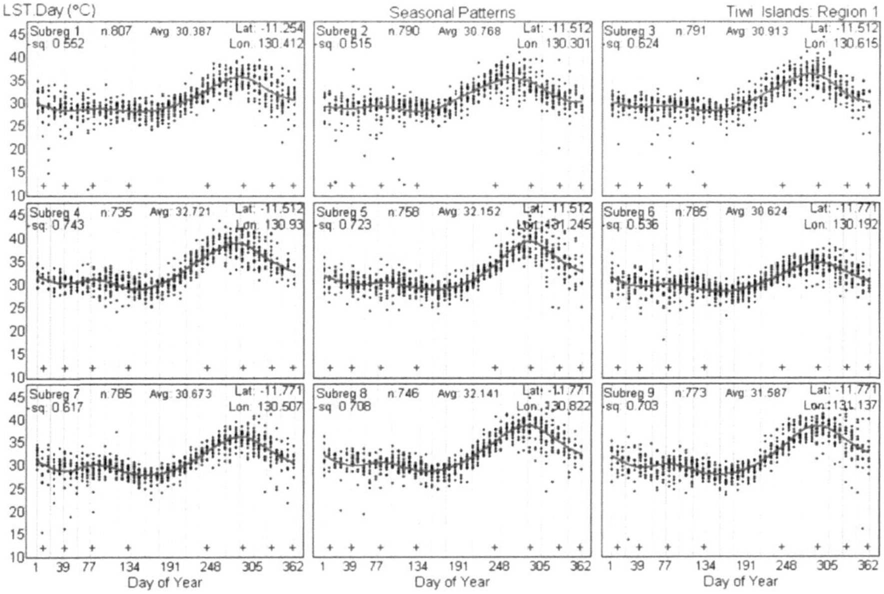

FIGURE 16.2 Seasonal patterns of the Tiwi islands. The horizontal axis is a day of the year (1 to 365). The vertical axis is 8-day LST in °C. The "Lat and Lon" display sub-region ordinate ("Lat" for latitude and "Lon" for longitude). The "Avg" displays the LST average from 2000–2019 in each sub-region. The "n" displays the observation number. Eight blue pluses are knots located on the Julian days 10, 35, 60, 90, 115, 310, 335, and 355. The black dots represent the LST data. The red curve is a smooth seasonal pattern from the LST average per day in the observed years.

Table 16.2 illustrates the mean of LST increase or decrease per decade and acceleration per decade for each sub-region in the Tiwi Islands. From 2000 to 2019, there were six regions experiencing a cooling trend and only three regions that experienced a warming trend. However, these changes were not substantial except in sub-region 1 (p: 0.000), the location of a dense forestry plantation area. Prior research conducted in an afforested area found that daytime LST decreased by about 1.1 ± 0.5°C (Peng et al., 2014). Some studies comparing the LST between built-up areas and vegetated areas also found that LSTs in built-up areas are warmer than vegetated areas (Bharath et al., 2013; Tan et al., 2020). A vegetative area tends to experience LST cooling (Bakar et al., 2016). In the city of Rosignano, Italy, LST was found to be decreasing as the government built more areas of urban tree vegetation (Barbierato et al., 2019). Evapotranspiration is the combined amounts of evaporation from land and plant transpiration that are incorporated in Earth's energy budget scheme.

Evapotranspiration has been considered to play an important role in maintaining LST stability in a region because plant transpiration releases oxygen that sinks carbon levels in the atmosphere. According to (Ramachandra & Bharath, 2019), the forest is the major carbon sink. This brief explanation helps portray the reasons behind cooling LST in sub-region 1.

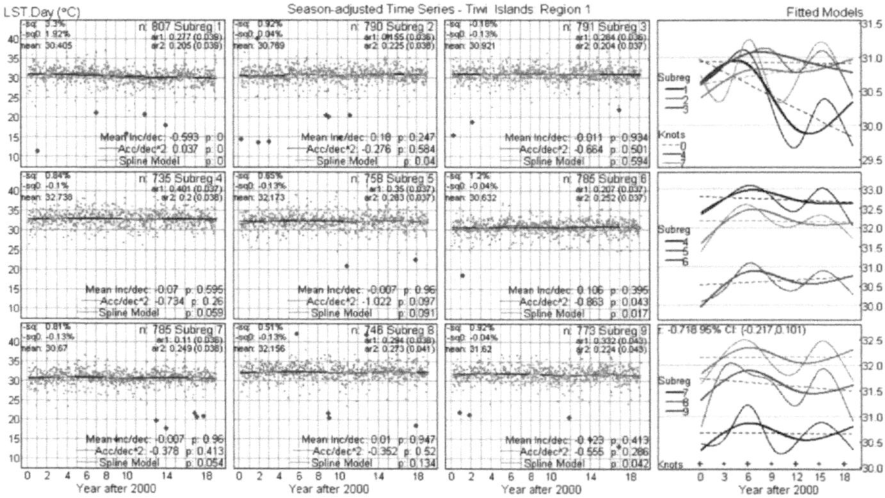

FIGURE 16.3 Season-adjusted time series of the Tiwi islands. The horizontal axis represents the years (2000 to 2019). The vertical axis represents 8-day LST in °C. The "n" displays the observation number. Mean inc/dec displays the degree Celsius of the LST increase or decrease for each sub-region. Acc/dec² is the acceleration per decade in squared degrees Celsius, indicating whether the LST trend is accelerating or decelerating. The rightmost panel displays the fitted model used in this study: 1) the dashed line indicates the linear pattern, 2) the solid bold line displays the model fitted with a 4-knots second-derivative spline to determine the acceleration per decade, and 3) the solid line displays the model fitted with a 7-knots second-derivative spline to examine the cyclic pulse of LST. The blue plus on the rightmost panel shows the knot's location.

Figure 16.4 illustrates that LSTs in the Tiwi Islands were stable. This is shown by the green-colored region. Our analysis shows that LSTs were stable in the eight sub-regions ($|z| \leq 1$) and decreased in one sub-region ($z < -1.96$), in the north-western part of Melville Island. As seen in Table 16.1, 90% of the Tiwi Islands are covered by closed forest land, while the north-western part of Melville Island is known for its extensive area of dry rainforest vegetation, particularly the coastal rainforest patch which extends over 2000 ha (Woinarski et al., 2003). Dry rainforests are floristically diverse,

TABLE 16.2 The Mean of LST per Decade in Each Sub-Region of the Tiwi Islands

Subregion	Longitude	Latitude	Mean Inc/dec (°C) [a]	Acc/dec °(C²) [b]	Sig.
1	130.412	–11.254	–0.593	0.037	0.000*
2	130.301	–11.512	0.180	–0.276	0.247
3	130.615	–11.512	–0.011	–0.664	0.934
4	130.93	–11.512	–0.070	–0.734	0.595
5	131.245	–11.512	–0.007	–1.022	0.960
6	130.192	–11.771	0.106	–0.863	0.395
7	130.507	–11.771	–0.007	–0.378	0.960
8	130.822	–11.771	0.010	–0.352	0.947
9	131.137	–11.771	–0.123	–0.555	0.413

[a]*Mean LST increases or decreases per decade in degrees Celsius.*

[b]*Acceleration per decade in squared degrees Celsius.*

**Significance at the 0.05 level.*

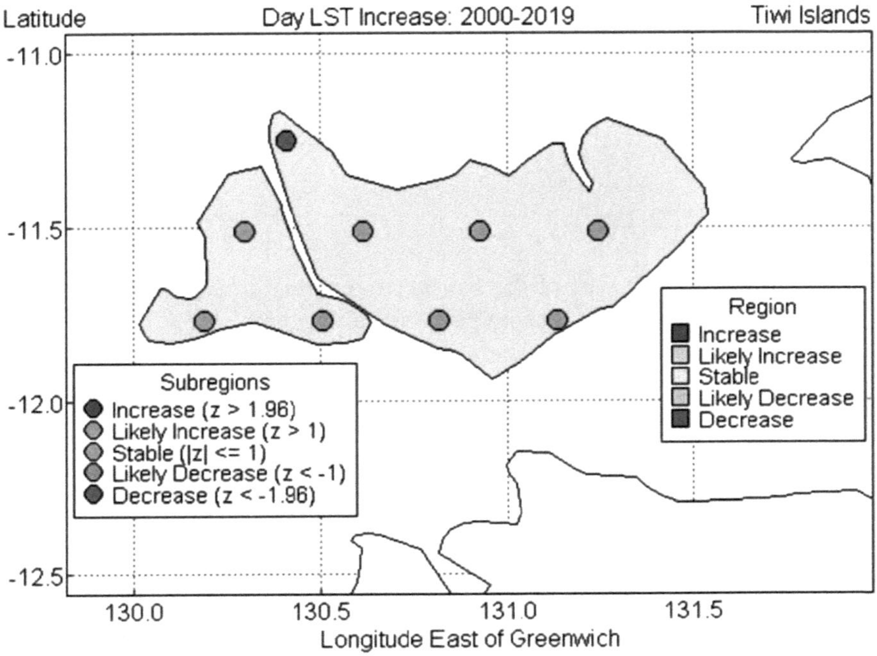

FIGURE 16.4 LST illustration in the Tiwi islands (2000–2019).

with moderate to closed canopies (covering 30–100%) and canopy heights ranging from 4 to 20 meters. Dry rainforests consist of drought-resistant plants that have a robust structure enabling them to cope with a warmer climate better than plants in wet rainforests. Furthermore, a potential adaptation method to enhance resilience is to diversify the forest by planting tree species that are more likely to thrive in a future climate (Hof et al., 2017). A larger area for vegetation and fewer built-up areas have effectively maintained LST stability in the Tiwi Islands. The selection of a group of plants in an area is also crucial for achieving stable energy transfer.

The global climate change issue is timeless and has been continuously researched since 1880. To determine how severe climate change is, several climate factors are examined, including ice sheet loss, sea level rise, world temperature, and others. The World Meteorological Organization (WMO) has confirmed that 2010 to 2019 was the warmest decade on record when compared to pre-industrial periods. We are aware that our study is limited because the dataset has only been accessible since the satellite's orbit in 2000 and that we did not compare LULC during the study period or examine its association with LST. Our study also only shows that the area where there is a significant cooling trend is a closed-dense forestry plant area. However, our analysis of LST changes in the Tiwi Islands shows that despite global warming, certain regions nevertheless have stable or even declining LST conditions.

16.4 CONCLUSION

The Tiwi Islands are part of the Northern Territory of Australia, known for their high biodiversity values and extensive vegetated area. The islands are only inhabited by a few populations centralized in Wurrumiyanga, Milikapiti, and Pirlangimpi. This study found that LST in the Tiwi Islands was stable ($|z| \leq 1$) from 2000 to 2019 but decreased by 0.057°C per decade (Z: –0.718; CI: –0.217–0.101) regionally. LST significantly decreased in one sub-region, which is a location of a dense forestry plantation. Forestry plantation and diversification help to reduce carbon levels in the atmosphere and lower the LST locally. A wider area for vegetation and fewer built-up areas have successfully maintained LST stability in the Tiwi Islands. Adaptation and mitigation measures should be adopted to successfully suppress and prevent climate change effects. However, further investigation is needed to reconfirm the findings of this study and its impact on a global scale. Other approaches are also needed to improve accuracy and estimation.

KEYWORDS

- **essential climate variable**
- **land surface temperature**
- **land use and land cover**
- **multivariate analysis**
- **remote sensing**
- **time series analysis**

REFERENCES

Al Kafy, A., Al-Faisal, A., Mahmudul Hasan, M., Sikdar, Md. S., Hasan Khan, M. H., Rahman, M., & Islam, R. (2020). Impact of LULC changes on LST in Rajshahi District of Bangladesh: A remote sensing approach. *Journal of Geographic Studies, 3*(1), 11–23. https://doi.org/10.21523/gcj5.19030102.

Alshaikh, A. (2015). Vegetation cover density and land surface temperature interrelationship using satellite data, a case study of Wadi Bisha, South KSA. *Advances in Remote Sensing, 4*(3), 248–262. http://dx.doi.org/10.4236/ars.2015.43020.

Bakar, S. B. A., Pradhan, B., Lay, U. S., & Abdullahi, S. (2016). Spatial assessment of land surface temperature and land use/land cover in Langkawi Island. *IOP Conference Series: Earth and Environmental Science, 37*, 012064. http://dx.doi.org/10.1088/1755-1315/37/1/012064.

Barbierato, E., Bernetti, I., Capecchi, I., & Saragosa, C. (2019). Quantifying the impact of trees on land surface temperature: A downscaling algorithm at city-scale. *European Journal of Remote Sensing, 52*(sup4), 74–83. http://dx.doi.org/10.1080/22797254.2019.1646104.

Bharath, S., Rajan, K. S., & Ramachandra, T. V. (2013). Land surface temperature responses to land use land cover dynamics. *Geoinformatics & Geostatistics: An Overview, 1*(4), 1–10, http://dx.doi.org/10.4172/2327-4581.1000112.

Buchhorn, M., Smets, B., Bertels, L., Lesiv, M., Tsendbazar, N. E., Masiliunas, D., Linlin, L., Herold, M., & Fritz, S. (2020). Copernicus Global Land Service: Land cover 100 m: Collection 3: Epoch 2019: Globe (Version V3.0.1) [Data set]. Zenodo. http://dx.doi.org/10.5281/zenodo.3939050.

Duveiller, G., Hooker, J., & Cescatti, A. (2018). The mark of vegetation change on Earth's surface energy balance. *Nature Communications, 9*(1), 1–12. https://doi.org/10.1038/s41467-017-02810-8.

Firoozi, F., Mahmoudi, P., Jahanshahi, S. M. A., Tavousi, T., Liu, Y., & Liang, Z. (2020). Modeling changes trend of time series of land surface temperature (LST) using satellite remote sensing productions (case study: Sistan plain in the east of Iran). *Arabian Journal of Geosciences, 13*(10), 1–14. https://doi.org/10.1007/s12517-020-05314-w.

Fonseka, H. P. U., Zhang, H., Sun, Y., Su, H., Lin, H., & Lin, Y. (2019). Urbanization and its impacts on land surface temperature in Colombo Metropolitan Area, Sri Lanka, from 1988 to 2016. *Remote Sensing, 11*(8), 1–18. https://doi.org/10.3390/rs11080957.

Freudenberger, L., Hobson, P. R., Schluck, M., & Ibisch, P. L. (2012). A global map of the functionality of terrestrial ecosystems. *Ecological Complexity, 12*, 13–22. https://doi.org/10.1016/j.ecocom.2012.08.002.

Gamerman, D., & Moreira, A. R. B. (2004). Multivariate spatial regression models. *Journal of Multivariate Analysis, 91*(2), 262–281. https://doi.org/10.1016/j.jmva.2004.02.016.

Haynes, M. W., Horowitz, F. G., Sambridge, M., Gerner, E. J., & Beardsmore, G. R. (2018). Australian mean land-surface temperature. *Geothermics, 72*, 156–162. https://doi.org/10.1016/j.geothermics.2017.10.008.

Hof, A. R., Dymond, C. C., & Mladenoff, D. J. (2017). Climate change mitigation through adaptation: The effectiveness of forest diversification by novel tree planting regimes. *Ecosphere, 8*(11), 1–29, https://doi.org/10.1002/ECS2.1981.

Jiang, J., & Tian, G. (2010). Analysis of the impact of land use/land cover change on land surface temperature with remote sensing. *Procedia Environmental Sciences, 2*, 571–575. https://doi.org/10.1016/j.proenv.2010.10.062.

Khandelwal, S., Goyal, R., Kaul, N., & Mathew, A. (2018). Assessment of land surface temperature variation due to change in elevation of the area surrounding Jaipur, India. *Egyptian Journal of Remote Sensing and Space Sciences, 21*(1), 87–94. https://doi.org/10.1016/j.ejrs.2017.01.005.

Lambin, E. F., & Meyfroidt, P. (2011). Global land use change, economic globalization, and the looming land scarcity. *Proceedings of the National Academy of Sciences of the United States of America, 108*(9), 3465–3472. https://doi.org/10.1073/pnas.1100480108.

MacLachlan, A., Biggs, E., Roberts, G., & Boruff, B. (2017). Urbanization-induced land cover temperature dynamics for sustainable future urban heat island mitigation. *Urban Science, 1*(4), 38. https://doi.org/10.3390/urbansci1040038.

Muro, J., Strauch, A., Heinemann, S., Steinbach, S., Thonfeld, F., Waske, B., & Diekkrüger, B. (2018). Land surface temperature trends as an indicator of land use changes in wetlands. *International Journal of Applied Earth Observation and Geoinformation, 70*, 62–71. https://doi.org/10.1016/j.jag.2018.02.002.

Mustafa, E. K., Co, Y., Liu, G., Kaloop, M. R., Beshr, A. A., Zarzoura, F., & Sadek, M. (2020). Study for predicting Land Surface Temperature (LST) using Landsat data: A comparison of four algorithms. *Advances in Civil Engineering, 2020*, 1–16. https://doi.org/10.1155/2020/7363546.

Pal, S., & Ziaul, S. (2017). Detection of land use and land cover change and land surface temperature in English Bazar urban center. *Egyptian Journal of Remote Sensing and Space Science, 20*(1), 125–145. https://doi.org/10.1016/j.ejrs.2016.11.003.

Peng, S. S., Piao, S., Zeng, Z., Ciais, P., Zhou, L., Li, L. Z. X., Myneni, R. B., Yin, Y., & Zeng, H. (2014). Afforestation in China cools the local land surface temperature. *Proceedings of the National Academy of Sciences of the United States of America, 111*(8), 2915–2919. https://doi.org/10.1073/pnas.1315126111.

Peng, X., Wu, W., Zheng, Y., Sun, J., Hu, T., & Wang, P. (2020). Correlation analysis of land surface temperature and topographic elements in Hangzhou, China. *Scientific Reports, 10*(1), 1–16. https://doi.org/10.1038/s41598-020-67423-6.

Phan, T., Kappas, M., & Tran, T. (2018). Land surface temperature variation due to changes in elevation in northwest Vietnam. *Climate, 6*(2), 28. https://doi.org/10.3390/cli6020028.

Ramachandra, T. V., & Bharath, S. (2019). Global warming mitigation through carbon sequestrations in the central Western Ghats. *Remote Sensing in Earth Systems Science, 2*(1), 39–63. https://doi.org/10.1007/s41976-019-0010-z.

Rasul, A., Ibrahim, S., Onojeghuo, A. R., & Balzter, H. (2020). A trend analysis of leaf area index and land surface temperature and their relationship from global to local scale. *Land, 9*(10), 1–17. https://doi.org/10.3390/land9100388.

Retamales-Muñoz, G., Durán-Alarcón, C., & Mattar, C. (2019). Recent land surface temperature patterns in Antarctica using satellite and reanalysis data. *Journal of South American Earth Sciences, 95*, 102304. https://doi.org/10.1016/j.jsames.2019.102304.

Shiff, S., Helman, D., & Lensky, I. M. (2021). Worldwide continuous gap-filled MODIS land surface temperature dataset. *Scientific Data, 8*(1), 1–10. https://doi.org/10.1038/s41597-021-00861-7.

Song, X. P., Hansen, M. C., Stehman, S. V., Potapov, P. V., Tyukavina, A., Vermote, E. F., & Townshend, J. R. (2018b). Global land change from 1982 to 2016. *Nature, 560*(7720), 639–643. https://doi.org/10.1038/s41586-018-0411-9.

Song, Z., Li, R., Qiu, R., Liu, S., Tan, C., Li, Q., Ge, W., Han, X., Tang, X., Shi, W., Song, L., Yu, W., Yang, H., & Ma, M. (2018a). Global land surface temperature was influenced by vegetation cover and PM2.5 from 2001 to 2016. *Remote Sensing, 10*(12), 1–18, https://doi.org/10.3390/rs10122034.

Suwanwong, A., & Kongchouy, N. (2016). Cubic spline regression model and GEE for land surface temperature trend using MODIS in the cloud forest of Khao Nan National Park Southern Thailand during 2000–2015. *Journal of Engineering and Applied Sciences, 11*(11), 2387–2395.

Tan, J., Yu, D., Li, Q., Tan, X., & Zhou, W. (2020). The spatial relationship between land-use/land-cover changes and land surface temperature in the Dongting Lake area, China. *Scientific Reports, 10*(1), 1–9. https://doi.org/10.1038/s41598-020-66168-6.

Wang, J., Xu, C., Hu, M., Li, Q., Yan, Z., & Jones, P. (2018a). Global land surface air temperature dynamics since 1880. *International Journal of Climatology, 38*, e466–e474. https://doi.org/10.1002/joc.5384.

Wang, S., Ma, Q., Ding, H., & Liang, H. (2018b). Detection of urban expansion and land surface temperature change using multi-temporal Landsat images. *Resources, Conservation and Recycling, 128*, 526–534. https://doi.org/10.1016/j.resconrec.2016.05.011.

Woinarski, J., Brennan, K., Cowie, I., & Kerrigan, R., & Hempel, C. (2003). Biodiversity conservation on the Tiwi Islands, Northern Territory: Part 1. Environments and plants; Northern territory government; Department of Infrastructure, Planning and Environment; Darwin, Australia.

Wongsai, N., Wongsai, S., & Huete, A. R. (2017). Annual seasonality extraction using the cubic spline function and decadal trend in temporal daytime MODIS LST data. *Remote Sensing, 9*, 1254. https://doi.org/10.3390/rs9121254.

Xu, H., Lin, D., & Tang, F. (2013). The impact of impervious surface development on land surface temperature in a subtropical city: Xiamen, China. *International Journal of Climatology, 33*(8), 1873–1883. https://doi.org/10.1002/joc.3554.

CHAPTER 17

Trends and Research Progress in Intelligible Models for Healthcare

NOR HAMIZAH MISWAN[1] and CHAN CHEE SENG[2]

[1]*Department of Mathematical Sciences, Faculty of Science and Technology, Universiti Kebangsaan Malaysia, Bangi, Selangor, Malaysia*

[2]*Department of Artificial Intelligence, Faculty of Computer Science and Information Technology, University of Malaya, Kuala Lumpur, Malaysia*

ABSTRACT

Hospital readmission is a major cost for healthcare systems worldwide. If patients with a higher potential of readmission could be identified at the start, appropriate plans could be implemented to reduce the risk of readmission. Therefore, it is important to predict the right target patients and provide interpretable insights of the model predicted. The current study develops a hospital readmission prediction framework using Logistic Regression and Artificial Neural Network (ANN), two commonly used Machine Learning classifiers. On top of that, Association Rule Mining (ARM), particularly the Apriori algorithm, was utilized to extract the hidden patterns and relationships between readmission factors and the output of the predicted model. The comparative performance of prediction models shows that ANN recorded promising predictive modeling performances across several performance measures. With respect to model interpretation using ARM, the overall dataset demonstrated that the main rules of readmission were associated with multiple past hospitalizations and hospital visits, in particular from the elderly age group. Finally, the current framework focuses on obtaining reliable prediction performance as well as providing interpretability insight into the predicted output of the modeling.

Engineering Design and Technical Applications of Physical Science: An Integrated Approach for Sustainable Development. Chin Hua Chia, Sarani Zakaria, Ruey Shan Chen, & Md. Shahariar Chowdhury (Eds.)

17.1 INTRODUCTION

Hospital readmission or re-hospitalization is defined as a patient's admission to a hospital, either to the same or a different hospital after being discharged, within a certain period of typically thirty days (Turgeman & May 2016). Hospital readmission is among the most critical concerns due to its high burden and financial implications on the healthcare system of many countries. Hence, predicting the right patient target for readmission would reduce the risk of readmission in the future and ensure resource allocation is performed in a cost-effective manner.

In the field of predictive modeling, machine learning (ML) models have emerged as a highly promising approach with advantages over traditional statistical approaches based on recent comparative studies (Artetxe et al., 2018). LR was the most widely used ML technique in the previous studies on hospital readmission prediction (Bradford et al., 2016; Futoma et al., 2015; Golas et al., 2018). It is the most well-established model for binary response variables (readmitted or not readmitted) and is still being used as the benchmark classifier. Apart from LR, the ANN model was also actively used in hospital readmission prediction due to its ability to learn the complexity of the non-linear pattern of readmission tasks (Jamei et al., 2017; Garcia-acre et al., 2017; Hung et al., 2020; Li et al., 2021). Lately, with the great progress of deep learning in healthcare, many researchers have adopted deep learning approaches that can extend the basic neural network into multiple hidden layers. Even though deep learning family models have great potential to boost overall performance, it remains a challenge to manage and interpret due to the large number of parameters involved.

Apart from establishing the prediction model, the interpretability method is an important concept that should be taken into account in healthcare studies (Stiglic et al., 2020). Interpretable ML models refer to methods or models in which the behaviors and predictions of ML systems can be understood by humans. Interpretability is essential in medical applications because of the underlying complexity of the phenomenon and the potential impact of making the wrong decision. When applying the models to some healthcare contexts that involve potentially life-threatening responses to infection, such as the risk prediction of sepsis, the predictive model is required to not only be of high performance (Caruana et al., 2015) but also be interpretable to provide insights into the ML systems. To date, there are very few attempts to implement interpretability methods, and most studies highlighted the need to include more interpretable models in future readmission studies (Cui et al.,

2018). Hence, this paper attempts to provide a compact interpretable machine learning model by combining the rule-based interpretation using ARM and the prediction model using ML classifiers to bridge the gap between the issues related to accuracy and interpretability within the ML community.

17.2 METHOD

The overall proposed interpretable machine learning framework for hospital readmission in this study is shown in Figure 17.1. Three phases are involved: data collection and pre-processing, prediction model, and model interpretation. This study focused on short-term hospital readmission length, particularly 30-day readmission since it was the most widely used readmission threshold in previous research (Artetxe et al., 2018). The dataset involved in this study is the patients admitted with heart failure, which was extracted from the Customer Healthcare Information System, developed and maintained by the Health Informatics Centre, Ministry of Health Malaysia. The dataset consisted of demographics, admission and discharge information, past hospital admissions, diagnoses, other associated causes during admission, and status of readmission. A detailed explanation of the dataset can be found in the previous study (Miswan et al., 2021).

Before data analysis, data pre-processing of the raw dataset was developed to ensure that the data were free from irrelevant records, unnecessary variables, and incomplete instances. Several missing instances were imputed using Missing Imputation of Chained Equation, which was among the frequently used methods in readmission studies (Hatipoglu et al., 2018), and the feature selection process was performed using Grey-LASSO, which has been established previously (Miswan et al., 2021). All the categorical variables were encoded as one-hot-encoded to capture different categories in a feature. The data were separated into a training set and a test set with a ratio of 7:3. In handling data imbalance, the under-sampling approach was used in this study as it was among the most utilized strategies in readmission tasks. The minority of the observations on both datasets comprised thousands of samples, which were sufficient for representing the whole dataset and superior in terms of computation time (Haixiang et al., 2017).

These pre-processed data were then used as the inputs for predictive modeling in Phase 2. Machine learning in hospital readmission has evolved in recent years and has shown outstanding performance over established statistical methods. Logistic regression (LR) and artificial neural networks

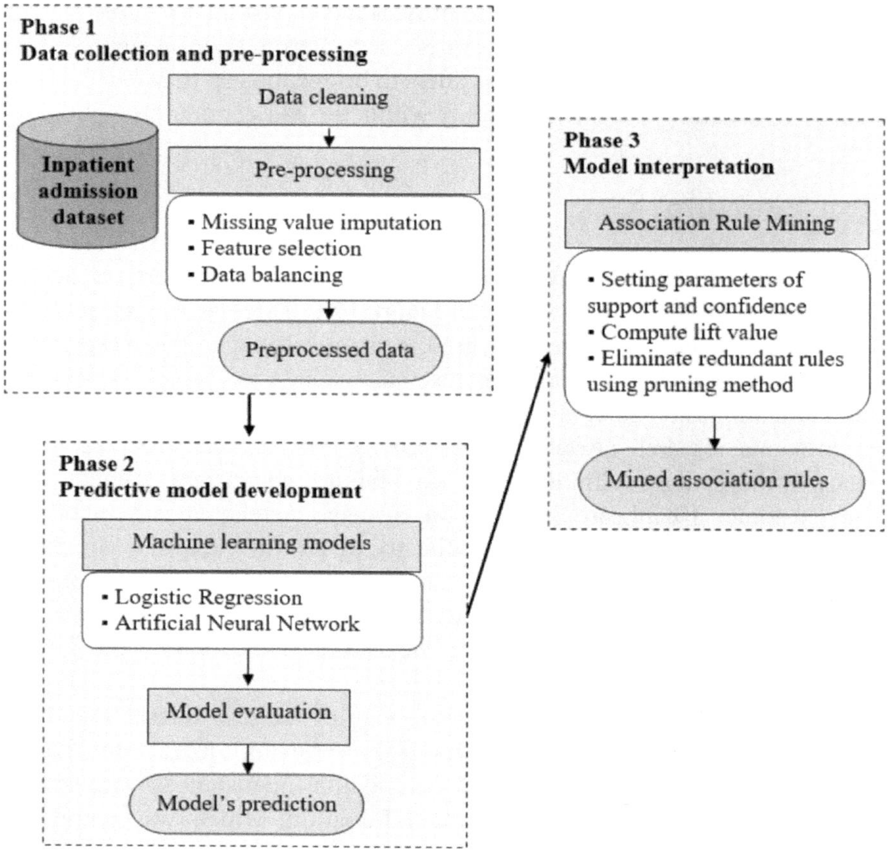

FIGURE 17.1 The interpretable machine learning framework for hospital readmission.

(ANN) were selected as predictive models due to their popularity in hospital readmission studies. In the model, let Y represent the discrete target value, where 1 indicates the patient was readmitted within 30 days, and 0 indicates that the patient was not readmitted within 30 days. Hence, the discrete target value was assumed to fall into exactly one K possible class $\{C_k\} k \in \{0,1\}$ and $X = \langle X_1, \ldots, X_d \rangle$ was the vector of predictor variables. For LR, it is commonly used for outcome variables measured on a binary scale. Here, the aim is to estimate the distribution of P(Y|X), where the model can be defined by equation (1) below,

$$P(Y = C_k | X) = g(\theta^T X) = \frac{1}{1 + e^{-\theta^T X}} \tag{1}$$

Where; $g\ (\theta^T X)$ is the sigmoid function and $\theta^T X = \sum_0^d \theta_i X_i$.

Next, ANN is used to perform high-level abstraction in the data with nonlinear transformation. The information is transferred from the input layer through multiple layers of the nonlinear activation function. It is then transferred to the output layers to produce the target class output, provided that the input could be noted as $a^{(0)}$ the first hidden layer $a^{(1)}$ and subsequent l layers operated based on equation (2) below,

$$\begin{aligned} a^{(1)} &= f\left(W^{(0)}a^{(0)} + b^{(0)}\right) \\ &\vdots \\ a^{(l+1)} &= f\left(W^{(l)}a^{(l)} + b^{(l)}\right) \end{aligned} \tag{2}$$

where; $W^{(l)}$, $a^{(l)}$ and $b^{(l)}$ represent the matric of weights, activations, and bias of layer l, respectively. The prediction model with optimal discrimination performances between LR and ANN would be selected as the reference to extract the interpretable insight in Phase 3.

In the final phase of the research framework, model interpretation is employed for the model's prediction of the classifiers that have optimal discrimination performances. In the current study, ARM was employed, motivated by the widespread application found in the market basket analysis. ARM provides the rule analysis to discover the underlying interactions and patterns among the features. The Apriori algorithm of ARM is frequently utilized in the healthcare domain (Li et al., 2008) to extract valuable information and assist physicians in treating patients (Stilou et al., 2001). On the other hand, the Apriori algorithm was used in healthcare to generate association rules between clinical events and various medications, tests, and other relations (Kulkarni & Mundhe, 2017; Sarıyer, 2019).

The rules in ARM usually represent the implication rules, such as $x_i \rightarrow x_j$, where both x_i, x_j, $\subseteq I$ and x_i, $\cap\, x_j = \phi$. The strength of the association rules is calculated based on two important measures: support and confidence. Support determines the percentage of transactions in T that contains $x_i \cup x_j$. Hence, the support measures the frequency of the rule applied to the whole dataset. Meanwhile, the confidence determines the percentage of transactions in T that contain both x_i and x_j. It measures the strength of the confidence that the rule holds. Besides, other measures such as lift value can also help in finding the important rules, where sometimes there are many redundant rules. The lift, which also refers to the measure of interestingness, could be defined as the confidence of $x_i \rightarrow x_j$ over the proportion of all transactions that contained x_j. Hence, the overall framework would provide explainable insight beyond the standard interpretability outputs that commonly focus on feature importance.

17.3 RESULTS AND DISCUSSION

To evaluate the performance of the models, several evaluation metrics including AUC and accuracy were used. The area under the receiver operating characteristic (ROC) curve (AUC) was a quantitative method used to compare the classifiers in terms of predictive performance, which reflected the trade-off between the rate of correct prediction of readmitted patients and the rate of incorrect prediction of readmitted patients (Ali et al., 2015). Accuracy shows the correct measurement for the predicted patient. The evaluation metrics on a testing set of several ML classifiers based on different feature selection algorithms were quantified, as shown in Table 17.1.

TABLE 17.1 The Evaluation Metrics of Ml Models for Hospital Readmission Prediction

ML Models	AUC	Accuracy
Logistic Regression	83.3%	92.0%
Artificial Neural Network	83.6%	92.7%

Based on the evaluation metrics in Table 17.1, both models recorded promising results in AUC and accuracy. ANN recorded the top promising predictive modeling performances due to its ability to capture the complexity of the data. The predicted output of the ANN model served as the output variables for rules generated using ARM. Hence, the mined rules generated were presented in Table 17.2, together with the rules generated using the original output.

From the predicted data of dataset D2, the top readmitted category recorded the elderly age group with one or two past visits in the last 180 days. On the other hand, the top non-readmitted category was represented by one or two external causes with zero other diagnoses. In comparison with the actual readmission output, the top important rules were generally associated with one or two past visits in the last 180 days, whereas non-readmitted patients were associated with one or two underlying causes and zero other diagnoses.

TABLE 17.2 Rules Generated Based on Actual and Predicted Output for Hospital Readmission

Dataset	Rules	Support	Confidence	Lift
Actual data	{Total visits within 180 days = 1–2 times} → {Readmit = Yes}	0.022	1.000	13.551
	{Past admission = 1–2 times} → {Readmit = Yes}	0.024	1.000	13.551
	{Total diagnoses = 1–2 times, Surgery experience = No, Gender = Male, Diagnosis E14 = No, Diagnosis J18 = No} → {Readmit = Yes}	0.010	0.193	2.616
	{Total diagnoses = None, Total causes = 1–2 times} → {Readmit = No}	0.010	1.000	1.080
	{Total causes = 1–2 times, Diagnosis I25 = No, Other diagnoses = No} → {Readmit = No}	0.010	1.000	1.080
Predicted model	{Age group = Elderly, Total visits within 180 days = 1–2 times} → {Readmit = Yes}	0.011	1.000	9.683
	{LOS = 1–5 days, Total visits within 180 days = 1–2 times} → {Readmit = Yes}	0.013	1.000	9.683
	{Past admission = 1–2 times, Discharge discipline = General medicine} → {Readmit = Yes}	0.018	1.000	9.683
	{Total diagnoses = None, Total external = 1–2 times} → {Readmit = No}	0.010	1.000	1.115

**Abbreviations for ICD codes diagnosis*: E14, Diabetes mellitus; J18, Bronchopneumonia; I25, Cardiovascular

17.4 CONCLUSION

Hospital readmission prediction has gained growing interest in the research community due to its relationship with hospital management costs. It is also useful to obtain explainable insights from the prediction model for proper healthcare planning. In this study, an interpretable machine learning framework was proposed by combining interpretability using ARM on top of the prediction phase using ML models. In the modeling process, LR and ANN were trained with pre-processed datasets using a combination of missing value imputation, feature selection, and data balancing techniques. Overall, the results from the proposed framework demonstrated that the main rules of readmission were associated with multiple past hospitalizations and hospital visits, particularly from the elderly age group. Meanwhile, non-readmitted patients were associated with zero past hospitalizations and hospital visits with a minimal number of other diagnoses.

ACKNOWLEDGMENT

The authors are thankful for the support from the Health Informatics Centre, Ministry of Health Malaysia for providing the data to support this study under the medical ethics of NMRR-18-2909-44625.

KEYWORDS

- **artificial neural network**
- **Hospital readmission**
- **machine learning**
- **machine learning**
- **model interpretation**

REFERENCES

Ali, A., Shamsuddin, S. M., & Ralescu, A. L. (2013). Classification with class imbalance problem. *International Journal of Advances in Soft Computing and its Applications, 5*(3), 1–30.

Artetxe, A., Beristain, A., & Grana, M. (2018). Predictive models for hospital readmission risk: A systematic review of methods. *Computer Methods and Programs in Biomedicine, 164*, 49–64.

Bradford, C., Shah, B. M., Shane, P., Wachi, N., & Sahota, K. (2017). Patient and clinical characteristics that heighten the risk for heart failure readmission. *Research in Social and Administrative Pharmacy, 13*(6), 1070–1081.

Caruana, R., Lou, Y., Gehrke, J., Koch, P., Sturm, M., & Elhadad, N. (2015). Intelligible models for healthcare: Predicting pneumonia risk and hospital 30-day readmission. In *Proceedings of the 21st ACM SIGKDD International Conference on Knowledge Discovery and Data Mining* (pp. 1721–1730).

Cui, S., Wang, D., Wang, Y., Yu, P. W., & Jin, Y. (2018). An improved support vector machine-based diabetic readmission prediction. *Computer Methods and Programs in Biomedicine, 166*, 123–135.

Futoma, J., Morris, J., & Lucas, J. (2015). A comparison of models for predicting early hospital readmissions. *Journal of Biomedical Informatics, 56*, 229–238.

Garcia-Arce, A., Rico, F., & Zayas-Castro, J. L. (2018). Comparison of machine learning algorithms for the prediction of preventable hospital readmissions. *The Journal for Healthcare Quality, 40*(3), 129–138.

Golas, S. B., Shibahara, T., Agboola, S., Otaki, H., Sato, J., & Nakae, T. (2018). A machine learning model to predict the risk of 30-day readmissions in patients with heart failure: A retrospective analysis of electronic medical records data. *BMC Medical Informatics and Decision Making, 18*(1), 44.

Haixiang, G., Yijing, L., Shang, J., Mingyun, G., Yuanyue, H., & Bing, G. (2017). Learning from class-imbalanced data: Review of methods and applications. *Expert Systems with Applications, 73*, 220–239.

Hatipoğlu, U., Wells, B. J., Chagin, K., Joshi, D., Milinovich, A., & Rothberg, M. B. (2018). Predicting 30-day all-cause readmission risk for subjects admitted with pneumonia at the point of care. *Respiratory Care, 63*(1), 43–49.

Hung, M., Hon, E. S., Lauren, E., Xu, J., Judd, G., & Su, W. (2020). Machine learning approach to predict the risk of 90-day hospital readmissions in patients with atrial fibrillation: Implications for quality improvement in healthcare. *Health Services Research and Managerial Epidemiology, 7*, 1–7.

Jamei, M., Nisnevich, A., Wetchler, E., Sudat, S., & Liu, E. (2017). Predicting all-cause risk of 30-day hospital readmission using artificial neural networks. *PloS One, 12*(7), e0181173.

Kulkarni, A. R., & Mundhe, D. S. D. (2017). Data mining technique: An implementation of association rule mining in healthcare. *International Advanced Research Journal in Science, Engineering, and Technology, 4*(7), 62–65.

Li, J., & Fu, A. W.-c., & Fahey, P. (2009). Efficient discovery of risk patterns in medical data. *Artificial Intelligence in Medicine, 45*(1), 77–89.

Li, W., Lipsky, M. S., Hon, E. S., Su, W., Su, S., He, Y., & Hung, M. (2021). Predicting all-cause 90-day hospital readmission for dental patients using machine learning methods. *BDJ Open, 7*(1), 1–7.

Miswan, N. H., Chan, C. S., & Ng, C. G. (2021). Hospital readmission prediction based on improved feature selection using grey relational analysis and lasso. *Grey Systems: Theory and Application, 11*(4), 796–812.

Sarıyer, G., & Öcal Taşar, C. (2020). Highlighting the rules between diagnosis types and laboratory diagnostic tests for patients of an emergency department: Use of association rule mining. *Health Informatics Journal, 26*(2), 1177–1193.

Stiglic, G., Kocbek, P., Fijacko, N., Zitnik, M., Verbert, K., & Cilar, L. (2020). Interpretability of machine learning-based prediction models in healthcare. *Wiley Interdisciplinary Reviews: Data Mining and Knowledge Discovery, 10*(5), e1379.

Stilou, S., Bamidis, P. D., Maglaveras, N., & Pappas, C. (2001). Mining association rules from clinical databases: An intelligent diagnostic process in healthcare. *Studies in Health Technology and Informatics, 84*(2), 1399–1403.

Turgeman, L., & May, J. H. (2016). A mixed-ensemble model for hospital readmission. *Artificial Intelligence in Medicine, 72*, 72–82.

CHAPTER 18

Preliminary Microgap Formation on Leadframe Mold Compound Interface of Encapsulated Electronic Components Subjected to Flood Colloidal Nanoparticles

A. JALAR,[1,2] M. A. BAKAR,[2] A. ATIQAH,[2] and A. A. OMAR[2]

[1]*Department of Applied Physics, Faculty of Science and Technology, Universiti Kebangsaan Malaysia, UKM Bangi, Selangor, Malaysia*

[2]*Institute of Microengineering and Nanoelectronics, Universiti Kebangsaan Malaysia, UKM Bangi, Selangor, Malaysia*

ABSTRACT

Consumer electronics are prone to failure due to floods. The increase in electronic products will cause environmental problems due to the unexpected amount of e-waste. Thus, there is a need to design and understand reliable electronic components that can withstand flood situations. This work aims to investigate the reliability of encapsulated electronic components when subjected to flood colloidal nanoparticles. A typical commercial single outline transistor (SOT023) was used as an electronic component sample. The SOT consists of copper wire bonding that is bonded to the Ag-plated Cu leadframe. The wire bonding and the leadframe were then encapsulated in an epoxy mold compound via a transfer molding process. The flood colloidal nanoparticles (FCNPs) used in this study were taken from Sungai Serai, Hulu Langat, Selangor. The samples were then subjected to the FCNPs environment. After being subjected to the FCNPs, the samples were observed using an optical microscope. Cross-section samples were prepared using a metallographic technique to capture the interface between the lead frame and the epoxy

Engineering Design and Technical Applications of Physical Science: An Integrated Approach for Sustainable Development. Chin Hua Chia, Sarani Zakaria, Ruey Shan Chen, & Md. Shahariar Chowdhury (Eds.)

mold compound. It was found that a microgap was formed at the external interface between the leadframe and the epoxy mold compound. The internal microgap size is approximately ranging from 1.5–1.9 μm. This finding shows that the formation of a microgap will create a path for any nanoparticles to ingress inside the electronic component. Proper encapsulation will prevent the formation of a microgap. Thus, this will create a reliable component that can withstand flood environments. The flood-reliable component will extend the lifespan of the electronic component, therefore minimizing the damage to electronic devices and equipment. This will directly benefit the environment by reducing the unexpected e-waste due to flash flood phenomena.

18.1 INTRODUCTION

The electronic era has been introduced since the development of electronic devices at a large scale and has tremendously increased in many aspects of human well-being and technology. The electronics application started as assisting in technological development and changing human lifestyle and quality of life. Electronic components are the backbone of electronic devices (Shi et al., 2020; Kang, Tok, & Bao, 2019; Fay, Jena, & Maki, 2019). The high reliability of electronic components promises a longer lifespan of the devices and thus ensures their functionality without unexpected failure.

From a semiconductor technology point of view, an electronic component consists of a die that is encapsulated before being assembled on a printed circuit board. Sometimes it is called a chip. A die consists of semiconductor materials with specific functionalities, whereas the encapsulation process is considered IC packaging or semiconductor packaging. In semiconductor packaging, the encapsulation process is called the back-end assembly process. Therefore, back-end assembly refers to the final stages of semiconductor processing. The backend assembly process typically involves the following steps: beginning with wafer sawing, the wafer is cut into individual die. Then, the die is attached to the lead frame and bonded to a package or substrate, typically using wire bonding. These wires establish the electrical connections between the die and the package. Next, the die and wire bonds are encapsulated in a protective material, such as plastic or ceramic, using a process called molding or encapsulation. This encapsulation protects the die and wire bonds from physical damage, moisture, and contaminants. To ensure the packaged ICs meet the required specifications, they then undergo rigorous testing including electrical, quality assurance, reliability, and endurance tests. Finally, the packaged ICs are ready for shipment for the next assembly process, such as surface mount technology (SMT) to position the packaged onto PCB. Figure 18.1 shows the important back-end assembly process.

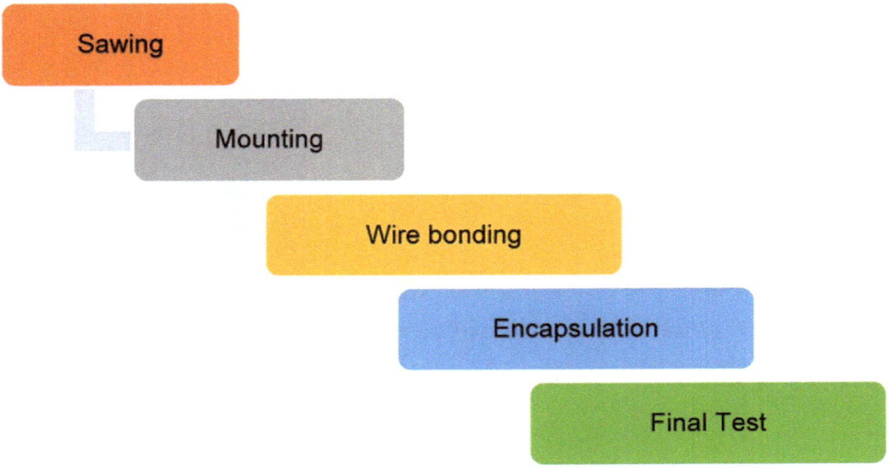

FIGURE 18.1 Important back-end assembly process.

For electronic components, encapsulation plays a vital role in protecting the internal structure, such as the interconnection and integrated circuit, from the environment (Ng, Abas, & Abdullah, 2020; Ardebili, Zhang, & Pecht, 2018). Figure 18.2 below shows the printed circuit board assembled with the electronic components. For example, corrosion can occur due to inadequate encapsulation, leading to delamination between the mold compound and the metal lead frame. As a result, the internal interconnection and integrated circuit may experience corrosion, leading to the failure of the electronic component. A recent study by Omar et al., (2022) reported the existence of an ingression route between the mold compound and leadframe during the production of the electronic component. Surprisingly, the gap between the mold compound was reported to be about 1.7 microns. This size is large enough for nanoparticles to ingress toward the center of the package and cause the ultimate failure of the component. The sample shown in Figure 18.3 is an example of wire bonding on a bond pad and leadframe before the encapsulation process. The epoxy mold compound is depicted in Figure 18.4, and the cross-section of a lead frame and epoxy mold compound after the encapsulation process is shown in Figure 18.5.

FIGURE 18.2 Example of encapsulated final product.

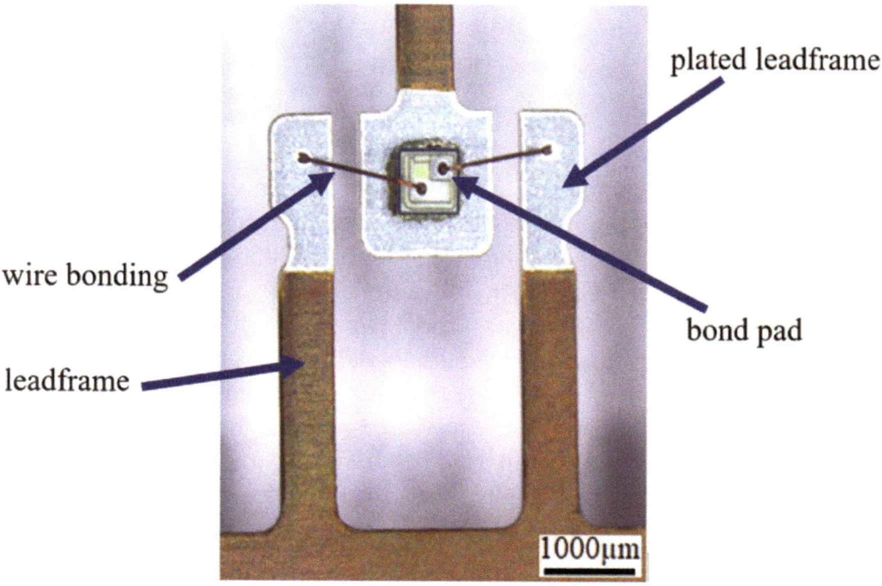

FIGURE 18.3 Example of wire bonding on a bond pad and lead frame before the encapsulation process.

Delamination in encapsulated electronic components is one of the main causes of failure. It could occur at the interface between the mold resin material and substrate, the die and the die attach material, or the die attach material and the substrate due to many contributing factors such as high-stress concentration, weak adhesion strength, or moisture from humid environments (Tan et al., 2021; Jalal, Ansari, & Atiqah, 2021; Garete, Li, & Taduran, 2020). To reduce this type of failure, a combination of Finite Element Analysis (FEA) and Design of Experiment (DOE) methods is often introduced to obtain optimized

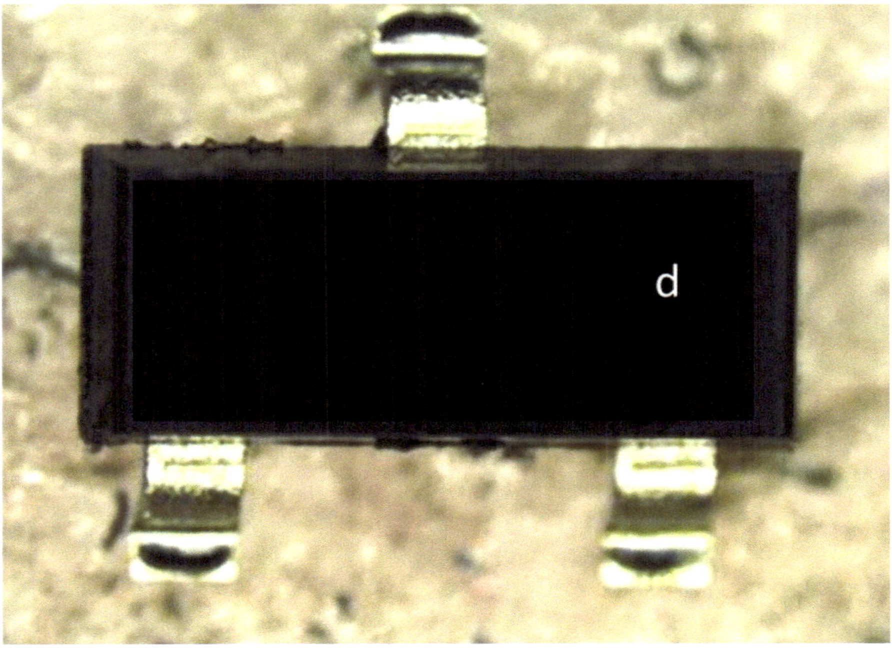

FIGURE 18.4 Epoxy mold compound.

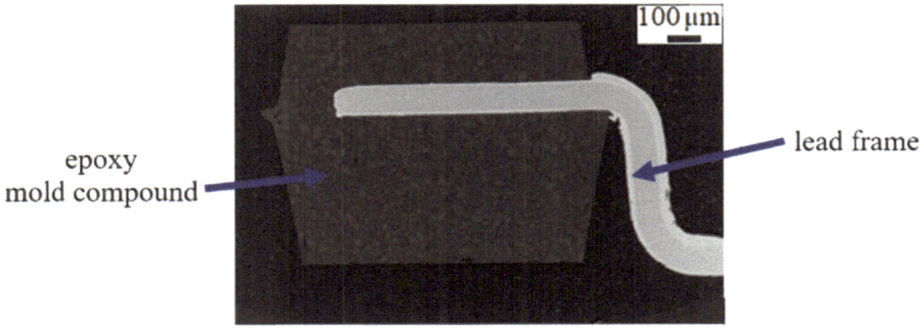

FIGURE 18.5 Cross section of a leadframe and epoxy mold compound after encapsulation process.

parameters (lead frame, molding compound, die-attach materials, and so on) to produce IC packages (Ching & Abdullah 2022). The literature available on delamination measurements of electronic packages mainly focuses on the internal stress-induced delamination of the entire package. Van Driel et al., (2008) developed combined finite element modeling with simulation-based optimization methods to reduce delamination-related reliability problems in exposed pad packages. Based on the simulated findings, accurate material models with design guidelines can be derived to provide generic insight into the mechanism of delamination for future work. Meanwhile, Stoyanov et al., (2020) reported a combined experimental and finite element modeling study to predict the interfacial stress limits that components endure before showing signs of delamination failure under solder conditions. To date, none of the studies analyze and report measurements on reliability-related issues caused by flood colloidal nanoparticles.

Electronic waste (e-waste) is a hazardous solid waste related to greenhouse gases (GHG) that causes environmental hazards. In the event of a flood, electronic equipment that was submerged is regarded as a total loss since the cost of repairs is high, and most often, it is more economical to get a new one. The damaged items end up as e-waste that significantly affects the environment. However, consumer electronic components are designed for normal everyday use and are not intended for flood resistance. Therefore, societal and environmental demands for reliable and affordable electronic components are essential to address the e-waste issue resulting from recent urban flooding. As such, the adverse effects of colloidal nanoparticles due to flooding that cause damage to electronic components will be thoroughly investigated. This work incorporates the degradation study of electronic encapsulation and is part of a big project to integrate reliability in designing electronic components to overcome accumulated e-waste due to flood colloidal nanoparticles (FCNPs). This would be reliable, affordable, and beneficial to society.

E-waste contributes approximately about 10% by weight to the overall flood waste and 40% to the damage value of electronic products (Leader et al., 2018). The damaged items end up as e-waste, significantly affecting the environment (Wu et al., 2015; Leader et al., 2018). E-waste is categorized as a hazardous solid waste stream due to the presence of toxic materials that could cause environmental hazards, especially if leached out and solubilized in soil, groundwater, and surface water. Ingestion of soluble substances could result in human health risks. Hence, disposal of e-waste in conventional landfills is not recommended. In Malaysia, e-waste is regulated under Code SW110, First Schedule, Environmental Quality (Scheduled Wastes) Regulations 2005 (MALAYSIA, 2010; Sulaiman et al., n.d.; Che Jamin & Mahmood, 2015). E-waste is generally divided into two main types according to its generation sources, i.e., e-waste generated from the industrial sector and household e-waste. "E-waste" refers to broken, non-working, or old/obsolete electric electronic appliances such as TVs, PCs, air conditioners, washing machines, and refrigerators. Code SW110 is defined as waste from electrical and electronic assemblies containing components such as accumulators, mercury switches, glass from cathode-ray tubes, and other activated glass or polychlorinated biphenyl capacitors, or contaminated with cadmium, mercury, lead, nickel, chromium, copper, lithium, silver, manganese, or polychlorinated biphenyl. The more electrical and electronic equipment is being produced, the more e-waste needs to be disposed of or managed properly. If e-waste is discarded without implementing an environmentally sound manner such as into the river, landfill, burning, or sent to the informal sector, e-waste may endanger our lives, affecting human health and causing deterioration of environmental quality.

Figure 18.6 shows the estimation of e-waste that Malaysia generates from 2016 until 2025. The Global E-waste Monitor Report 2017 presented by the United Nations University (UNU) estimated the global e-waste generation as 44.7 million tons in 2016. The raw material value of this fraction was estimated to be 55 billion euros (about 60 billion dollars) (Baldé et al., 2017). The StEP (Solving the e-waste problem) Annual Report (2015/2016) estimated the global e-waste generation in 2015 as 43.8 million tons. The countries in Southeast Asia and China significantly contributed to the e-waste generation (i.e., 40.7% e-waste generation in Asia in 2016) and this is supported by the growth of printed circuit boards (PCBs) manufacturing, which was 10.8–14.4% per annum (Cui & Anderson, 2016; Baldé et al., 2017). Based on research, estimations show Malaysia will generate 24.5 million units of e-waste in 2025 as shown in Figure 18.6.

In this research work, a colloidal nanoparticle flood-reliable electronic component is analyzed using experimental works subjected to flood colloidal nanoparticles. The study on the reliability of encapsulated electronic components when subjected to flood colloidal nanoparticles will be carried out and characterized by using optical microscopy and scanning electron microscopy. The colloidal nanoparticle flood-reliable electronic component will reduce the e-waste affected by a flood, thus contributing to a better environment.

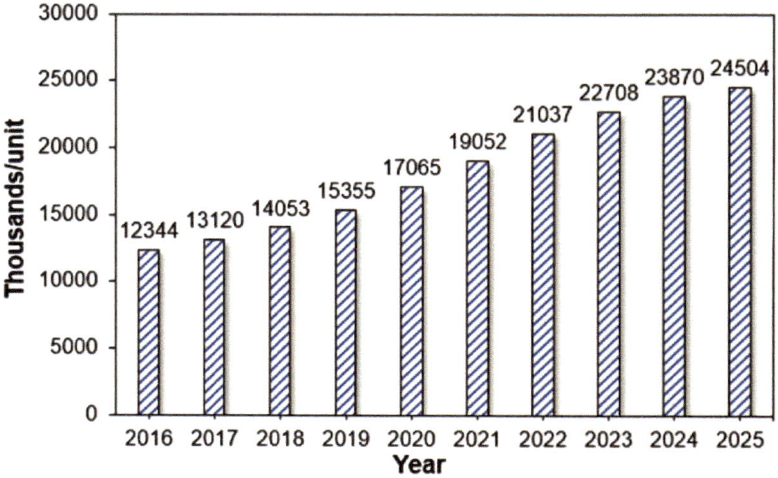

FIGURE 18.6 Total of E-Waste in Malaysia including air conditioner, refrigerator, washing machine, mobile phone, personal computers and television from 2016 until 2025.

Source: Adapted from Malaysian Department of Environment Water, 2020

18.2 METHODOLOGY

A typical and commercial SOT023 package was used in this study. The SOT023 package is a small outline transistor package commonly used for consumer electronic devices. This package has three leads and is very popular for low-power transistors, such as small signal transistors and low-voltage regulators. The dimensions of the SOT023 package were 2 mm x 1 mm x 1.25 mm. The package has three leads in a triangular arrangement. The center lead is the collector, while the two outer leads are the emitter and the base.

The sample of flood colloidal nanoparticles used in this study originated from a flood. To imitate the flood condition with nanoparticles, riverbank soil was collected and mixed with Hulu Langat River water located in Sungai Lui, Hulu Langat, Selangor. A total of 1 kg of sample was collected and mixed with 20 liters of water, then stirred thoroughly for 15 minutes using a hand mixer.

To study the structural integrity of the SOT023, optical and SEM techniques were used to determine the cross-section of the SOT023. The sample's cross-section was prepared using metallographic techniques. The sample is placed in a mounting cup, and a sample clip is used to hold it in place. A mixture of epoxy and resin with a ratio of 2:1 was mixed and stirred homogeneously. The mounting mixture was poured into the cup containing the sample and left to harden for approximately 4 hours. The sample was then removed from the mounting cup for the next step, which involved grinding and polishing. The sample was ground using SiC abrasive papers with grit sizes ranging from 800 to 4000, at a slow rpm of approximately 30 to 50. After the grinding process was completed, the sample underwent a polishing process using Dp-Nap polishing cloth and diamond spray with particle sizes of 6 μm and 1 μm. The sample was cleaned using tepol and deionized water before being viewed under an optical microscope and FESEM for internal structure analysis.

Another point that needs careful consideration in analyzing the samples in this study is the fact that the epoxy mold compounds first need to be removed so Cu wire and internal leadframe are directly accessible by analysis tools visually and physically. While there are many methods and approaches available for this purpose, challenges to the preservation of the area of interest must be kept as a top priority to ensure the validity of the analysis outcome. For this reason, no chemicals are allowed in preparing the samples for analysis. Standard approaches of chemical de-cap with Nitric (HNO_3) and Sulfuric (H_2SO_4) acid or any other methods involving chemicals of any nature are prohibited as they may introduce or remove elements from the target area that could lead to invalid observations. Furthermore, they may also compromise the surface morphology of the samples, particularly at the Cu wedge or stitch bond as well as the Cu leadframe especially as Cu is known to react notoriously with Nitric acid. Therefore, for this study, the method of decapsulation used is primarily laser ablation, with mechanical decapsulation being a secondary option.

For microscopic analysis, a metallurgical microscope model Leica S8AP0 was used, with a magnification of 10x and bright field lighting attached to a JEOL Scanning Electron Microscope (SEM), model JSM-6010LV.

18.3 RESULTS AND DISCUSSION

Electronic equipment/devices are controlled by electronic components, which are normally assembled onto the printed circuit board as shown in Figure 18.7. Taking a closer look at the component, it is built from semiconductor/electronics materials, connected to the leadframe (such as copper), and encapsulated with mold compound, a type of epoxy-based material with other materials (Omar et al., 2022; V. W. J. Tan et al., 2018; Kim, Raleigh, & Saiyed, 2019). The reliability of the component depends on the 'tightness' of the encapsulation (Haybat et al., 2021). A small opening or gap (Figure 18.8) (sometimes referred to as delamination) will cause many complications and make it prone to failure (V. W. J. Tan et al., 2018; Kim, Raleigh, & Saiyed, 2019; Subramanian et al., 2018).

The upper and lower parts of the EMC show a weldline from the transfer molding process of encapsulation. A microgap with a width size ranging from approximately 1.5–1.9 μm occurs between the lead frame/lead and EMC (Figure 18.8). The occurrence of a microgap is not consistent and its formation depends on many factors such as during manufacturing and service. The size of the FCNPs is much smaller compared to the occurrence of the microgap. Thus, the formation of the microgap will allow the FCNPs to ingress and interrupt the internal structure of the electronic component.

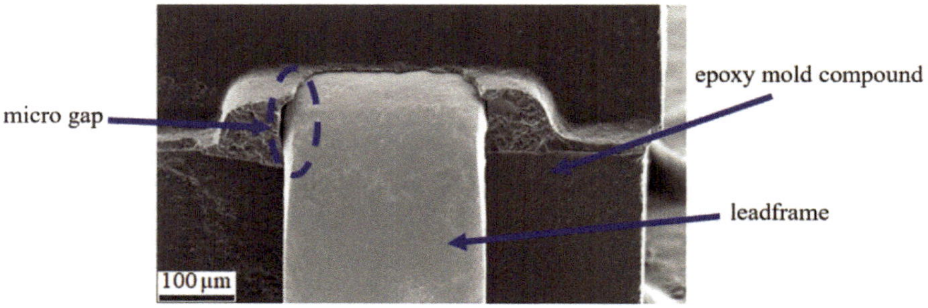

FIGURE 18.7 A micro gap was observed between the lead frame and epoxy mold compound.

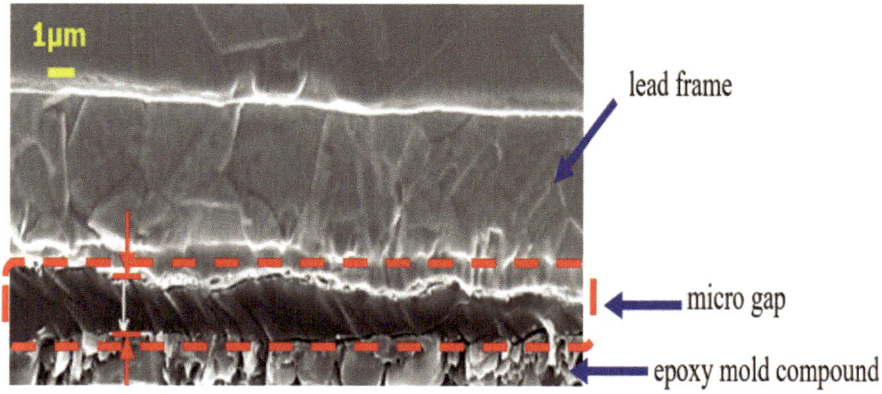

FIGURE 18.8 Occurrence of microgap between mold compound and metal lead frame.

Figure 18.7 shows an example of a major electronics component printed circuit board. The interface between epoxy and metal defines the quality and reliability of electronic components. Floods contain water and other contaminants, including colloidal nanoparticles commonly called 'mud.' From a water perspective, electronic components are normally considered for water and humidity resistance and measures have already been established to overcome this issue. However, manufacturers do not typically consider the impact of colloidal nanoparticles on the reliability of the component during the design phase. Different types of colloidal nanoparticles exist in floodwater, varying in size and shape. These nanoparticles can react with the interface, leading to degradation and potentially causing separation or gaps in the structure. Understanding the interaction between colloidal nanoparticles and epoxy is essential for designing electronic component encapsulation that can minimize damage from floods. Manufacturers can create highly reliable encapsulation for electronic components, especially for military and aviation applications (Subramanian et al., 2018; Ildarovich et al., 2021; Lancaster & Keswani, 2018). However, the manufacturing cost is higher due to specific materials and processing. For instance, certain electronic components in aviation and military applications still use some percentage of lead in the interconnection, which is not intended for common applications. Therefore, when producing electronic devices for ordinary people, the production cost must be taken into consideration to encourage people to buy them. Thus, socio and techno-economic studies must be carried out to get clearer input from society, so that the product is affordable and reliable to meet and comply with flood issues.

For ordinary consumer electronics, it is acceptable to have such a micro gap in the electronic component. This is indicated by IPC-7621 (code standard and practice in the electronic industry), which allows a certain size of the micro gap for consumer electronics. However, this practice is not favorable for harsh environments. For example, during floods, corrosive, and harsh environments.

The main factor of damaged components due to floods is FCNP containment which can cause failure through the epoxy-metal interface of the encapsulation. If this issue is resolved through engineering and manufacturing design, encapsulation that withstands FCNPs from a flood can be produced. The cost of encapsulation estimated for these components is very small compared to the overall cost of electronic equipment. Thus, repairing electronic equipment will not involve parts and ensure total low repairing costs. This will reduce disposing of damaged equipment resulting in an increase of unscheduled e-waste as well as spending the extra cost on new unplanned purchasing.

18.4 CONCLUSION

This work reports on preliminary observations on the electronic component that is prone to failure when exposed to the flood that contains flood colloidal nanoparticles (FCNPs). The microgap with a size of about 1.5–1.9 μm existed at the interface of the leadframe and epoxy mold compound. This size is much bigger than the FCNP size. Thus, this microgap possibly allows the FCNPs to interrupt the "tightness" of encapsulation leading to exposure of the internal structure. Addressing this issue will have a possible impact on societal, environmental, and economic aspects.

ACKNOWLEDGMENT

This work was supported by the Ministry of Higher Education, Malaysia under the Fundamental Research Grant Scheme (FRGS/1/2020/TK0/UKM/01/3).

KEYWORDS

- **design of experiment**
- **electronic consumer**
- **e-waste**
- **finite element analysis**
- **flood colloidal nanoparticles**
- **greenhouse gases**
- **microgap**
- **mold compound encapsulation**

REFERENCES

Ardebili, H., Zhang, J., & Pecht, M. G. (2018). *Encapsulation Technologies for Electronic Applications*. William Andrew.

Baldé, C. P., Forti, V., Gray, V., Kuehr, R., & Stegmann, P. (2017). *The Global E-Waste Monitor 2017: Quantities, Flows and Resources*. United Nations University, International Telecommunication Union, and International Solid Waste Association.

Che Jamin, N., & Mahmood, N. Z. (2015). Scheduled waste management in Malaysia: An overview. *Advanced Materials Research, 1113*, 841–846.
Ching, L. R., & Abdullah, M. Z. (2022). A review of mold flow and finite element analysis simulation of Chip Scale Packaging (CSP) for Light Emitting Diode (LED). *Journal of Advanced Research in Fluid Mechanics and Thermal Sciences, 99*(1), 158–173.
Cui, H., & Anderson, C. G. (2016). Literature review of hydrometallurgical recycling of Printed Circuit Boards (PCBs). *Journal of Advanced Chemical Engineering, 6*(1), 142–153.
Driel, W. D. van, van Gils, M. A. J., Fan, X., Zhang, G. Q., & Ernst, L. J. (2008). Driving mechanisms of delamination-related reliability problems in exposed pad packages. *IEEE Transactions on Components and Packaging Technologies, 31*(2), 260–268.
Fay, P., Jena, D., & Maki, P. (2019). *High-Frequency GaN Electronic Devices*. Springer.
Garete, A. J., Li, Z., & Taduran, A. (2020). Epoxy molding compound development for improved MSL1 delamination resistance in plastic encapsulated clip bond power package. In *2020 IEEE 22nd Electronics Packaging Technology Conference (EPTC)* (pp. 90–94). IEEE.
Haybat, M., Guenther, T., Kulkarni, R., Sahakalkan, S., Grözinger, T., Rothermel, T., & Zimmermann, A. (2021). Characterization of hermetically sealed metallic feedthroughs through injection-molded epoxy-molding compounds. *Applied Mechanics, 2*(4), 976–995.
Ildarovich, Y. I., Rustemovich, U. R., Nailevich, S. S., Alekseevich, Z. I., Igorevich, L. L., Maratovich, S. A., & Urakseev, M. (2021). To the question of the approach to the aviation modern and advanced electrical machines electronics units design formation. In *2021 International Conference on Electrotechnical Complexes and Systems (ICOECS)* (pp. 480–485). IEEE.
Jalal, S. N. S., Ansari, M. N. M., & Atiqah, A. (2021). Effect of molding cure time on high-density quad-flat-no lead sawn package. In *IOP Conference Series: Materials Science and Engineering, 1092*, 12040. IOP Publishing.
Kang, J., Tok, J. B. H., & Bao, Z. (2019). Self-healing soft electronics. *Nature Electronics, 2*(4), 144–150.
Kim, Y., Raleigh, C., & Saiyed, S. (2019). Design study to prevent mold delamination for over-molded lead frame package. In *2019 18th IEEE Intersociety Conference on Thermal and Thermomechanical Phenomena in Electronic Systems (ITherm)* (pp. 485–489). IEEE.
Lancaster, A., & Keswani, M. (2018). Integrated circuit packaging review with an emphasis on 3D packaging. *Integration, 60*, 204–212.
Leader, A., Gaustad, G., Tomaszewski, B., & Babbitt, C. W. (2018). The consequences of electronic waste post-disaster: A case study of flooding in Bonn, Germany. *Sustainability, 10*(11), 4193.
MALAYSIA, ELECTRONIC EQUIPMENT IN. (2010). *Guidelines for the Classification of Used Electrical and Electronic Equipment in Malaysia*. Department of Environment, Malaysia.
Malaysian Department of Environment Water (2020). E-waste Management in Malaysia. What is e-waste? https://ewaste.doe.gov.my/index.php/what-is-e-waste/
Ng, F. C., Abas, M. A., & Abdullah, M. Z. (2020). Filling efficiency of flip-chip underfill encapsulation process. *Soldering & Surface Mount Technology, 32*(1), 10–18.
Omar, A. A., Jalar, A., Bakar, M. A., & Hamid, K. A. (2022). Sulfur ingression on the encapsulated lead frame of the discrete semiconductor package. *IEEE Transactions on Components, Packaging and Manufacturing Technology*. 544–550.
Shi, Q., Dong, B., He, T., Sun, Z., Zhu, J., Zhang, Z., & Lee, C. (2020). Progress in wearable electronics/photonics—Moving toward the era of artificial intelligence and the Internet of Things. *InfoMat, 2*(6), 1131–1162.

Stoyanov, S., Bailey, C., Stewart, P., Parker, M., & Roulston, J. F. (2020). Experimental and modeling study on delamination risks for refinished electronic packages under hot solder dip loads. *IEEE Transactions on Components, Packaging and Manufacturing Technology, 10*(3), 502–515.

Subramanian, N. R., Dexter, R., Joel, F., & Mohamad, Y. (2018). Evaluating delamination in molded packages with process simulation. In *2018 IEEE 20th Electronics Packaging Technology Conference (EPTC)* (pp. 811–814). IEEE.

Sulaiman, S. S., Hashim, N., Kamaruddin, H., Ibrahim, I., & Dian, A. M. (n.d.). Safeguarding the natural environment: Laws on e-waste management in Malaysia. *European Proceedings of Social and Behavioral Sciences, 70,* 274–280.

Tan, J. S., Lee, W. W., Atiqah, A., Hamid, K. A., Jalar, A., & Bakar, M. A. (2021). The effect of electrolytic deflash current on delamination of epoxy mold compound during the tin plating process. In *2021 IEEE Regional Symposium on Micro and Nanoelectronics (RSM)* (pp. 73–75). IEEE.

Tan, V. W. J., Au, Y. K., Low, B. Y., Eu, P. L., & Tan, L. C. (2018). Mold compound and copper wire selection for quad-flat packages with high-density lead frame in automotive applications. In *2018 IEEE 38th International Electronics Manufacturing Technology Conference (IEMT)* (pp. 1–5). IEEE.

Wu, Q., Leung, J. Y. S., Geng, X., Chen, S., Huang, X., Li, H., & Lu, Y. (2015). Heavy metal contamination of soil and water in the vicinity of an abandoned e-waste recycling site: Implications for dissemination of heavy metals. *Science of the Total Environment, 506,* 217–225.

CHAPTER 19

Deciphering the Comparative Pharmacological Insights of *Thysanolaena maxima, Diploclisia glaucescens,* and *Pterospermum semisagittatum*: Chemico-Biological Investigations

RAIHAN UZ ZAMAN,[1,2,3] NAZIM UDDIN EMON,[3,4] MINHAJUL ISLAM,[3,5] and RASHADUZ ZAMAN[3,6]

[1]*Institute for Biomedicine and Glycomics, Griffith University, Gold Coast campus, Queensland, Australia*

[2]*Faculty of Traditional Thai Medicine, Prince of Songkhla University, Hat Yai, Songkhla, Thailand*

[3]*Department of Pharmacy, Faculty of Biological Science, International Islamic University Chittagong, Chittagong, Bangladesh*

[4]*Department of Public Health, School of Science and Technology, Bangladesh Open University, Gazipur, Bangladesh*

[5]*Department of Pharmaceutical Science, North South University, Dhaka, Bangladesh*

[6]*Department of Renewable Resources, University of Alberta, Edmonton, Canada*

Engineering Design and Technical Applications of Physical Science: An Integrated Approach for Sustainable Development. Chin Hua Chia, Sarani Zakaria, Ruey Shan Chen, & Md. Shahariar Chowdhury (Eds.)

ABSTRACT

Thysanolaena maxima (Family: Poaceae), *Pterospermum semisagittatum* (Family: Sterculiaceae), and *Diploclisia glaucescens* (Family: Menispermaceae) are three well-known medicinal plants in Bangladesh owing to their medicinal values. The study was conducted to unveil the antidiarrheal, analgesic, anti-inflammatory, anthelmintic, and anticoagulant activities of the methanol extract of these plants. The *in vivo* antidiarrheal assessment was conducted by castor oil-induced diarrheal test, while the analgesic and anti-inflammatory studies were carried out by acetic acid-induced pain and formalin-induced inflammation tests, respectively. The *in vitro* anticoagulant and anthelmintic studies were performed by clot lysis and worms' mortality measurement tests, respectively. In addition, an *in-silico* study of phytoconstituents was conducted by virtual molecular docking analysis. The methanol extracts of the selected plants showed substantial ($p < 0.05$, 0.01, 0.001) analgesic, anti-inflammatory, anthelmintic, and anticoagulant activity. MEPS 400 mg/kg showed the maximum antidiarrheal, analgesic, and anti-inflammatory potencies. In the clot lysis test, the METM, MEPS, and MEDG (100 μL) showed 27.07%, 27.27%, and 15.13% of clot lysis, respectively. METM yielded the highest anthelmintic effect and took 05:20 ± 0.89 to 23:42 ± 1.74 minutes paralysis time and 25:10 ± 0.88 to 1:03:54 ± 1.87 minutes death time. In addition, the selected phytoconstituents showed strong binding affinity to the receptors and ranged from –3.6 kcal/mol to –16.2 kcal/mol. The outcomes suggest strong antidiarrheal, analgesic, anti-inflammatory, anthelmintic, and anticoagulant potency of the selected plants, and the phytoconstituents have high binding affinity to the selected receptors.

19.1 INTRODUCTION

Diarrhea is a very common and notable concern in the contemporary public health sector. The scientific evidence extracted from *in vivo* and *in vitro* studies has shown that this gastrointestinal disease is generally associated with food indigestion, intestinal complications, infections, and incompatible medications followed by some deadly implications (Zhang et al., 2013). The diarrheal patients are afflicted with acute or toxic destructive impulses resulting in physiological, anatomical, and biological disease conditions as well as experiencing an exhausted feeling of pain (Bouhassira, 2008).

Generally, some common effector mechanisms and mediators beget gastrointestinal inflammation giving baneful outcomes such as multiple types of exudation, tissue breakdown, and tissue degeneration (Medzhitov, 2008). The pathophysiological episodes behind both inflammatory and non-inflammatory diarrheal diseases are mainly affected by pathogenic toxins. In the case of inflammatory diarrhea, the gastrointestinal cells are damaged, or cellular proteins are triggered resulting in the secretion of inflammatory mediators by the cytotoxins of pathogens like *Clostridium difficile, Shigella,* or entero-invasive *Escherichia coli* (EIEC). On the other hand, in non-inflammatory diarrhea, the gastrointestinal cells are actively seized by the endogenous enterotoxins of other pathogens; for example, heat-stable or labile toxins of enterotoxigenic *Escherichia coli* (ETEC) and cholera toxins. However, most of the diarrheal treatments intend to reduce discomforts and dysfunctions related to gastric motility rather than dealing specifically with this pathophysiologic incident (Rudra et al., 2020). Management of pain may not be effective in the case of chronically ill patients without any proper ideal treatment, and inadequate management may result in severe back pain, recurrent migraines, rheumatoid arthritis, etc., (Adnan et al., 2019). In some instances, non-steroidal anti-inflammatory drugs (NSAIDs) exhibit promising efficacy by restraining the activities of cyclooxygenase-1 and cyclooxygenase-2 enzymes, which in turn reduce the biosynthesis of several inflammatory agents. However, some major side effects like hepatotoxicity, renal disorder, peptic ulcer, etc., have been manifested due to prolonged use of some marketed NSAIDs. For this reason, the contemporary research sectors demand the exploration and establishment of novel phytochemicals and bioactive compounds for the establishment of better substitutes (Alam et al., 2021).

From the ancient era to the recent ones, plant extracts have exhibited significant therapeutic roles in global healthcare (Yadav, Kumar, & Siwach, 2006) and have also served humankind with remedial effects by healing physical torment (Mukul, Uddin, & Tito, 2007). The use of safe and cost-efficient herbal substances in diarrheal, microbial, and oxidative treatments has garnered much attention, as phytochemicals are more likely to provide significant nutrients, minimize the unwanted effects of synthetic treatments, and be safe for extended therapy (Kuna et al., 2019). Moreover, 80% of the world's population utilizes various plants for medicinal purposes, as reported by the WHO, due to reasons such as safety profiles, fewer side effects, and beneficial outcomes (Kamboj, 2000). However, the high-tech revolution and scientific diversification have paved the way to uncover the previously hidden potential of phytochemicals and efficacious pharmacological candidates, reinforcing opportunities. Therefore, today's pharmaceutical and other related companies are more likely to emphasize the discovery of novel natural lead compounds. Additionally, to identify the safety and efficacy of natural entities with fewer side effects, both pre-clinical and clinical trials are carried out (Hrckova & Velebny, 2012). Various natural plant substances as well as phyto-compounds are being assessed for significant anthelmintic properties (Chen et al., 2015). Previously found scientific evidence showed the cost-effective and promising uses of anthelmintics in the treatment of livestock parasitic diseases over the decades (Eguale, Tadesse, & Giday, 2011). Thus, the present research focuses on the constructive growth of novel, potential, and safe anti-inflammatory, anthelmintic, antidiarrheal, and anti-nociceptive drugs with very few side effects from natural plant sources based on the evidence of indigenous and ethno-pharmacological applications (Patrick-Iwuanyanwu et al., 2012). *Thysanolaena maxima* (Roxb.), Kuntze, a perennial forest grass plant belonging to the *Poaceae* family, is commonly found in the craggy regions of Bangladesh, Nepal, Bhutan, Thailand, and India, and the native people of these countries have been utilizing various parts of this plant from ancient times. Notably, the conventional medical practitioners of Meghalaya, India used to treat eye infections with the floral head puree of *T. maxima* (Hoque et al., 2020). Also, the Dimasa tribes of Assam, India used to treat digestion problems and GI gaseous problems with the raw tender leaves and flower buds (Rout, Sajem, & Nath, 2012). It is synonymously known as *Thysanolaena latifolia* and is generally familiar as "Tiger grass" (Hoque et al., 2020). Chemical isolation from the florets of this plant has displayed some significant compounds like isoswertisin, ergosterol peroxide, luteolin, benzoic acid, isoquercetin, 7α-hydroxy sitosterol, β-sitosterol, leonuriside A, 7R, 8R-4-O-methylsyringylglycerol, caffeic anhydride, jaboticabin acid, syringic acid, butyl protocatechuate, etc., (Shrestha et al., 2016). Again, *Pterospermum semisagittatum* Buch.-Ham. Ex Roxb, another medium to large tree belonging to the Sterculiaceae family, is commonly found in the Indian subcontinent where it is known as "laona assawar" or "asswar" (Khan et al., 2003). Scientific research on the family of this plant has revealed the presence of several flavonoids, terpenes, alkaloids, phenylpropanoids, cyclopropenoid fatty acids, sterols, and their glycosides including frangufoline, chamaedrine, waltherine A, waltherine B, waltherine C, adouetine X, melanovine A, melanovine B, adouetine, melofoline, scutianine B, scutianine C, integerrenine, franganine, waltherione A, waltherione B, melovinone, anorldianine 27-N-oxide, melosatin A, melosatin B, melochinone, (7S,8R)-dihydrodehydrodiconiferyl alcohol-90-O-b-D-glucopyranoside, mansonone C, D-H, luteolin, luteolin 7-O-glucoside (Cynaroside), luteolin-7-b-O-neohesperidoside, 10-hydroxy-4,7-megastigmadien-3-one-9-O-b-D-glucopyranoside, and 9–10-hydroxy-4,7-megastigmadien-3-one-9-O-b-D-glucopyranoside, and 9-hydroxy-4,7-megastigmadien-3-one-10-O-b-D-glucopyranoside, quercetin, quercetin 3-O-glucoside, quercetin 3-O-galactoside, apigenin, apigenin-7-b-O-neohesperidoside, kaempferol-3-O-b-D-glucoside, kaempferol-3-O-b-D-rutinoside, 3-O-a-L-arabinopyranoside, vitexin, angustifolia, etc., This species exerts beneficial medicinal effects in liver disease, digestive tract disorders, malaria, pulmonary disorders, rheumatic pain, tumors, cancer, burning sensation, heart palpitations, etc., *Diploclisia glaucescens* (B1.) Diels, another large woody perennial plant belonging to the Menispermaceae family, is scarcely found in the forest areas of the Chittagong Hill Tracts, the mid-country region of South India, and Sri Lanka (Jayasinghe, Wannigama, & MacLeod, 1998). The botanical characteristics of this plant include 20 cm wide stems at maximum; coriaceous, oval, or suborbicular-shaped leaves with rounded or acute apex; and yellowish tiny flowers (Jayasinghe, Wannigama, & MacLeod, 1998). Upon chemical evaluation of this plant extract, several compounds of significant classes have been demonstrated, such as serianic acid, stigmasterol, twenty-five saponins including phytolaccagenic acid, diploclisin (Jayasinghe et al., 2003), phytoecdysteroids including ecdysterone (Yekun et al., 1999), octadecanoic acid-9-yne, undecanoic acid, acetylenic acid, triacontanoic acid, β-sitosterol (Jayasinghe, Hara, & Fujimoto, 2007), (-) syringarenol (lignin) (Association, 2011), diploclidin, vibo-quercitol, alkaloid stepharine, proaporphine (Jayasinghe, 1992), glaucescine, and phenyl glycosides (Xiaochun et al., 2003). In addition, from the aqueous extraction evaluation of *D. glaucescens*, various amino acids, proteins, flavonoids, phenols, alkaloids, tannins, and saponins were also found (Cahlíková et al., 2011). Among these, the compound ecdysterone showed promising pharmacological activity in mice by increasing muscle insulin signaling via modulating acylcarnitine profile and mitochondrial oxidative phosphorylation complexes (Panda et al., 2009). Other research studies of this plant have also revealed evidence of possessing various pharmacological activities, such as spermicidal, insecticidal (Yekun et al., 1999), cytotoxicity, molluscicidal (Jayasinghe et al., 2003), antibacterial activities (Cahlíková et al., 2011), and ocetadecanoic acid-9-yne showed strong PAI-1 bioactivity (Jayasinghe, Hara, & Fujimoto, 2007). Moreover, it has also been reported that the leaves of this plant are effective in treating venereal diseases, urethritis, inflammation of the gall bladder, rheumatic arthritis, snake bites, pettishness, and diarrhea in Rangamati (Bandara et al., 1989).

In this current study, the methanol extracts of *Thysanolaena maxima* (METM), *Pterospermum semisagittatum* (MEPS), and *Diploclisia glaucescens*

(MEDG) leaves were investigated on mice for the antidiarrheal, analgesic, anti-inflammatory, anthelmintic, and anticoagulant activities as no scientific evidence was revealed on this before. Additionally, to observe the ligand binding interactions of pharmacologically active substances of *Thysanolaena maxima, Pterospermum semisagittatum*, and *Diploclisia glaucescens* with diarrhea, pain, inflammation, helminthic, and blood coagulation-related receptors; computer-aided *in silico* molecular docking studies were carried out.

19.2 MATERIALS AND METHODS

19.2.1 DRUGS AND CHEMICALS

The methanol and ethanol were bought from Panreac (Barcelona, Spain) and Carlo Erba, Strada Rivoltana (Rodano, Italy) respectively. Other necessary chemicals were taken from Sigma Aldrich (St. Louis, MO). All the chemicals were at their highest analytical grade.

19.2.2 COLLECTION AND IDENTIFICATION OF THE PLANT

Adequate amounts of fresh mature leaves of *Thysanolaena maxima* and *Diploclisia glaucescens* were obtained from Proshanti, Kaptai, Rangamati, Chittagong, Bangladesh, while *Pterospermum semisagittatum* leaves were collected from Mohamaya, Mirsarai, Chittagong, Bangladesh. All the samples were identified by Mr. Sajib Rudra, a taxonomist from the Department of Botany, University of Chittagong, Bangladesh. The voucher specimens are: *Thysanolaena maxima* (accession number: CTGUH SR 2799), *Pterospermum semisagittatum* (accession number: CTGUH SR 2801), and *Diploclisia glaucescens* (accession number: CTGUH SR 2800) and were deposited in the University of Chittagong Herbarium Center.

19.2.3 EXTRACTION OF PLANT MATERIAL

The collected leaves were cleaned and dried well under natural shade at room temperature (23 ± 0.5°C). Then, the dried leaf samples were ground to a coarse powder with a mechanical grinder. After that, about 500 g of powder was soaked in methanol and left for two days in a rotary shaker followed by filtration with Whatman filter paper, which yielded an adequate amount of crude extract successfully. Finally, the filtrate was evaporated in a rotary evaporator at 40°C (RE200, Bibby Sterling, UK) and collected. The yield value of methanol extracts of *Thysanolaena maxima* (METM), *Pterospermum semisagittatum* (MEPS), and *Diploclisia glaucescens* (MEDG) were 11 g, 13 g, and 15 g, respectively. The METM, MEPS, and MEDG were stored at 4°C for further experimental use.

19.2.4 EXPERIMENTAL ANIMALS

Seven weeks older Swiss albino mice weighing 23–30 g were acquired from the animal research division of the International Centre for Diarrheal Disease and Research, Bangladesh (ICDDR, B) for experimental purposes. The mice were kept in polycarbonate cages at room temperature (25 ± 2°C) and 55–60% humidity with a 12-hour daylight cycle resembling standard laboratory conditions. They were provided with an adequate amount of pellet animal diet and tap water. The mice were habituated for 14 days to adapt to the laboratory environment prior to the experiments. For each method, mice were divided into 8 groups (n = 5) where Group I received 1% Tween 80 (10 mL/kg) and Group II received a specific standard drug for each experiment in specific dosages. Whereas groups III–V were given METM, MEPS, and MEDG at 200 and 400 mg/kg of body weight orally, respectively.

19.2.5 IN VIVO STUDIES

19.2.5.1 ACUTE TOXICITY TEST

For the acute oral toxicity test, a standard laboratory setup according to the "Organization for Environmental Control Development" guidelines (OECD: Guidelines 420; Fixed-Dose Method) was ensured. Mice were kept starved for 4 hours. Then, the experimental mice (n = 5) were given a single dose (2000 mg/kg, body weight) of METM, MEPS, and MEDG orally. The mice were observed for unexpected responses such as behavioral changes, itching, rash, swelling, allergic symptoms, and mortality over the next 72 hours (Zaman et al., 2015).

19.2.5.2 CASTOR OIL-INDUCED DIARRHEA

For this test, a formerly documented procedure had been followed (Alam et al., 2021). Here, groups of mice (n = 5) were starved for 18 hours with only access to water. The dosing of drugs was mentioned in the 'Experimental animals' section. About 1 hour after the dosing, followed by extract feeding by oral gavage, 0.5 mL of castor oil was administered to each mouse orally to induce diarrhea. Then each of them was placed separately in a transparent cage lined with blotting paper. The blotting paper was changed every hour for up to four hours. After 4 hours of observation, the occurrence of diarrhea, fecal volume, weight of wet feces, and total fecal content were determined.

19.2.5.3 ACETIC ACID INDUCED PAIN

To assess the analgesic activity of METM, MEPS, and MEDG, the writhing test induced by acetic acid was performed (Emon, Jahan, & Sayeed, 2020). Mice were divided into 8 groups (n = 5) where Group I and II were given tween-80 (10 mL/kg of body weight orally) and acetylsalicylic acid (10 mg/kg of body weight intraperitoneally) and were considered as negative and positive controls, respectively. Whereas, groups III-IV, V-VI, VII-VIII were given METM, MEDG, and MEPS 200 and 400 mg/kg of body weight orally. After that, acetic acid solution (0.6% v/v) was injected intraperitoneally following the administration of test samples. Then, 5 min after the inoculation of the acetic acid solution, the occurrence of writhing was observed and counted over 25 min. Here,

$$\%\text{ of inhibition } = \frac{\textit{mean writhing numbers by control} - \textit{mean writhing numbers by test samples}}{\textit{mean writhing numbers by control}} \times 100$$

19.2.5.4 FORMALIN INDUCED INFLAMMATION

A formerly established protocol (Alam et al., 2021) was followed for the formalin-induced paw-licking experiment. Here, the experimental samples (METM, MEPS, and MEDG) were compared to ibuprofen (10 mg/kg), usually used as an anti-inflammatory drug. Around 1% Tween 80 was used as the negative control, and 200 and 400 mg/kg of body weight of METM, MEDG, and MEPS samples were weighed individually to conduct the experiment. The mice were divided into 8 groups (n=5) where Group-I and II were given 1% Tween-80 (10 mL/kg b.w; p.o.) and ibuprofen (10 mg/kg b.w; p.o) respectively. Group III-IV, V-VI, VII-VIII were treated with the METM, MEDG, and MEPS 200 and 400 mg/kg of body weight p.o. dose, respectively. After injecting 2.5% formalin into the sub-plantar of the mice's left hind paw with a dose of 20 μL/kg, they were transferred to clear watch boxes for observation over 30 minutes. The early phase (0–5 min) and late phase (15–30 min) time span were considered, respectively as the neurogenic phase and inflammatory phase. Finally, outcomes of pain reduction ratios were reported using the following formula:

$$\%\text{ of inhibition} = \frac{\textit{mean licking time by control group} - \textit{mean licking time by test samples}}{\textit{mean licking time by control group}} \times 100$$

19.2.6 *IN VITRO* STUDIES

19.2.6.1 SERUM-CLOT LYTIC ACTIVITY

A previously established methodology was adopted to carry out the serum-clot lytic activity test (Emon et al., (n.d.); Emon et al., 2021). In this experiment, for each plant extract sample of METM, MEPS, and MEDG, 0.5 ml of fresh whole blood sample was collected from healthy individuals (n=10). It was ensured that the individuals did not consume anticoagulants or any oral contraceptives for the most recent 10 days. Then, the blood samples were placed into previously weighed (W1) 15 sterile Eppendorf tubes and kept for incubation at 37°C for 45 minutes for clot formation. After clot formation, without muddling the clot, serum was separated from each tube and the Eppendorf with clot was weighed (W2). So, the clot weight was determined using the following equation:

$$\text{Clot weight} = (\textit{Weight of the Eppendorf with clot} - \textit{weight of empty Eppendorf}).$$

After that, for clot lysis, the METM, MEDG, and MEPS samples were individually added to each tube at a concentration of 100 μL (100 mg/10 mL). Additionally, streptokinase and distilled water were also added to the tube with the samples for positive and negative controls, respectively. Then the Eppendorf tubes were kept for incubation at 37°C for 90 min, and the thrombolysis occurred. The tubes were then weighed again after clot lysis to determine the percentage of clot lysis. The experiments were repeated three times with the same blood sample from five volunteers. The level of thrombolysis was calculated using the following equation:

$$\text{Clot lysis (\%)} = \frac{\textit{Weight of the Eppendorf with clot after lysis}}{\textit{Weight of the clot}} \times 100$$

19.2.6.2 ANTHELMINTIC ACTIVITY

The research protocol of anthelmintic activity: Simply put, the anthelmintic activity was conducted following the earlier report (Sajib Rudra et al., 2020). In this study, twelve petri dishes were used, each containing 6 Tubifex tubifex worms. Samples of METM, MEDG, and MEPS were applied to the petri dishes at different concentrations (10 mg/mL, 5 mg/mL, 2.5 mg/mL). Albendazole (0.5 mg/mL) was used as the standard drug. The paralysis of the worms was confirmed by zero movement, and the death of the worms was confirmed by no reaction even after vigorous shaking or plunging into warm water. The 'time for paralysis' and 'time for death' were then recorded, which later helped determine the anthelmintic activity of the different samples.

19.2.7 IN SILICO COMPUTER-AIDED VIRTUAL SCREENING

- **Software Tools:** For computer-aided analysis, the following software, devices, and databases were used: Auto-Dock Vina, UCSF Chimera, DrugBank, Discovery Studio Visualizer 2020 (BIOVIA), MGL Instruments, DrugBank, PubChem, and Protein Data Bank (PDB) (Adeniji et al., 2020).
- **Selection of the Target Proteins:** A precise review of the contemporary scientific investigation has primarily accomplished the assortment of the conceivable targets for diarrhea, pain, inflammation, helminthic infections, and coagulation docking. The selected targets were reconfirmed from the PDB (Protein Data Bank).
- **Preparation of Target Proteins:** Before preparing the target proteins, the specific receptors responsible for stimulating diarrhea, inflammation, helminthic infections, pain, and blood clotting were extracted, such as M3 muscarinic acetylcholine receptor (PDB: 5ZHP) (Liu et al., 2018), Cyclooxygenase-2 (PDB: 6COX) (Kurumbail et al., 1996), tubulin-colchicine (PDB: 1SAO) (Ravelli et al., 2004),

cyclooxygenase-1 (PDB: 2OYE) (Harman et al., 2007), and tissue plasminogen activator (PDB: 1A5H) (Renatus et al., 1997) from the RCSB Protein Data Bank (PDB). These receptors have been observed conferring similar aspects in *Rattus norvegicus, Escherichia virus T4, Bos taurus, Rattus norvegicus*, and *Homo sapiens*, respectively. Moreover, these structures have already been used for various recent research purposes and noted in several putative publications. Hence, the findings can be reproduced by utilizing appropriate measurements. For preparing the target proteins, the protein ligands were separated from their complexes using Discovery Studio Visualizer (BIOVIA). Then, with molecular docking, the proteins were optimized and assembled via UCSF Chimera. Removal of water molecules from the structure was performed followed by the addition of necessary polar hydrogen atoms. Finally, after assessing the Gasteiger charge, the proteins were converted into PDBQT format and saved via the AutoDock Vina docking program (Masters, Eagon, & Heying, 2020).

- **Virtual Filtering:** For this purpose, a virtual screening tool namely PyRx-AutoDock Vina docking software has been used (Masters, Eagon, & Heying, 2020). With this software, the ligands such as isoquercetin, butyl protocatechuate, luteolin, benzoic acid, undecanoic acid, serjanic acid, stigmasterol, phytolaccagenic acid, waltherione A, ecdysterone, diclofenac Na, albendazole, Ibuprofen, loperamide, aspirin were minimized and optimized. Then these ligands were allowed to interact with the selected receptor targets namely: 5ZHP, 2OYE, 6COX, 1SAO, and 1A5H. The grid selection allows the ligand to freely move through the assigned values in planes (x, y, and z). The coordinating variables x, y, and z were acquired from the corresponding PDB crystals by using the Discovery Studio Visualizer, i.e., 5ZHP, 2OYE, 6COX, 1SAO, and 1A5H. Following the previous publications, the predominant size for the grid was 60 Å x 60 Å x 60 Å. Finally, BIOVIA Discovery Studio Visualizer 2020 visualized the docking analysis (ligand-protein interactions).

19.2.8 STATISTICAL ANALYSIS

The data analysis was done by one-way analysis of variance (ANOVA). For determining the significant differences between the test and control groups, Dunnett's test was done using GraphPad Prism, Version 9.0.0

(GraphPad Software Inc., San Diego, CA, USA). All the values were demonstrated as mean ± standard error of the mean (SEM). The obtained *p*-values ($*p < 0.05$, $**p < 0.01$, and $***p < 0.001$) were noted as statistically significant.

19.3 RESULTS

19.3.1 EFFECT OF METM, MEPS, AND MEDG ON ACUTE TOXICITY TEST

No behavioral changes, itching, rash, swelling, dermal infection, ear and eye infections, convulsions, diarrhea, elevated heart rate, elevated breathing, coma, or death were observed after 72 hours of oral administration of extracts.

19.3.2 EFFECT OF METM, MEPS, AND MEDG ON CASTOR OIL-INDUCED DIARRHEA

The crude methanol extracts of *Thysanolaena maxima, Diploclisia glaucescens*, and *Pterospermum semisagittatum* exhibited dose-dependent antidiarrheal potency. With the administration of loperamide (2 mg/kg) as a standard drug, the diarrheal episodes and weight of wet feces were reduced predominantly in an extremely significant ($p < 0.001$) manner (diarrheal episodes: 2.16 ± 0.30, weight of wet feces: 0.20 ± 0.04). Among the extracts, MEPS 400 mg/kg yielded the maximum depletion of diarrheal episodes (4.26 ± 0.37; $p < 0.001$) and weight of wet feces (0.42 ± 0.01; $p < 0.001$). The summary of the findings has been demonstrated in Table 19.1.

19.3.3 EFFECT OF METM, MEPS, AND MEDG ON ACETIC ACID-INDUCED PAIN

The pain of mice that were generated by acetic acid has been significantly reduced ($p < 0.5$, 0.01, 0.001) with the treatment of METM, MEPS, and MEDG (200 and 400 mg/kg, b.w; p.o). With the treatment of METM, MEPS, and MEDG (400 mg/kg), acetic acid-induced pain was decreased to 43.93%, 59.23%, and 52.48% subsequently, where the standard drug possessed 75.66% of pain inhibition. The comprehensive data has been explored in Figure 19.1.

TABLE 19.1 Castor Oil Induced Inhibition Reports of METM, MEPS, MEDG and Control Samples in Mice Model

Dose Treatment	Number of wet feces	Total number of feces	Total weight of wet feces (gm)
Tween 80 (mL/kg)	$10.17 \pm 0.79^{\#}$	$12.67 \pm 0.80^{\#}$	$1.37 \pm 0.02^{\#}$
Loperamide (2 mg/kg)	$1.16 \pm 0.16^{***}$	$2.16 \pm 0.30^{***}$	$0.20 \pm 04^{***}$
METM 200 mg/kg	$7.78 \pm 0.57^{**}$	$8.66 \pm 0.33^{**}$	1.12 ± 0.13
METM 400 mg/kg	$5.23 \pm 1.63^{***}$	$5.78 \pm 0.33^{***}$	$0.65 \pm 0.17^{***}$
MEPS 200 mg/kg	$6.26 \pm 0.57^{***}$	$9.66 \pm 1.2^{*}$	$0.70 \pm 0.06^{***}$
MEPS 400 mg/kg	$3.33 \pm 0.33^{***}$	$4.26 \pm 0.37^{***}$	$0.42 \pm 0.01^{***}$
MEDG 200 mg/kg	$5.88 \pm 0.57^{***}$	$5.00 \pm 0.57^{***}$	$0.69 \pm 0.13^{***}$
MEDG 400 mg/kg	$4.33 \pm 0.88^{***}$	$7.33 \pm 0.33^{***}$	$0.81 \pm 0.10^{***}$

$^{***}p<0.001$, $^{**}p<0.01$, $^{*}p<0.05$. All values are Mean ± SD and statistically analyzed using Two Way Analysis of Variance (ANOVA) followed by Dunnett's multiple comparison test, n=5. METM = methanol extract of Thysanolaena maxima, MEPS = methanol extract of Pterospermum semisagittatum and MEDG = methanol extract of Diploclisia glaucescens.

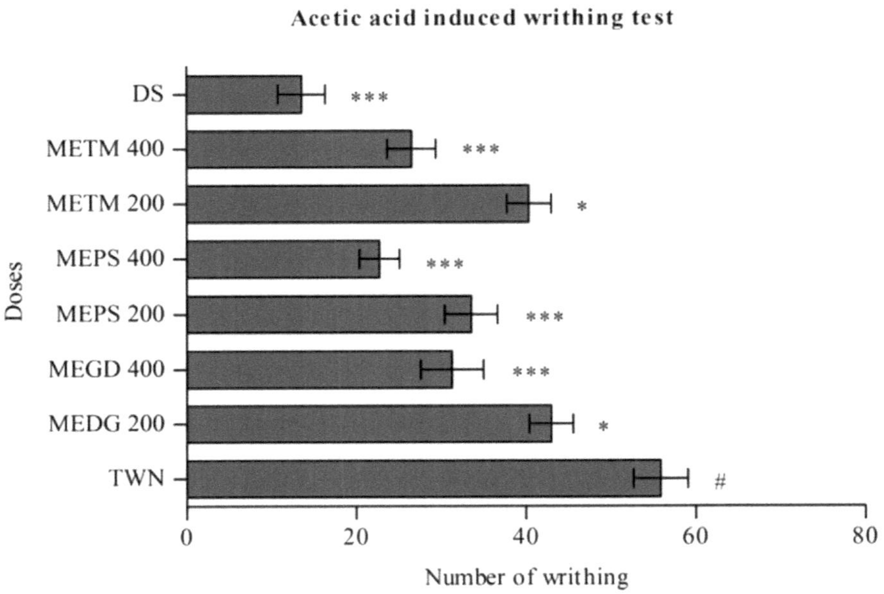

FIGURE 19.1 Analgesic activity of METM, MEPS, MEDG. $^{***}p<0.001$, $^{**}p<0.01$, $^{*}p<0.05$. All values are Mean ± SD and statistically analyzed using Two Way Analysis of Variance (ANOVA) followed by Dunnett's multiple comparison test, n=5. METM = methanol extract of *Thysanolaena maxima*, MEPS = methanol extract of *Pterospermum semisagittatum* and MEDG = methanol extract of *Diploclisia glaucescens*. DS = diclofenac sodium, TWN = 1% tween 80.

19.3.4 EFFECT OF METM, MEPS, AND MEDG ON FORMALIN-INDUCED INFLAMMATION

In the formalin-induced test, the MEPS exhibited a substantial anti-inflammatory response. Compared to other extracts, the peak percentage of inhibition of paw licking during the early and late phases has been shown by MEPS at 400 mg/kg. The percentage inhibition of formalin-induced paw licking by METM, MEPS, and MEDG (400 mg/kg) is 49.64%, 60.38%, and 21.80% respectively. Additionally, METM, MEPS, and MEDG (200 mg/kg) showed 20.98%, 48.93%, and 14.22% of inhibition, while the standard drug (ibuprofen at 10 mg/kg) exhibited 71.03% inhibition of formalin-induced inflammation. The comparative data are presented in Figure 19.2.

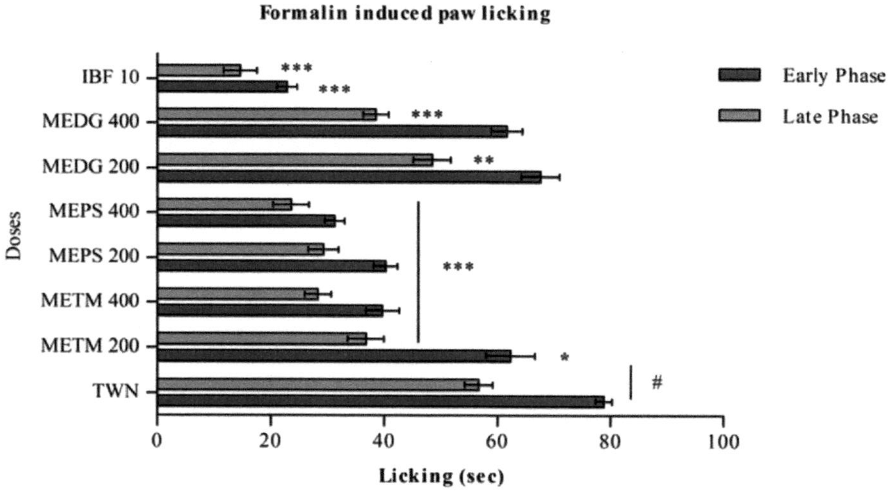

FIGURE 19.2 Anti-inflammatory activity of METM, MEPS, MEDG. $^{***}p<0.001$, $^{**}p<0.01$, $^{*}p<0.05$. All values are Mean ± SD and statistically analyzed using Two Way Analysis of Variance (ANOVA) followed by Dunnett's multiple comparison test, n=5. METM = methanol extract of *Thysanolaena maxima*, MEPS = methanol extract of *Pterospermum semisagittatum* and MEDG = methanol extract of *Diploclisia glaucescens*, IBF = Ibuprofen, TWN = 1% tween 80.

19.3.5 EFFECT OF METM, MEPS, AND MEDG ON SERUM-CLOT LYTIC ACTIVITY TEST

The clot lytic activity of the extracts showed effects in very trace amounts. In this study, the METM, MEPS, and MEDG (100 μL) exhibited a significant percentage of clot lysis (27.27%, 27.07%, and 15.13% respectively; $p < 0.001$) activity where the standard control (positive and negative) yielded 59.68% and 3.59% of clot lysis. The findings are presented in Figure 19.3.

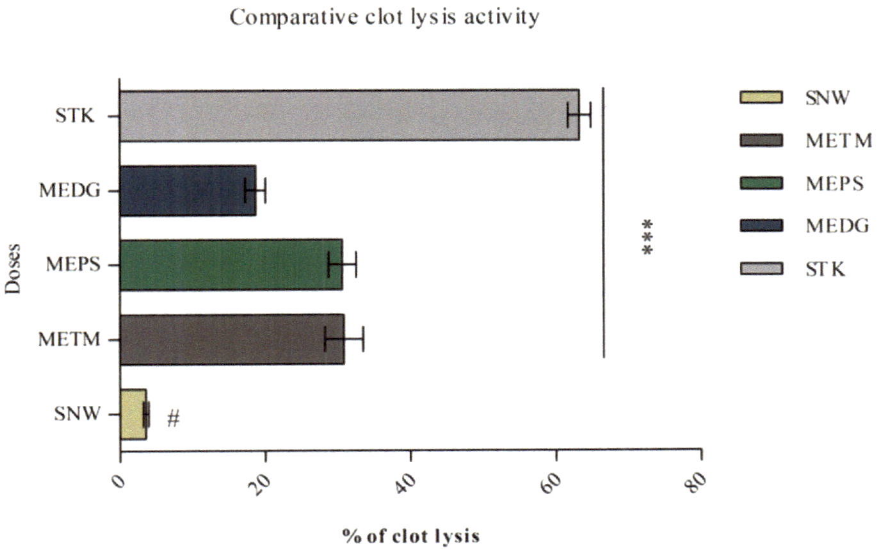

FIGURE 19.3 The clot lytic activity of METM, MEPS and MEDG. $^{***}p<0.001$, $^{**}p<0.01$, $^{*}p<0.05$. All values are Mean ± SD and statistically analyzed using One Way Analysis of Variance (ANOVA) followed by Dunnett's multiple comparison test, n=10. METM = methanol extract of *Thysanolaena maxima*, MEPS = methanol extract of *Pterospermum semisagittatum* and MEDG = methanol extract of *Diploclisia glaucescens*, STK = streptokinase, SNW = Saline water.

19.3.6 EFFECT OF METM, MEPS, AND MEDG ON ANTHELMINTIC ASSAY

METM, MEPS, and MEDG developed dose-dependent anthelmintic activity. For the doses of 20 mg/kg, the time taken by METM, MEPS, and MEDG to paralyze the worms was 05:20 ± 0.89, 12:22 ± 1.62, and 11:56 ± 0.97 minutes, respectively. The paralysis and the death time for the standard drug (albendazole 5 mg/kg) were 3.48 ± 0.73 and 31.00 ± 1.71 minutes, respectively. The data has been briefly presented in Table 19.2.

TABLE 19.2 Anthelmintic Activity of METM, MEPS, MEDG and Control Samples

Dose Treatment	Paralysis time (min)	Death time (min)
ALDZ 5 mg/kg	3.48 ± 0.73	31.00 ± 1.71
METM 10 mg/kg	08:12 ± 1.58	25:10 ± 0.88
METM 20 mg/kg	05:20 ± 0.89	30:40 ± 1.32
MEPS 10 mg/kg	16:22 ± 0.65	42:15 ± 2.44
MEPS 20 mg/kg	12:22 ± 1.62	30.42 ± 2.36
MEDG 10 mg/kg	23:42 ± 1.74	1:03:54 ± 1.87
MEDG 20 mg/kg	11:56 ± 0.97	40:10 ± 1.53

METM = methanol extract of Thysanolaena maxima, MEPS = methanol extract of Pterospermum semisagittatum and MEDG = methanol extract of Diploclisia glaucescens. ALDZ = albendazole.

19.3.7 IN SILICO VIRTUAL SCREENING

The results of antidiarrheal, analgesic, anti-inflammatory, anthelmintic, and anticoagulant docking have been presented in Table 19.3 and Figure 19.4. In antidiarrheal docking, among the interactions of the M3 muscarinic acetylcholine receptor (PDB ID: 5ZHP) and the selected compounds, stigmasterol showed the highest binding affinity to the 5ZHP and scored –7.9 kcal/mol. The ranking of the docking score is as follows: stigmasterol > loperamide > serjanic acid > waltherione A > phytolaccagenic acid > luteolin > ecdysterone > isoquercitrin > undecanoic acid > butyl protocatechuate. Besides, in analgesic and anti-inflammatory docking, ecdysterone unveiled the highest binding score when it interacted with the cyclooxygenase-1 (PDB: 2OYE) and cyclooxygenase-2 (PDB: 6COX) receptors. Ecdysterone binds to the COX-1 receptor via a series of bonds: unfavorable bump (his207) and conventional hydrogen bond (glu454). The interaction of COX-2 and ecdysterone occurred through conventional hydrogen bonding (amino acid residues: asn375, gly225). The ranking of the binding score for the selected compounds and tubulin-colchicine is: waltherione A > benzoic acid > luteolin > isoquercitrin > serjanic acid > ecdysterone > albendazole > undecanoic acid > butyl protocatechuate. The binding scores for the waltherione A and the albendazole are –8.4 kcal/mol and –6.8 kcal/mol, respectively. Moreover, phytolaccagenic acid and stigmasterol did not bind to the tubulin-colchicine receptor. In anticoagulation docking analysis, waltherione A exhibited the highest binding affinity when docking was employed for the selected compounds and tissue plasminogen activator (PDB: 1A5H). Waltherione A obtained a docking score of –8.0 kcal/mol. Furthermore, waltherione A binds

TABLE 19.3 Molecular Docking Score of the Selected Phytocontituents with the M3 Muscarinic Acetylcholine Receptor (PDB: 5ZHP), Cyclooxygenase-2 (PDB: 6COX), Tubulin-Colchicine (PDB: 1SAO), Cyclooxygenase-1 (PDB: 2OYE) and Tissue Plasminogen Activator

		Docking Score				
		Anti-diarrhea	**Analgesic**	**Anti-inflammatory**	**Anthelmintic**	**Clot-lytic**
Compounds	PubChem CID	5ZHP	2OYE	6COX	1SA0	1A5H
Isoquercitrin	5280804	-5.8	-8.5	-9.4	-7.7	-7.6
Benzoic acid	243	-	-7.6	-7.7	-8.2	-7.0
Luteolin	5280445	-6.3	-9.2	-8.7	-7.9	-7.5
Butyl Protocatechuate	21888677	-4.9	-	-5.9	-5.7	-5.7
Ecdysterone	5459840	-6.3	**-12.6**	**-16.2**	-7.2	-7.2
Waltherione A	11474862	-7.6	-8.2	-9.5	**-8.4**	**-8.0**
Serjanic acid	12315473	-7.7	-3.6	10.0	-7.5	-7.5
Phytolaccagenic Acid	13878342	-7.3	-11.1	-9.0	-	-7.4
Undecanoic acid	8180	-4.5	-	-5.0	-5.5	-5.0
Stigmasterol	5280794	**-7.9**	-4.0	-9.9	-	-7.0
Standard drugs (Loperamide/ Diclofenac sodium/Ibuprofen/ Albendazole/Aspirin)	3955/5018304/3672/2082/2244	-7.7	-7.1	-6.6	-6.8	-5.2

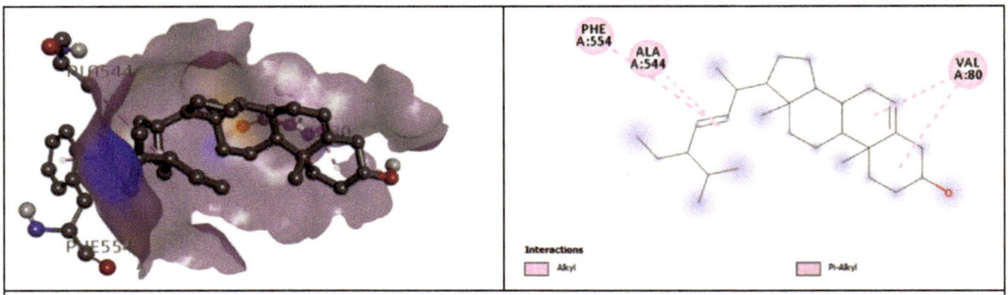

Stigmasterol-5ZHP

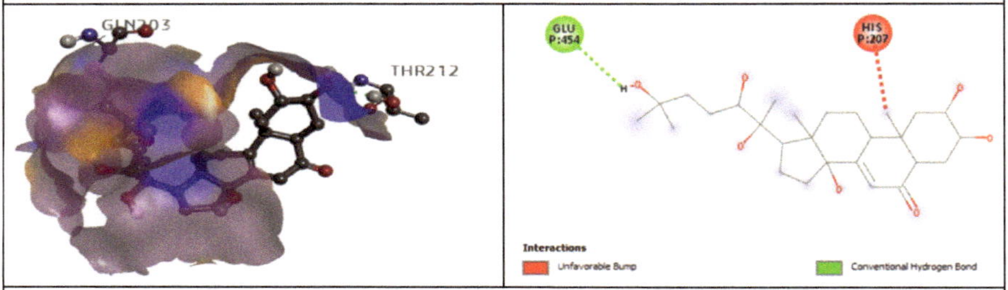

Ecdysterone-2OYE

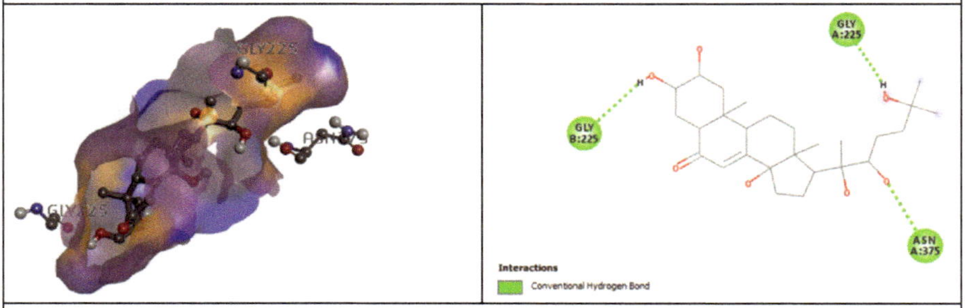

Ecdysterone-6COX

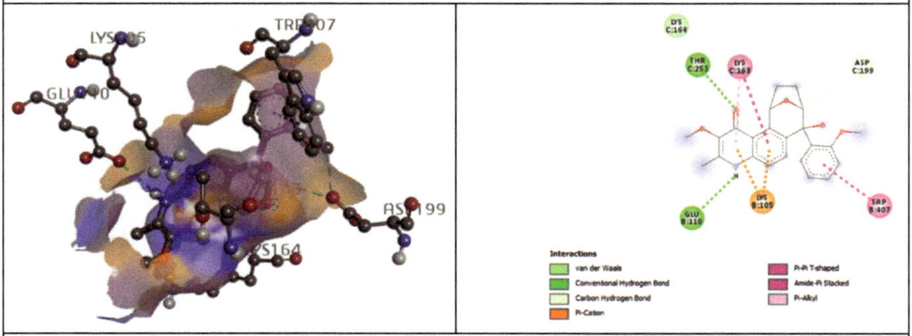

1AS0- Waltherione A

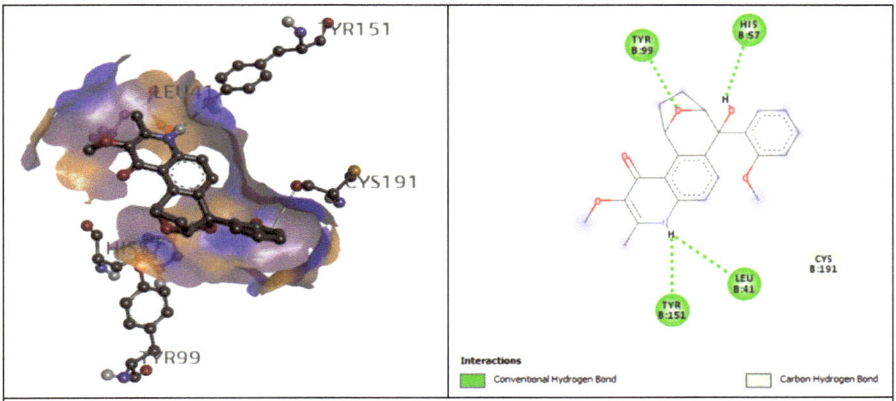

1A5H- Waltherione A

FIGURE 19.4 Best (2D and 3D) interaction of selected proteins and ligands (Stigmasterol-5ZHP, Ecdysterone-2OYE, Ecdysterone-6COX, 1AS0- Waltherione A, 1A5H- Waltherione A).

to the tissue plasminogen activator through a group of amino acids, and the amino acid binds to the compounds through several bonds: conventional hydrogen bond (tyr151, tyr99, his57, leu41) and carbon-hydrogen bond (cys191). The standard drug (aspirin) leads to a binding score of –5.2 kcal/mol with the tissue plasminogen activator.

19.4 DISCUSSION

Scientists claim that about 4,000 to 10,000 species of different extinct therapeutic sanative plants, which may be found locally, nationally, regionally, or transnationally, can confer a significant contribution to the betterment of public health, economics, and living standards (Kala, Dhyani, & Sajwan, 2006). Phytochemicals or plant-derived marketed drugs can be a good alternative to antidiarrheal drugs such as Zangrado (Miller et al., 2000), Seirogan (Kuge, Shibata, &. Willett, 2003), Kampo (Oi et al., 2002), tormentil root (Subbotina et al., 2003), and Tong-Xie-Ning (Gang et al., 2006), which are used in the treatment of diarrhea. Some medicinal herbs, namely, *Camellia sinensis* (Hrckova & Velebny, 2012), *Psidium guajava* (Yadav, Kumar, & Siwach, 2006), *Croton lechleri* (Alam et al., 2020), and *Mentha piperita* (Mukul, Uddin, & Tito, 2007), also provide notable antidiarrheal activity (Rawat, Singh, & Kumar, 2017). Based on ethno-pharmacological use, plant sources have been utilized and evaluated against diarrhea and microbial invasion resulting from their intrinsic therapeutic potentialities (Williamson, Okpako, & Evans, 1996). In this research study, 200 and 400 mg/kg body weight doses of METM, MEPS, and MEDG samples were taken into consideration and administered orally to the animal models. As a result, a dose-dependent impediment of the frequency of diarrheal feces was seen with statistically significant p-values ($p < 0.05$, 0.01, and 0.001). Besides, impairment between peristalsis of the intestine and/or GI absorption may lead to diarrhea (Gidudu et al., 2011). Castor oil can be used as a very handy laxative option which can exert a laxative effect successively by stimulating fluid secretion, inhibiting absorption of water and electrolytes, decreasing absorption of active Na^+ and K^+ ions, and lessening Na^+, K^+, -ATPase in the small intestine and colon via hydrolyzation and irritant actions of ricinoleic acid, which is liberated by pancreatic acid in the upper small intestine (Shiferie & Shibeshi, 2013; Mbagwu& Adeyemi, 2008). Besides, ricinoleic acid releases prostaglandins which initiate diarrhea via regulating GIT and stimulating motility secretion (Hu et al., 2009; Agunu et al., 2005). A recent study on molecular and cellular mechanisms suggests that the laxative effect of castor oil may be exerted by the liberation of ricinoleic acid, which subsequently activates the EP3 prostanoid receptor triggering the intestinal and uterine muscle cells. Hence, one of the mechanisms of antidiarrheal activity of METM, MEDG, and MEPS might be the depletion of prostaglandin secretions (Tunaru et al., 2012). In addition, plant extracts possessing flavonoids, alkaloids, and/or other bioactive phytoconstituents may also exert antidiarrheal activity as these phyto-compounds restrict the production of prostaglandins and autacoids via modification of cyclooxygenase 1 (COX-1), cyclooxygenase 2 (COX-2), and lipo-oxygenase (LOX) production, which also justifies the reported antidiarrheal activities of this study (Mishra, Seth, & Maurya, 2016).

Peripheral pain has been detected as a manifestation of acetic acid-induced writhing response (Meite et al., 2009). Additionally, the pain sensory neuron is triggered via the invocation of different biochemical agents responsible for pain such as PGI2, PGE2, PGF2α, etc., and gives the perception of pain to a greater extent through capillary permeability (Zahid, Rizwani, & Shareef, 2014, Amico-Roxas et al., 1984). The respective extracts, i.e., METM, MEPS, MEDG (400 mg/kg), and MEPS 200 mg/kg have shown significant pain inhibitors, which may be attributed to possible mechanisms of affecting prostaglandin synthesis like Aspirin and other NSAIDs do in the formalin-induced licking test. It has been stated that formalin-induced pain possesses two phases, namely the early phase (0–5 minutes) and the late phase (15–20 minutes) (Kolesnikov et al., 2004). When C-fiber neurons are activated, the early phase is augmented, and sequentially the late phase is stimulated via the discharge of inflammatory agents that result from the early phase (Shibata et al., 1989). Though peripherally acting drugs only perform effectively at the late phase via prostaglandin synthesis inhibition, centrally acting drugs can be effective in both phases.

Thrombosis or blood clot formation process can be denoted as a series of events including deposition of platelets, tissue factor, and fibrin in the bruised areas of the blood vessels or endothelial cell surface (Furie & Furie, 2008). Here, the cascade mechanism is initiated by the activated platelets causing platelet-platelet bonds which later bind with leukocytes and ultimately result in the complicated process of plaque formation and growth (Das et al., 2013). On the other hand, plasmin is a natural antithrombotic that is converted from plasminogen and causes fibrinolysis. So, to exert thrombolytic activity an agent should be able to desecrate the fibrinogen and fibrin. For example, streptokinase converts bacterial plasminogen to plasmin and causes the enzymatic breakdown of fibrin (Pantzar, Ljungh, &. Wadström, 1998; Bhowmick, Sarwar, Rahman Dewan, Das et al., 2014). The phytoconstituents are believed to possess a remarkable role in the thrombolytic process (Alam et al., 2020). The extracts (100 μL) of *Thysanolaena maxima, Pterospermum semisagittatum*, and *Diploclisia glaucescens* showed substantial thrombolytic activity. Phytoconstituents of the plant extracts are expected to be responsible for their thrombolytic effects (Rudra, Sawon, Emon, Alam et al., 2020). However, with the progression of further research on cell viability tests and *in vivo* studies, it may ensure a significant function in treating mostly outgrowing cardiovascular disorders.

The anthelmintic activity of METM, MEPS, and MEDG at different concentrations on *Tubifex tubifex* by causing paralysis and death was compared with the reference drug albendazole. Albendazole generally suppresses the helminth β-tubulin polymerization affecting microtubule-dependent roles like glucose uptake (Bhowmick, Sarwar, Rahman Dewan, Das et al., 2014). The phytochemical screening outcomes of METM, MEPS, and MEDG extracts have demonstrated the presence of tannins, alkaloids, and steroids, and hence the anthelmintic activity of extracts can be expected due to the presence of alkaloids and tannins (Shrestha et al., 2016; Jayasinghe et al., 2003; Jayasinghe, Hara, & Fujimoto, 2007; Jayasinghe, 1992). Thus, it can be assumed that one feasible anthelmintic mode of action of the extracts may be owing to tannins, which are able to cause the death of the parasite by binding with free proteins of glycoprotein in the parasitic cuticle (Hussain, Sonkar, Ahmad Md & Wahab, 2012). Another conceivable anthelmintic mode of action may be attributed to alkaloids, which are able to cause paralysis by affecting the central nervous system (CNS) of the parasite (Chander, Sri, Sravanthi, & Susmitha, 2014).

Molecular docking, a computer-aided molecular modeling technique, is able to build an idea of the anticipated mode of action and binding pattern of a ligand to the targeted binding pockets of different proteins. Therefore, it is extensively utilized in the estimation of the bioactivity of various phytoconstituents (Khan, Nazir, Raiz, Saleem et al., 2019). Hence, to justify and validate the experimental findings of conducted biological studies, molecular docking was performed in this research work. Here, a docking study was conducted for ten specified phytochemicals of *Thysanolaena maxima, Diploclisia glaucescens*, and *Pterospermum semisagittatum* against five targets for antidiarrheal, analgesic, anti-inflammatory, thrombolytic, and anthelmintic studies, respectively, to establish a better insight into their prospective biological activity along with the validation of the outcomes found in *in vivo* and *in vitro* studies. In the antidiarrheal molecular docking analysis, the ranges of docking scores were within –4.9 to –7.9 kcal/mol, where several amino acid residues interacted with selected compounds via alkyl and pi-alkyl bonds. In the case of analgesic and anti-inflammatory docking analysis, the ranges of docking values were within –3.6 to –12.6 kcal/mol and –5.0 to –16.2 kcal/mol, respectively. The outcomes were compared with the respective standard drugs, and the phytoconstituents showed a more potent receptor-ligand binding affinity. The docking results are near and likely better than the standard drug's binding affinity toward the selected receptors. Overall, the comparative analysis unveils that METM, MEPS, and MEDG are highly potential to reduce castor oil-induced diarrhea. Again, MEPS showed better inhibition in acetic acid-induced analgesia. Indifferently, MEPS obtained the highest inhibition of formalin-induced inflammation in mice. MEPS and METM showed significant clot lytic activity than the negative control but in very trace amounts compared to the standard drug. Similarly, both of them showed better paralysis and death time against *Tubifex tubifex* worms. The phytochemical constituents of the plants studied here are expected to exhibit the pharmacological outcomes as demonstrated.

19.5 CONCLUSION

The bioassays of METM, MEPS, and MEDG affirmed the safety concern of methanol extracts of *Thysanolaena maxima, Pterospermum semisagittatum,* and *Diploclisia glaucescens* as well as substantial efficacy as antidiarrheal, analgesic, anti-inflammatory, anthelmintic, and thrombolytic agents. In addition, *the in silico* study also supports stigmasterol, ecdysterone, and waltherione A as effective phytocompounds for diarrhea, pain, inflammation, blood coagulation, and abdominal worm management by considering their best docking score. However, further studies are still required to reveal better insights into their pharmacological activities as well as the proper mode of action.

ACKNOWLEDGMENTS

We are thankful to the Department of Pharmacy, International Islamic University Chittagong, for technical and laboratory support.

KEYWORDS

- **Antidiarrheal Activity**
- **Plant Extraction**
- **Pharmacological Formulation**
- **Anti-inflammatory Properties**
- **Anthelmintic Activity**

REFERENCES

Adeniji, S. E., et al. (2020). Quantitative structure-activity relationship model, molecular docking simulation, and computational design of some novel compounds against DNA gyrase receptor. *Chemistry Africa, 3*(2), 391–408.

Adnan, M., et al. (2019). Evaluation of anti-nociceptive and anti-inflammatory activities of the methanol extract of *Holigarna caustica* (Dennst.) Oken leaves. *Journal of Ethnopharmacology, 236*, 401–411.

Agunu, A., et al. (2005). Evaluation of five medicinal plants used in diarrhea treatment in Nigeria. *Journal of Ethnopharmacology, 101*(1–3), 27–30.

Alam, S., et al. (2020). Pharmacological and computer-aided studies provide new insights into *Millettia peguensis* Ali (Fabaceae). *Saudi Pharmaceutical Journal. 28*(12), 1777–1790.

Alam, S., et al. (2021). Antidiarrheal, antimicrobial and antioxidant potentials of methanol extract of *Colocasia gigantea* Hook. f. leaves: Evidenced from in vivo and in vitro studies along with computer-aided approaches. *BMC Complementary Medicine and Therapies, 21*(1), 119.

Amico-Roxas, M., et al. (1984). Gangliosides antinociceptive effects in rodents. *Archives Internationales de Pharmacodynamie et de Thérapie, 272*(1), 103–117.

Association, A. D. (2011). Ecdysterone enhances muscle insulin signaling by modulating acylcarnitine profile and mitochondrial oxidative phosphorylation complexes in mice fed a high-fat diet withdrawn. *Diabetes, 60*(5), 1645.

Bandara, B. R., et al. (1989). Diploclisin, a bidesmosidic triterpenoid saponin from *Diploclisia glaucescens*. *Phytochemistry, 28*(10), 2783–2785.

Bhowmick, R., et al. (2014). In ws1-inflammatory potential in Swiss albino mice and in vitro thrombolytic activity of hydroalcoholic extract from *Litsea glutinosa* leaves. *Biological Research, 47*(1), 1–8. https://doi.org/10.1186/0717-6287-47-56.

Bouhassira, D. (2008). Définition et classification des douleurs neuropathiques. *La Presse Médicale, 37*(2), 311–314.

Cahlíková, L., et al. (2011). Ecdysterone and its activity on some degenerative diseases. *Natural Product Communications, 6*(5), 1934578X1100600527.

Chander, P. A., Sri, H. Y., Sravanthi, N. B. M., & Susmitha, U. V. (2014). In vitro anthelmintic activity of *Barleria buxifolia* on Indian adult earthworms and estimation of total flavonoid content. *Asian Pacific Journal of Tropical Disease, 4*, S233–S235.

Chen, Y., et al. (2015). Antinociceptive and anti-inflammatory activities of *Schefflera octophylla* extracts. *Journal of Ethnopharmacology, 171*, 42–50.

Das, A., et al. (2013). Investigation of in vitro thrombolytic potential of ethanolic extract of *Momordica charantia* fruits: An anti-diabetic medicinal plant.

Eguale, T., Tadesse, D., & Giday, M. (2011). In vitro anthelmintic activity of crude extracts of five medicinal plants against egg-hatching and larval development of Haemonchus contortus. *Journal of Ethnopharmacology, 137*(1), 108–113.

Emanuel, E. J. (2013). Reconsidering the Declaration of Helsinki. *The Lancet, 381*(9877), 1532–1533. https://doi.org/10.1016/S0140-6736(13)60970-8.

Emon, N. U., et al. (2021). Antidepressant, anxiolytic, antipyretic, and thrombolytic profiling of methanol extract of the aerial part of Piper nigrum: *In vivo, in vitro*, and *in silico* approaches. *Food Science & Nutrition, 9*(2), 833–846.

Emon, N. U., et al. (n.d.). In vitro comparative cardioprotective activity of methanol extract of *Caesalpinia digyna* (Rottl.) stems and *Senna sophera* (L.) Roxb. stems.

Emon, N. U., Jahan, I., & Sayeed, M. A. (2020). Investigation of antinociceptive, anti-inflammatory, and thrombolytic activity of *Caesalpinia digyna* (Rottl.) leaves by experimental and computational approaches. *Advances in Traditional Medicine, 20*(3), 451–459.

Furie, B., & Furie, B. C. (2008). Mechanisms of thrombus formation. *New England Journal of Medicine, 359*(9), 938–949.

Gang, W., et al. (2006). Tong-xie-ning, a Chinese herbal formula, in treatment of diarrhea-predominant irritable bowel syndrome: A prospective, randomized, double-blind, placebo-controlled trial. *Chinese Medical Journal, 119*(24), 2114–2119.

Gidudu, J., et al. (2011). Diarrhea: Case definition and guidelines for collection, analysis, and presentation of immunization safety data. *Vaccine, 29*(5), 1053–1059.

Harman, C. A., et al. (2007). Structural basis of enantioselective inhibition of cyclooxygenase-1 by S-alpha-substituted indomethacin ethanol amides. *Journal of Biological Chemistry, 282* (38), 28096–28105.

Hoque, N., et al. (2020). Cytotoxic metabolites from *Thysanolaena maxima* Roxb. Available in Bangladesh. *Clinical Phytoscience, 6*(1), 1–10.

Hrckova, G., & Velebny, S. (2012). *Pharmacological Potential of Selected Natural Compounds in the Control of Parasitic Diseases*. Springer Science & Business Media.

Hu, J., et al. (2009). Antidiarrhoeal and intestinal modulatory activities of Wei-Chang-An-Wan extract. *Journal of Ethnopharmacology, 125*(3), 450–455.

Hussain, A., Sonkar, A. K., Ahmad, Md. P., & Wahab, S. (2012). In-vitro anthelmintic activity of *Coleus aromaticus* root in Indian adult earthworm. *Asian Pacific Journal of Tropical Disease, 2*, S425–S427.

Jayasinghe, J. H. M. U. L. B. (2015). Chemistry and bioactivity of some Sri Lankan menispermaceae and lauraceae (Doctoral dissertation).

Jayasinghe, L., et al. (2003). A pyridine ring-containing ecdysteroid from *Diploclisia glaucescens*. *Tetrahedron Letters, 44*(49), 8769–8771.

Jayasinghe, U., Hara, N., & Fujimoto, Y. (2007). (2-Nitro ethyl) phenyl and cyanophenyl glycosides from the fruits of *Diploclisia glaucescens*. *Natural Product Research, 21*(3), 260–264.

Jayasinghe, U., Wannigama, G., & MacLeod, J. (1998). Glucuronides of Diploclisia glaucescens. *Journal of the Chemical Society of Pakistan, 20*(2), 131–137.

Kala, C. P., Dhyani, P. P., & Sajwan, B. S. (2006). Developing the medicinal plants sector in northern India: Challenges and opportunities. *Journal of Ethnobiology and Ethnomedicine, 2*(1), 1–15.

Kamboj, V. P. (2000). Herbal medicine. *Current Science, 78*(1), 35–39.

Khan, M., et al. (2005). Neolignan and megastigmane glycosides from the leaves of *Pterospermum semisagittatum*. *Die Pharmazie – An International Journal of Pharmaceutical Sciences, 60*(1), 72–74.

Khan, S., et al. (2003). Three megastigmane glycosides from the leaves of *Pterospermum semisagittatum*. *Pharmaceutical Biology, 41*(7), 512–515.

Khan, S., et al. (2019). Phytochemical profiling, in vitro biological properties and in silico studies on *Caragana ambigua* stocks (Fabaceae): A comprehensive approach. *Industrial Crops and Products, 131*, 117–124. https://doi.org/10.1016/j.indcrop.2019.01.044.

Kolesnikov, Y., et al. (2004). Evaluation of the tail formalin test in mice as a new model to assess local analgesic effects. *Brain Research, 1029*(2), 217–223.

Kuge, T., Shibata, T., & Willett, M. S. (2003). Multiple-dose escalation, safety, and tolerability study of wood creosote, the principal active ingredient of Seirogan, an herbal antidiarrheal medication, in healthy subjects. *The Journal of Clinical Pharmacology, 43*(3), 284–290.

Kuna, L., et al. (2019). Peptic ulcer disease: A brief review of conventional therapy and herbal treatment options. *Journal of Clinical Medicine, 8*(2), 179.

Kurumbail, R. G., et al. (1996). Structural basis for selective inhibition of cyclooxygenase-2 by anti-inflammatory agents. *Nature, 384*(6610), 644–648.

Liu, H., et al. (2018). Structure-guided development of selective M3 muscarinic acetylcholine receptor antagonists. *Proceedings of the National Academy of Sciences, 115*(47), 12046–12050.

Masters, L., Eagon, S., & Heying, M. (2020). Evaluation of consensus scoring methods for AutoDock Vina, smina, and idock. *Journal of Molecular Graphics and Modelling, 96*, 107532.

Mbagwu, H., & Adeyemi, O. (2008). The anti-diarrhoeal activity of the aqueous extract of *Mezoneuron benthamianum* Baill (Caesalpiniaceae). *Journal of Ethnopharmacology, 116* (1), 16–20.

Medzhitov, R. (2008). Origin and physiological roles of inflammation. *Nature, 454*(7203), 428–435.
Meite, S., et al. (2009). Antidiarrheal activity of the ethyl acetate extract of *Morinda morindoides* in rats. *Tropical Journal of Pharmaceutical Research, 8*(3), 201–206.
Miller, M. J., et al. (2000). Treatment of gastric ulcers and diarrhea with the Amazonian herbal medicine sangre de grado. *American Journal of Physiology-Gastrointestinal and Liver Physiology, 279*(1), G192–G200.
Mishra, A., Seth, A., & Maurya, S. K. (2016). Therapeutic significance and pharmacological activities of antidiarrheal medicinal plants mentioned in Ayurveda: A review. *Journal of Intercultural Ethnopharmacology, 5*(3), 290–310.
Mukul, S. A., Uddin, M. B., & Tito, M. R. (2007). Medicinal plant diversity and local healthcare among the people living in and around a conservation area of Northern Bangladesh. *International Journal of Forest Usufructs Management, 8*(2), 50–63.
Ndebele, P. (2013). The Declaration of Helsinki, 50 years later. *JAMA, 310*(20), 2145–2146. https://doi.org/10.1001/jama.2013.281316.
Oi, H., et al. (2002). Identification in traditional herbal medications and confirmation by synthesis of factors that inhibit cholera toxin-induced fluid accumulation. *Proceedings of the National Academy of Sciences, 99*(5), 3042–3046.
Panda, S., et al. (2009). Thyroid inhibitory, antiperoxidative, and hypoglycemic effects of stigmasterol isolated from *Butea monosperma*. *Fitoterapia, 80*(2), 123–126.
Pantzar, M., Ljungh, Å., & Wadström, T. (1998). Plasminogen binding and activation at the surface of *Helicobacter pylori* CCUG 17874. *Infection and Immunity, 66*(10), 4976–4980.
Patrick-Iwuanyanwu, K., et al. (2012). Evaluation of acute and sub-chronic oral toxicity study of baker cleansers bitters—a polyherbal drug on experimental rats. *Excli Journal, 11*, 632.
Rahmatullah, M., et al. (2009). An ethnomedicinal survey of Dhamrai sub-district in Dhaka District, Bangladesh. *American Eurasian Journal of Sustainable Agriculture, 3*(4), 881–888.
Ravelli, R. B., et al. (2004). Insight into tubulin regulation from a complex with colchicine and a stathmin-like domain. *Nature, 428*(6979), 198–202.
Rawat, P., Singh, P. K., & Kumar, V. (2017). Evidence-based traditional anti-diarrheal medicinal plants and their phytocompounds. *Biomedicine & Pharmacotherapy, 96*, 1453–1464.
Renatus, M., et al. (1997). Structural mapping of the active site specificity determinants of human tissue-type plasminogen activator. Implications for the design of low molecular weight substrates and inhibitors. *Journal of Biological Chemistry, 272*(35), 21713–21719.
Rout, J., Sajem, A. L., & Nath, M. (2012). Medicinal plants of the North Cacher Hills district of Assam are used by the Dimasa tribe. *Indian Journal of Traditional Knowledge, 11*(3), 520–527.
Rudra, S., et al. (2020). Biological investigations of the methanol extract of *Tetrastigma leucostaphylum* (Dennst.) Alston ex Mabb. (Vitaceae): In vivo and in vitro approach. *Journal of Advanced Biotechnology and Experimental Therapeutics, 3*(3), 216–224. https://doi.org/10.5455/jabet.2020.d127.
Rudra, S., et al. (2020). Evaluation of various solvent extracts of *Tetrastigma leucostaphylum* (Dennst.) Alston leaves, a Bangladeshi traditional medicine used for the treatment of diarrhea. *Molecules, 25*(21), 4994.
Sajib Rudra, M., et al. (2020). Biological investigations of the methanol extract of *Tetrastigma leucostaphylum* (Dennst.) Alston ex Mabb. (Vitaceae): In vivo and in vitro approach. *Journal of Advanced Biotechnology and Experimental Therapeutics, 3*, 216–224.

Shibata, M., et al. (1989). Modified formalin test: Characteristic biphasic pain response. *Pain, 38*(3), 347–352.
Shiferie, F., & Shibeshi, W. (2013). In vivo antidiarrheal and ex-vivo spasmolytic activities of the aqueous extract of the roots of *Echinops kebericho* Mesfin (Asteraceae) in rodents and isolated Guinea-pig ileum. *International Journal of Pharmacy and Pharmacology, 2*(7), 110–116.
Shrestha, S., et al. (2016). Phytochemical constituents from the florets of tiger grass Thysanolaena latifolia from Nepal. *Journal of Asian Natural Products Research, 18*(2), 206–213.
Subbotina, M. D., et al. (2003). Effect of oral administration of tormentil root extract (*Potentilla tormentilla*) on rotavirus diarrhea in children: A randomized, double-blind, controlled trial. *The Pediatric Infectious Disease Journal, 22*(8), 706–711.
Tunaru, S., et al. (2012). Castor oil induces laxation and uterus contraction via ricinoleic acid activating prostaglandin EP3 receptors. *Proceedings of the National Academy of Sciences, 109*(23), 9179–9184.
Williamson, E. M., Okpako, D. T., & Evans, F. J. (1996). *Pharmacological Methods in Phytotherapy Research: Volume 1: Selection, Preparation, and Pharmacological Evaluation of Plant Material*. John Wiley & Sons Ltd.
Xiaochun, H., et al. (2003). Crystal structure study of (–) syringarenol, a lignan isolated from *Diploclisia glaucescens*. *Natural Product Research and Development, 15*(1), 1–4.
Yadav, J., Kumar, S., & Siwach, P. (2006). Folk medicine is used in gynecological and other related problems by the rural population of Haryana. *Indian Journal of Traditional Knowledge. 5*(3), 323–326.
Yekun, Y., et al. (1999). An acetylenic acid from *Diploclisia glaucescens*. *Acta Botanica Yunnanica, 21*(2), 265–268.
Zahid, H., Rizwani, G. H., & Shareef, H. (2014). Standardization of Hibiscus Schizopetalus (Mast) Hook according to World Health Organization (WHO) guidelines. *Journal of Medicinal Plants Research, 8*(22), 802–810.
Zaman, R., et al. (2015). In vitro clot lysis activity of different extracts of *Mangifera sylvatica* Roxb. leaves. *Research Journal of Medicinal Plant, 9*(3), 135–140.
Zhang, J., et al. (2013). Anti-diarrheal constituents of *Alpinia oxyphylla*. *Fitoterapia, 89*, 149–156.

Index

D

E

I

M

T

U